International Conference on Mechanical and Energy Technologies (ICMET 2019)

IOP Conference Series: Materials Science and Engineering
Volume 748

Greater Noida, India
7 - 8 November 2019

ISBN: 978-1-7138-2657-6
ISSN: 1757-8981

Printed from e-media with permission by:

Curran Associates, Inc.
57 Morehouse Lane
Red Hook, NY 12571

Some format issues inherent in the e-media version may also appear in this print version.

Institute of Physics
Dirac House, Temple Back
Bristol BS1 6BE UK

Phone: 44 1 17 929 7481
Fax: 44 1 17 920 0979

techtracking@iop.org

Additional copies of this publication are available from:

Curran Associates, Inc.
57 Morehouse Lane
Red Hook, NY 12571 USA
Phone: 845-758-0400
Fax: 845-758-2633
Email: curran@proceedings.com
Web: www.proceedings.com

TABLE OF CONTENTS

IOP Conf. Series: Materials Science and Engineering **748** (2020) 011001 doi:10.1088/1757-899X/748/1/011001

Preface

International Conference on Mechanical and Energy Technologies (**ICMET-2019)** has been organized during 7-8 Nov, 2019 by the Department of Mechanical Engineering at Galgotias College of Engineering & Technology, Greater Noida, Uttar Pradesh, India. The motive of the conference is to bring together the scientists, professors, graduates and engineers to discuss and deliver ideas in concern to the recent advancement in Mechanical Engineering and Energy Technologies.

In the conference emerging concepts and developments in the field of Mechanical Engineering and Energy have been submitted. All the accepted papers have been submitted to strict peer-review by 2-4 expert referees, and selected based on originality, significance and clarity for the purpose of the conference. The conference program is extremely rich, profound and featuring high-impact presentations of selected papers and additional late-breaking contributions. We sincerely hope that the conference would not only show the participants a broad overview of the latest research results on related fields, but also provide them with a significant platform for academic connection and exchange.

To conclude, we would like to express our sincere gratitude to all the committee members and organizers for their hard work, precious time for the conference. Our deepest thanks also go to the volunteers and staffs for their long-hours work and generosity they've given to the conference. The last but not least, we would like to thank each and every of the authors, speakers and participants for their great contributions to the success of ICMET-2019.

ICMET-2019
Organizing Committee
Greater Noida, India

Peer review statement

All papers published in this volume of *IOP Conference Series: Materials Science and Engineering* have been peer reviewed through processes administered by the proceedings Editors. Reviews were conducted by expert referees to the professional and scientific standards expected of a proceedings journal published by IOP Publishing.

IOP Conf. Series: Materials Science and Engineering **748** (2020) 012001 doi:10.1088/1757-899X/748/1/012001

Energy Production from Mechanical Stress, Thermal and Wind Power

Rishabh Maurya, Praveen Kumar Maduri, Kushagra Singh, Apurva Soni
Galgotias College of Engineering and Technology, Greater Noida, India

Email-**rishabhmaurya011@gmail.com[1]**

Abstract - The two main factors for living, water and non-renewable energy are being over used these days which will ultimately lead to contamination and depletion of natural resources. Ground water is getting contaminated due to pollution while fossil fuels are being used in excess for generation of electricity. To overcome these issues author proposes a system which not only uses available natural resources to generate electricity but also helps in conserving rain water. A compact design is proposed which utilizes solar and wind energy along with rain water harvesting. System is designed in such a way that the energy produced does not depend on the ratio of availability of resources thus conserving the two main factors for living water and energy.

Key Words: Water harvesting, energy, renewable resources, electricity.

1) INTRODUCTION

The major part of industries depends upon non-renewable energy resources. To generate electricity, non-renewable energy is widely used. Burning non-renewable energy resources not only pollute environment but also harmful for us [1]. To overcome this problem renewable energy are used. Photovoltaic (PV) cell which are used to harness energy from sunlight are widely used as a renewable source of energy. Wind turbines with unique design to harness maximum wind energy are also designed [2]. One more source for harnessing energy is through rain ,and to harness rain energy piezoelectric materials are used but the problem is they are not feasible to use for generation of electricity[3].The problem with all these renewable sources is they are climate dependent as if there is no sunlight then PV cells will not work and with windmills is if air is not flowing then no wind energy can be harnessed. In present technique hybrid system are developed which can harness two energy at the same time[4].An integration of different processes can be designed so if one system stops working other will harness energy .This is done so it can work in such a way that it never stops harnessing the energy from the environment [2]. Economies are widely affected by the way energy is harnessed and distributed. Renewable energy sources are freely available and do not get exhausted with time. Many techniques have been developed to harness energy from rain, sunlight and wind [3,4]. Methods have been thought like hybrid system which integrate much energy together but those techniques get expensive and also switching between the energy is also not reliable to use. In this design a compact model is developed which efficiently integrates energy and stores it in a power source like battery. A major concern about the consumption of water and the quantity that is let also arises, which can be solved with the oppositely facing umbrella design. It is difficult with present technique to harvest rain water and at the same time filter it for drinking purpose. In areas of congestion where solar panels or windmills cannot be placed, a system is needed to harvest energy from these areas. The main problem with solar panels is that they not only use a large piece of land but also left the land for no other use hence leading to land pollution. So, to decrease this kind of pollution government has planned to give subsidy to user for selling electricity to them generated from solar panels that are fixed on their roofs. The product designed is also used by such users and also massive electricity can also be generated when we are harnessing energy from more than one resource at the same time creating a smarter way of harnessing energy. Traditional ways like making a tank for conserving rainwater are widely used which are not suitable and does only one job of conserving water and rainwater is now not suitable for drinking as it has some acidic chemicals altogether, so an efficient filter which can purify all those chemicals which are harmful for drinking is necessary. People of south region faced scarcity of water which tells a need for designs to harness energy as well as conserving water which will lead to a healthy and suitable form of living.

2) PROBLEM STATEMENT

Urban and rural both areas are facing a major concern regarding water and energy consumption. As due to increase in pollution which occur due to burning of non-renewable resources, so industries cannot rely more on non-renewable resources. These resources not only pollute water, land and air but also affects living. There is a transition seen from using non-renewable sources to renewable resources which do not degrade the environment and also create a healthy living. Many techniques like drip irrigation system in agriculture, roof water harvesting system have been developed but the feasibility of these systems is not enough with fast growth of technology the classical ways to harness energy from conventional resources cannot be used for long so a new system is created to solve such issues.

a) A solar cell harvesting system

Solar cells are fixed in ground due to which they utilize much of the land and also no other system can be fixed to harness energy, in this design solar cells are placed at a height in a circular manner so they take the shape of umbrella .when the sun rays hit the cells electrons starts moving .There are many types of photovoltaic cells like monocrystalline ,polycrystalline and concentrated photovoltaic cell which has the highest efficiency .Each of the method solar cell has their own advantage. They are categorized on the basis of purity of material used, efficiency of cell, cost and size. The three parameters necessary for design is the cost, size of cell and efficiency. Polycrystalline have efficiency around 15%, they are less expensive and take moderate space for energy production. When a photon hits a cell the current generated is not regular and if the batteries are directly charged then it may damage or overcharge it. Efficiency of the solar panels also depends on the way batteries are charged. To overcome this problem, we use charge controller which reduces the risk of overcharging and only provides a regular voltage as an output.

b) Harvesting energy from rainfall

Due to motion of water energy can be harvested, piezoelectric material is used for this task, and piezoelectric material is fixed on the circumference of the design. The task is to recover maximum energy from impact of rain drop. The impact experienced by piezoelectric material is inelastic in nature. The velocity of the raindrop gets constant at some point. Piezoelectric material work on a simple principle that when the material is pressed an electric field is generated when generates a potential difference which ultimately leads to flow of current. piezoelectric comes in many forms like plastic, ceramic and crystal. Polyvinylidene fluoride polymer (PVDF) is used in this design. Piezoelectric material generates an AC voltage so to store this energy in a battery it should be converted into DC and also the generated voltage is weak, so a voltage multiplier circuit is used to convert AC to DC and amplify the desired voltage.

c) Wind Energy harvesting system

A compact design is made which will be fixed just beneath the umbrella; the design has an outward curve fixed on the whole area of the wind system design. The important part is, it is very light and harness maximum wind energy.

A dynamo is used to harness wind energy. The shape of this system is similar to an umbrella and also sections are made which will experience a force due to which the wind system will move .The output of the dynamo is AC in nature which cannot be used to charge a battery, So to do that voltage multiplier is used which not only converts AC to DC but also amplifies the voltage, the circuit is a combination of diodes and capacitor which do this job. The way of structure, the size of dynamo will depend on the size of product. A system of gears can be used to increase the number of rotations of the shaft so that the more the number of rotations, higher the voltage obtained. As this system is also used in homes so a small system will be needed and to generate a considerable amount of voltage this should be done.

d) Integrating the three energies

In figure 1.0 it is shown that after harnessing the energy they all added by the integrating circuit, this mechanism has diodes in it which not only allows combining the three currents but also protects one source current to move to other sources. This can further be explained as when a current is coming from piezoelectric material, dynamo and solar cell they should be added in such a way that current coming from solar panel does travel to dynamo side else it may damage the dynamo and piezoelectric.

e) Regulation of current

After the integration of the three energy the net current cannot be directly used to charge the battery it may cause severe damage to the battery or may overcharge it. To reduce this risk charge controller is used. This controller acts like a valve which only allows charge to pass when desired charge is collected and also controls the current. The advantage is that it charges the battery faster if the battery is empty, if the battery is about to fill then it slows down the charging process hence increasing the lifespan of battery .It automatically shut downs when battery is fully charged hence reducing the risk of overcharging .It also has ports which will be connected to load as shown in figure1.0 to use the energy of the battery.

f) Water harvesting system

The benefit of opposite facing umbrella design is that it not only has solar panels and piezoelectric on it which generates electricity but also has a system which collects water. Due to global warming it is becoming difficult to predict rainfall many a time and also rainwater can't be drunk directly as it has many acids and chemicals. So, a filter specially designed to filter rainwater is designed and the electricity supply is given by the batteries for filter to work. The rainwater first hit the surface of the umbrella where solar panels and piezoelectric are fixed .In the middle of the umbrella a hole is given through which rainwater can pass .The surface of the umbrella is inclined upward which covers maximum area for collection of water, the water passes through a filter for purification and then it is stored in a tank for drinking and other purposes. When this system is used in street, the water collected can also be used to water plants in case of no rainfall for a longer period of time. The complete structure is so compact that it can be used in homes and in areas where the weather is in there extreme meaning heavy rainfall in shorter period and longer summer. This system can also provide shade in the time of heavy sunlight thus making it multipurpose.

IOP Conf. Series: Materials Science and Engineering **748** (2020) 012001 doi:10.1088/1757-899X/748/1/012001

3) BLOCK DIAGRAM

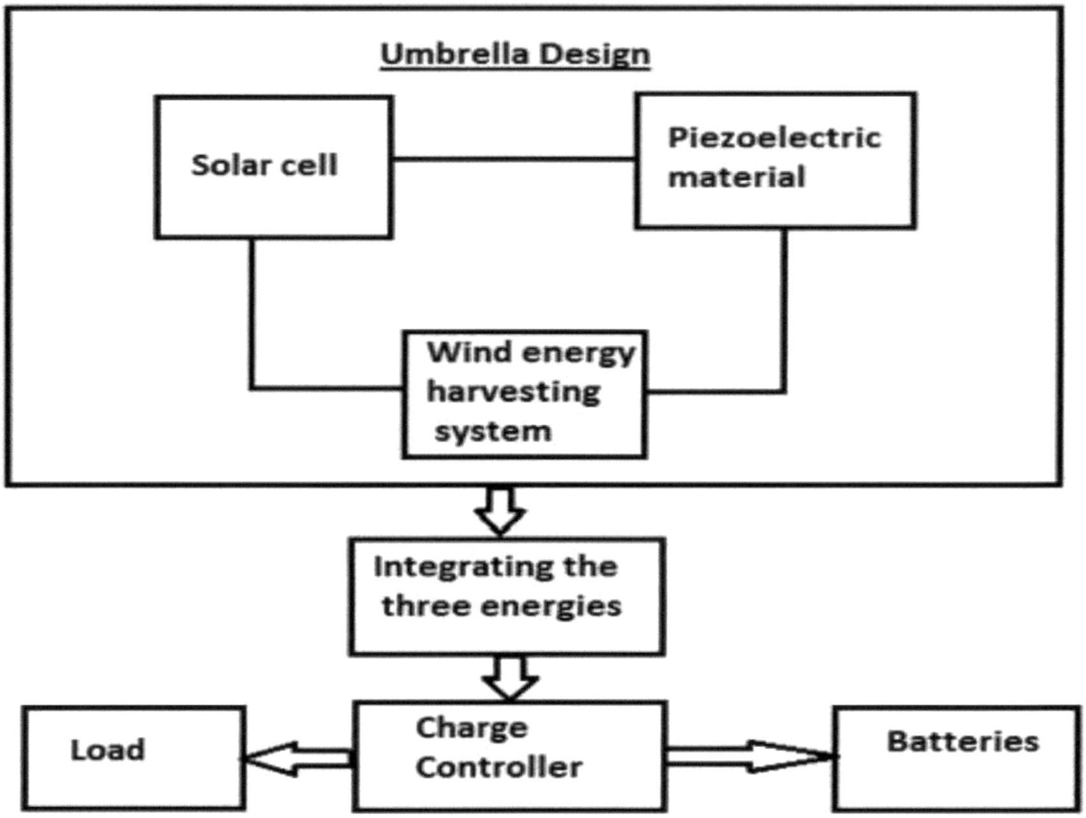

Fig (1.0) Block Diagram of proposed system

Block Diagram Description-
As in Fig (1.0) it is shown that umbrella design has all the three resources in it, where all the three are attached to each other. Going outside this design, joining the three-energy block is shown, after this joining charge controller is used to regulate and transfer that energy to battery and if excessive energy is available then energy meter calculates that energy and then transfer it to the main grid.

4) WORKING PRINCIPLE

This design consists of different energy sources and energy regulation, conversion system. The first energy resource is sunlight, which is harnessed by solar panel when photons hit the p-n junction of solar panel, they easily enter the junction hence produces electricity. There are many forms of solar panels available of which monocrystalline has highest efficiency because it is made of the highest grade of silicon. Whereas to harvest energy through rainfall piezoelectric material is used (PVDF). it works on the principle that when a mechanical stress is applied to this material the output is a flow of charge hence leading to flow of electricity but the energy obtained is less in magnitude. The challenge is to consume maximum energy from the impact of rain drop. Through the impact of rain in PVDF material the kinetic energy which get constant at some point for rain is to be converted in generating an electric field leading to a potential drop across the terminal. For generating electricity from mechanical motion in case generating electricity from wind, dynamo is used. When a moving coil is placed in the influence of magnetic field and potential drop is obtained at the terminals. The umbrella shaped design has fan like outer section which rotate in the influence of wind .this system

IOP Conf. Series: Materials Science and Engineering **748** (2020) 012001 doi:10.1088/1757-899X/748/1/012001

is further connected to gear system whose shaft is connected to a dynamo which generates electricity .When we have all three energies, the only need is to integrate them or add them so they can become one. A combination of diode is used to do that. When energy is collected the only part, left is to use this energy to charge the battery and if energy is in excessive form then transfer it to grid system. charge controller is used which only allows charge to flow when needed and also charges all battery with equal speed which leads to more lifespan of batteries, charge controller also has the ports which transfer excessive energy to grid system [6].

5) CHALLENGES AND FUTURE SCOPE:

As the conventional sources are unlimited in form so, new techniques can be developed to harness them. Transparent solar cell is one such example, these can be used in smart homes. and also, can be used in mobile phone screen which generate energy through the light coming out of the screen and also from outer environment. Further, present piezoelectric material generates very less output voltage, so increasing the efficiency will correspond more applications. Large system for solar panel and wind mill are widely used as a renewable source of energy but no such system is made to harness energy from rain at a very large scale. This project initiative can help to make it possible at a large scale. Charge controller are widely used in solar panels to regulate current and to charge batteries. Charge controller widely helps in controlling the charge for efficient charging but a new circuit is needed to boost the charging time and also to increase the life of battery.

6) CONCLUSION

Thus by implementing this idea the increase demand of energy and water resource can be fulfilled more efficiently for healthy living leading to a brighter and safe future.

7) REFERENCES

[1] Amundsen E S 1999 Environmental taxes on exhaustible resources *European journal of political economy.*

[2] Chavan L P , Nivale S, Shinde S, kamble D, Gaikwad 2016 A Design and Development of Novel *Vertical Axis Wind Turbine IJLTEMAS.*

[3] Wangde Y, Mendhe P, Soparker D, Gupta V, Prof. Dange J 2019 Voltage generation using piezo-Crystals *IJTSRD.*

[4] Lazarov V, Notton G, Zarkov Z, Bochev I 2015 Hybrid power systems with renewable energy sources *Eleventh international conference on electrical machines ELMA.*

[5] Glynne-Jones P, Beeby SP, James EP, White NM 2001 The modeling of a piezoelectric vibration powered generator for Microsystems *11th conference on solid state sensor and actuator.*

[6]_ Chakraborty S., Sinha A., Dutta S., Biswas N., A Review Paper of Wind Energy, *National Conference on Advances and Research in Electrical System Technology.*

[7]Shahzad U 2015 The need for renewable energy sources *International journal of Information technology and electrical engineering.*

[8] Reddy L M 2017 Comparative study on charge controller technique for solar PV cells *1st international conference on power engineering, computing and control.*

Applications of Automation and Robotics in Agriculture Industries; A Review

Amith A Kulkarni,[1] Dhanush P, [2] Chetan B S, [3] Thamme Gowda C S, [4] Prashant Kumar Shrivastava* [5]
[1,2,3,4,5]. Department of Mechanical Engineering, Vidyavardhaka College of Engineering, Mysuru, Karnataka [1]Amith.kulkrni.54@gmail.com, [2] Dhanushp147@gmail.com, [3]. Chetanbs050@gmail.com, [4]. thammegowda.cs@vvce.ac.in,
Corresponding Author Email: er.prashant1986@gmail.com*

Abstract: In this present work, we have discussed about the application of automation and robotics in the various stages of the agriculture industries like irrigation of horticulture land, parks, gardens, and golf courses efficiently and reduce the wastage of water. The uses of automation and smart systems also help in food safety by the selection of appropriate chemicals. The application of multitasking robots is also useful to make it fast and maintain quality. In the smart farming also helps to maintain the humidity, temperature and irrigation process by using the sensors and apps. The aim of this study to identify the opportunities and scope for future automation and the Internet of Things (IoT) in the field of agriculture industries

Keywords- Automation, Robotics, Artificial Intelligence, Internet of Things, Tele-psychiatry

1. INTRODUCTION

Nowadays, automation plays a major role in various type of industries to minimize the human efforts and increase productivity and achieve customer demand and fulfill customer satisfaction. Today, automation used in the different types of industries like Automobile, electronics, electrical, medical, defense and agriculture industries to improve the quality of products and reduce human errors, cost and time. Most of the industries are affected by the control and un-control factors. But for the industrial point of view, there are some limited un-control factors that affect the process because most of the process takes place under manufacturing plants. In the case of the agriculture industry, the output from the agriculture industries mainly based on the different types of aspects that effects on the output directly or indirectly like uncertain prediction of rain, storm, hot and cold waves etc. The conventional agriculture practices is not suitable for this kind of situations. The un-conventional agriculture practices with automation with IoT has been reduce the effects of uncertain factor and help to understand the weather condition that is responsible for increased agriculture output.

2. LITERATURE SURVEY

There are many technologies being introduced in the agricultural field and it is not yet being used in the daily agricultural activities as they are very expensive hence a dependable system with low cost and low power usage was introduced known by the name as UREA spraying arrangement with automated vehicle used for collecting data all by remote sensing hence it very low cost-effective, now by the recent trend of miniaturization of electric components there is an easy way of implementing these, the system captures green field pictures and products hence this type of model was introduced with high-performance quadrotor with brushless DC motor . The UAV can be programmed by an Arduino through network, the UAV has multispectral resolution and it has temperature cameras used for various climate change conditions, the programming of the UAV can be done in such a way that the system handling can be auto-corrected whenever there is an error as the feedback sensors provide regular data acquired by the sensors such as the green field images the UAV consists of accelerometers to calculate the acceleration forces and a gyroscope applied for measurement of angular velocity and combination of these provide independent degrees of freedom tracking system and the brushless dc .motor to drive the system as it has various advantages over the other motors in terms of life span, efficiency, and has overall higher performance over any other motors and propellers which are used to generate aerodynamic lift force and a speed controller unit which can alter the speed of the motor and should also act as the dynamic brake for the system and moreover a power source to run the system and a radio frequency control system used for various purposes such as voltage regulation and steering control and also take-off and landing. And camera and sensors for data acquisition hence by the combined function of all these parts the UAV can be used for various applications like crop field monitoring and monitoring of growth of plants and moreover the spray system to various fields based on the data the manual operator can drive the UAV to spray the various fertilizers over the required fields with high ease as the design is ecological this type of design can be utilized for the enhancement of the future developments of the various applications of agricultural areas [1].

Moreover, in agriculture where the farmers have to handle various heavy materials such as equipment's for the farming and heavy vegetables, fruits during harvesting time as this process requires hard intensive labor work hence a manipulator is used to handle such heavy material handling process, the agricultural-based manipulator is different from the other manipulator for the following reasons 1) they have different locomotive function 2) working condition and environment is different 3)they have different characteristics as they handle plants as there are two different types of work in the agriculture, they are the selective work and simultaneous work, hence the material handling is designed to be versatile, the design can be done in all the four possible ways that are the polar, Articulated, parallel, and cylindrical type coordinate systems which can be determined based on the end application.

The second factor to be considered for the design the analysis of the forces such as the kinematic forces because the manipulator will be having mobile platform and heavy material handling in agriculture as the material or target will be either scattered or handled horizontally hence the robot are designed based on the manipulability of condition, and based on the observations the parallel type manipulator are used for material handling of the agricultural applications [2].

However, the distribution of the chemical droplets on the crop field is very important for the crop development hence a sprayer technique with leveling function performed with the help of four-rod linkage self-leveling suspension and electro-hydraulic auto-leveling technique, as per in the agricultural field providing an better improvement for optimum plant development and use of chemicals is a elusive jobs as the agrochemical contains high concentrated chemicals which can serious environmental hazards hence we require a proper accurate and precise machine for the homogeneous spray applications, the quality of equal distribution of the spray is depended on the factor such as the distance between the nozzles and the crops, in regular sprayer machines the problems they face are due to vibrations, hence to overcome these problems a flexible sprayer boom system which is designed to eliminate the vibrations and the leveling should be properly done. And this can be achieved by minimize the vibration by the suspension and smoothing the tending boom with the hydraulic controller system.

For the better system two spray system with linked suspension and electro hydraulic techniques have been applied for spray. Apart from that it has been included infrared sensors, hydraulic sensors, and cylinder and controller system. Based on the application the mathematical model and ANSYS simulation can be carried out at the initial based on the outputs of the mathematical model the controller can be designed for the closed-loop system hence the system will show good correspondence to the requirements and the results which will lead to better spraying quality [3].

Moreover, walking tractors also known as Power wheel that is used for the plowing, tiling, and ground preparation in the small fields, using this type of machinery can overcome many drawbacks from the regular farming conventional machines the overall system of power wheel is considered kinematics, dynamics, and engine dynamics for independent action and the responses are the right and left clutch control signals, data from the inertial measurement unit (IMU) involved to power source permits signal and the analogue matching with integral least squares methods have been used for the validation of the model..

This steering arrangement are responsible for different types of mechanical advantages by this arrangement tractor perform in different types of conditions. Hence these type of power tillers can be used for both dry and wet cultivation, the remote operation of the power wheel controlled via wireless network. The IMU which is mounted at the center senses the angle, longitudinal and lateral acceleration, in the power wheel the power transmit to the wheels by a diesel engine, hence the whole system is driven which leads to the high efficient model for the cultivation of crops for any environment [4].

Thereafter, vision-based control system has been used in agriculture robot for providing the directing through the crop field from the concept of open architecture by using the machine vision system, at higher rate of efficiency which can be highly productive in the field of agriculture, and highly advantageous in material handling, and the combination of autonomous robot along with the navigation can be used to develop a highly intelligent system and it mainly comprises of both cabins. The power supply source, a portable computer and an electronic controller faced in the front cabin and a camera is fixed at front of it for the path detection and storage of the material the rear cabin is used such as water, pesticides, etc. and the encoders can be used for driving the speed of the motor based on the wheel drive used [5,6].

Which can be a front-wheel or rear-wheel drive and different mechanisms can be obtained for the driving of the mechanism of the wheel all of this can be controlled by the electronic control unit as it is the decision-making system based on the data achieved from the data obtained from the images of the camera, the control unit checks the data of the image if the organizing area and the application area to obtain planning data for the further action, this type of system can also adapt to the eco-friendly changes, body position and data of situation and inclinations if any and handle any emergency system. Moreover, to reduce the greenhouse effects in agriculture industries automation and IoT play an important role [7–9].

For the row guidance a machine vision system is used which has camera to identify the crop plants and based on that they find a suitable precise and established navigation data by the binary output that is also includes the heading angle and the offset from the image based on the coordinate system and the algorithm for the row guidance, the algorithm is written to ladder structure, for the edge recognition and judgement, then image binarization for the suitable for direction, in which the greyscale image is transformed to the binary image for guidance of row, hence based on the machine vision system the efficient autonomous robot can be used for the agricultural application [10].

However, the radio frequency identification is used in many areas of agriculture as it has various configurations such as non-interaction, highly appreciation , high data storage capacity, secure entrance, and easily compatible with any system, RFID based the seed quality grading system based on four subsystems naming 1)seed grade supervision technique in the industries, in this system the breeders can use a tag on which the experiment should be conducted and then taking a portable interrogator to note the information like the irrigation schedule, fertilizer schedule, pests schedule, maturing schedule, and situation for seeding schedule in RFID tag as these data are most essential for the breeders as they maintain the quality in subsequent procedures and then comes the 2) The RFID tag on the seed contains data such as the transport means, time, batches, initial location, final location, store environmental procedures etc. This data is most important for tracking the problems when the quality incident occurs 3)seed management and supervision system, in this system the data in RFID tags and seed packages are highly encrypted and can't be trespassed by the unauthorized and allows only the authorized supervision personnel to read the data 4)seed planting service system in this system as the RFID tag contains all the data and application program updated by the supervisor it allows the farmer to easily identify the quality of the product and also about its storage requirements, planting guidance, etc. which becomes very helpful for the farmers, hence to store all these data a suitable type of RFID tag, the RFID tag is selected based on the frequency range as it ranges from 125KHz to 245GHz based on the end application, and another factor to be considered is the anti-collision of the tags because there will be multiple tags hence care is taken such as using different algorithms for different RFID tag hence the tag collision is reduced and use of RFID is benefitted for the farmers [11]

Amongst various developed technologies for the agriculture the use of IoT(internet of things) is also one the emerging technologies which help in reducing the hard conventional farming method, the technology uses graphic related touch screens with high resolutions for multitasking along with using the external sensors, the system uses an embedded system to control the overall system, the controller unit consists of ARDUINO microcontroller which is from 8 bits to 32 bits, which is used to store the programs for the specific task and then the for the given program the Arduino controller controls the output devices connected to the output terminal of the controller and it can also be controlled with wireless network such as the RFID technology, the IoT is an interconnection of various objects along with wireless network driven by a power source for accomplishing various tasks along using various sensors [12–17].

Moreover, machine vision system for the row guidance mainly based on the mathematical modeling this model analyze the images and segment crops in images to extract target crops the machine uses a least-square method based on the targets of the crop , it houses a navigation technology based on the GPS or machine vision used for the agricultural vehicles the algorithm houses a various working principles for various operations in the machine which uses a a binarization edge system created by Bayes classification technique to division plants of the various steps and soil accurately for the cotton fields and a Hough transformation method was used by merchant find crop racket from the ground pictures and a separation of the 24 bit picture into a binary picture and a stochastic design method applied by slaughter to classify the cotton plants position between the weeds and several other methods.

To remove the intermittent generosity difference of the picture concentration and decrease the properties of fractional shadow and weeds, combining these algorithms a new approach to row recognition was introduced which was robust algorithm and was capable to mathematical morphological picture dispensation algorithms is done for the row guidance as it works in 3 stages 1)pretreatment of the image and simplification and then 2)identification extraction 3)discrimination of segmentation based on these process the machine vision system gives the data which is sent to the control unit and it, in turn, controls the DC motors which drive the wheels and controls the required position and inclination of the vehicle and guides the robot through environmental agricultural applications with high accuracy and precision [18].

Recently, due to increment in the population, the agricultural demands have not been able to meet the demand of the customer, hence to meet the production demand, the production system needs to be optimized it has been observed that inclemency of the environment creation of the organizations named greenhouse gases which controls the properties and features with completing a superior creation in a reduced space and improving harvests and enlightening quality. Recently automation under greenhouse gas has occurred in a confident way, one of it has related to the robotics that supports to achieve the cultivation tasks and helps to reduce the impact on the environment. As the automaton in agriculture starts from preparation of soil and sowing and development of the plant and harvest and then post-harvest and then managing these tasks has higher significance to complete the satisfactory yields and these factors control the environmental variables to maximize the benefits. The use of pesticides with recent new technologies improves the conditions of the use of pesticides as they are the danger to human health and the environment and are costly hence care should be taken to see the rigorous use of inputs through robotic platforms to supervise the work. Hence it allows finding the line of research in agriculture production under controlled condition hence it allows the management of environmental hazards so it highly recommended using automation [19].

Among various natural resources that are consumable water is also one of the most important Factors hence it is required to use an adequate amount of water and also the consideration of all the input resources of water such as rain and irrigation and productions such as the disappearance, transpiration, and drainage. To determine the water balance for the crop cultivation and the adequate amount of .water is decided by the soil sensors, agro-climatic stations these sensors provide approximate measurements these methods provide a typical approach for the measurement of evapotranspiration devices are used for the measurements for the variations in water pots and the data obtained by these equipment's and extra information are applied for agricultural irrigation system with decision making process and the data are regularly gathered for some intervals and based on the data various mechanisms can be used integration of automation, sensor technologies and other various mechatronic fields such as the use of the digital photography for the irrigation of lettuce and controlling these parameters via a computer to related parameters based on the analysis based on the color analysis can also be achieved along with observing of irrigation supervision by soil moisture sensors is simultaneously performed [20].

However, the use of ICT (information and communication technology) in the agriculture field as that can empower and provide information for the farmer about the various crops, weather, and production farming methods. The ICT will provide the information about the suitable quality of seed and which fertilizer is suitable for the particular seeds. Apart from that, it will provide the information about the plant nutrition and required water according to soil and seeds. The ICT fulfill the information gap and also provide the information how to increase the agriculture production by using this information. The used of automation and IoT with ICT are providing the different types of sensors based application system that will provides you the related information and automated machines reduce the man power and will help to increase the agriculture production [21–23].

3. RESULTS AND DISCUSSION

Agricultural system modernization is one of the important factors for the todays economy to develop, because the country is fully dependent on the factor that the today resource management in agriculture is most complicated, such as the water resource, the agriculture uses 70% of the fresh water hence water management is the most critical part in the agriculture to accomplish, hence to achieve this there are certain areas in agriculture where it is dependent such as the crop section for the soil, in this factor a PH sensor is used to measure the acidity in the solution, an IOT gate way Is used to connect the field devices with the wireless internet networks which could be WAN(wide area network) or it be a remote controlled device, the device can be integration of IOT with cloud computing, sensors and equipment's combined by the cloud computing along with raspberry pi sensor which is a module to which various sensors such as the temperature, humidity, are interfaced and also other will be provision for the GSM established programmed irrigation arrangement for the effective usage of properties and crop organization , and when it comes to the operation of the system initially the user registration is compulsory for the verification and then the admin checks the data provided by the user and allows to control the system, raspberry pi kit take the sensor send the information to Google spread sheet attached to the raspberry pi kit , the Google spread sheets conserving real time info and a data associated to crop protected on the sheets. This process is distributed in modules this proposed system is best suitable for the crop selection based on the soil quality and unwanted water wastage things are completely eliminated [24].

4. CONCLUSIONS

The applications of automation and smart systems in the field of irrigation that will help to make the irrigation process efficient and reduce the wastage of water. The applications of automation and smart systems also helps in food safety to find out the appropriate chemicals and pesticide for agriculture purpose. The automated and smart systems also useful in weed selection and control. The multitasking robots will help to agriculture activities and process and also complete the task within time and maintain the quality of the product and minimize the human efforts. The hybrid farming systems will help us to provide information about the humidity, temperature and water level.
By the better utilization of automation and IoT in field of agriculture we can simplify the agriculture process and it help us to select the suitable weed, crop and pesticides. On the other hand automation and IoT system makes agriculture process more predictable and provide the solutions. These technologies will help to reduce the human efforts and increase the production.

ACKNOWLEDGEMENT

The authors express sincerely heartiest thanks to Accendere Knowledge Management Services Pvt. Ltd for providing the assistant in manuscript preparation. The authors also grateful to the management of Vidhyavardhak College of Engineering, Mysuru, Karnataka, India for financial assistant to carry out this research work.

REFERENCES

[1] Meivel S, Dinakaran K, Gandhiraj N and Srinivasan M 2016 Remote sensing for UREA Spraying Agricultural (UAV) system *2016 3rd International Conference on Advanced Computing and Communication Systems (ICACCS)* vol 1 (IEEE) pp 1–6

[2] Sakai S, Iida M and Umeda M 2002 Heavy material handling manipulator for agricultural robot *Proceedings 2002 IEEE International Conference on Robotics and Automation (Cat. No. 02CH37292)* vol 1 (IEEE) pp 1062–8

[3] Sun J and Miao Y 2011 Modeling and simulation of the agricultural sprayer boom leveling system *2011 Third International Conference on Measuring Technology and Mechatronics Automation* vol 2 (IEEE) pp 613–8

[4] Piyathilaka L and Munasinghe R 2010 Modeling and simulation of power tiller for autonomous operation in agricultural fields *2010 The 2nd International Conference on Computer and Automation Engineering (ICCAE)* vol 5 (IEEE) pp 743–8

[5] Gat G, Gan-Mor S and Degani A 2016 Stable and robust vehicle steering control using an overhead guide in greenhouse tasks *Computers and Electronics in Agriculture* **121** 234–44

[6] Acaccia G M, Michelini R C, Molfino R M and Razzoli R P 2003 Mobile robots in greenhouse cultivation: inspection and treatment of plants *Memories. Paper presented in 1st International Workshop on Advances in Services Robotics. Bardolino, Italia* (Citeseer)

[7] Balloni S, Caruso L, Conti A, Schillaci G, Longo D and Muscato G 2008 Preliminary Study for the Development of an Electrical Autonomous Vehicle for Safe Agricultural Chemicals Distribution Inside Greenhouses *Atti su CD-rom del Congresso Internazionale "Innovation Technology to Empower Safety, Health and Welfare in Agriculture and Agro-food Systems", Ragusa, Italy* pp 15–7

[8] Vakilian K A and Massah J 2017 A farmer-assistant robot for nitrogen fertilizing management of greenhouse crops *Computers and electronics in agriculture* **139** 153–63

[9] Van Henten E J, Hemming J, Van Tuijl B A J, Kornet J G, Meuleman J, Bontsema J and Van Os E A 2002 An autonomous robot for harvesting cucumbers in greenhouses *Autonomous Robots* **13** 241–58

[10] Xue J and Xu L 2010 Autonomous agricultural robot and its row guidance *2010 International Conference on Measuring Technology and Mechatronics Automation* vol 1 (IEEE) pp 725–9

[11] Li H, Wang H, Shang Z, Li Q and Xiao W 2010 Application of RFID in agricultural seed quality tracking system *2010 8th World Congress on Intelligent Control and Automation* (IEEE) pp 3073–7

[12] Mohanraj I, Ashokumar K and Naren J 2016 Field monitoring and automation using IOT in agriculture domain *Procedia Computer Science* **93** 931–9

[13] Prathibha S R, Hongal A and Jyothi M P 2017 IOT Based monitoring system in smart agriculture *2017 International Conference on Recent Advances in Electronics and Communication Technology (ICRAECT)* (IEEE) pp 81–4

[14] Baranwal T and Pateriya P K 2016 Development of IoT based smart security and monitoring devices for agriculture *2016 6th International Conference-Cloud System and Big Data Engineering (Confluence)* (IEEE) pp 597–602

[15] Suma N, Samson S R, Saranya S, Shanmugapriya G and Subhashri R 2017 IOT based smart agriculture monitoring system *International Journal on Recent and Innovation Trends in computing and communication* **5** 177–81

[16] Cambra C, Sendra S, Lloret J and Garcia L 2017 An IoT service-oriented system for agriculture monitoring *2017 IEEE International Conference on Communications (ICC)* (IEEE) pp 1–6

[17] Gondchawar N and Kawitkar R S 2016 IoT based smart agriculture *International Journal of advanced research in Computer and Communication Engineering* **5** 838–42

[18] Jinlin X and Weiping J 2010 Vision-based guidance line detection in row crop fields *2010 International Conference on Intelligent Computation Technology and Automation* vol 3 (IEEE) pp 1140–3

[19] Tangarife H I and Díaz A E 2017 Robotic applications in the automation of agricultural production under greenhouse: A review *2017 IEEE 3rd Colombian Conference on Automatic Control (CCAC)* (IEEE) pp 1–6

[20] Ruiz-Canales A and Ferrández-Villena M 2015 New proposals in the automation and remote control of water management in agriculture: Agromotic systems *Agricultural Water Management* 1–3

[21] Singh S, Chana I and Buyya R 2015 Agri-Info: cloud based autonomic system for delivering agriculture as a service *arXiv preprint arXiv:1511.08986*

[22] Mendez G R and Mukhopadhyay S C 2013 A Wi-Fi based smart wireless sensor network for an agricultural environment *Wireless sensor networks and ecological monitoring* (Springer) pp 247–68

[23] Arroqui M, Mateos C, Machado C and Zunino A 2012 RESTful Web Services improve the efficiency of data transfer of a whole-farm simulator accessed by Android smartphones *Computers and Electronics in agriculture* **87** 14–8

[24] Shingarwade N D and Suryavanshi S C 2017 Performance Evaluation of Cloud Based Farm Automation *2017 International Conference on Computing, Communication, Control and Automation (ICCUBEA)* (IEEE) pp 1–6

A Brief Review on Biomass Converted Useful Materials

[1] Akshay M N, [2] Anand V, [3] Abhilash S, [4] Deepak G, [5] Prashant Kumar Shrivastava*
[1], Department of Mechanical Engineering, Vidyavardhaka College of Engineering, Mysuru, Karnataka
[1] akshaymn1998@gmail.com, [2] anand72045@gmail.com, [3], abhilashnarayan18@gmail.com, [4], deepak.g@vvce.ac.in,
Corresponding Author Email: er.prashant1986@gmail.com*,

Abstract In this work we have discussed the different types of biomass converted materials like pyrolysis of biomass, biomass-waste, gasification of harmful gases, carbon-based supercapacitor, silica, macadamia shell waste, lithium-sulfur battery, and rice husk, etc. from the different industries and different types of conversion approaches. There are different types of conversion techniques are available for biomass conversion like CVD, hydrothermal process, thermolysis, pyrolysis, combustion, and chemical treatments. The aim of this study to identify the various types of biomass converted materials from different types of industrial waste and also find out the better way to convert into a useful material.

Keywords- Graphene, Waste, Bio precursors, Biomass, Glucose, Rice husk, Hemp

1. INTRODUCTION

Nowadays, industrial waste is the fastest growing waste stream, and one of the most technically challenging for resource recovery given the complex mix of precious and semi-precious and strategic metals, plastics, glass, toxic elements, and contaminants. Previously industries are majorly responsible for atmospheric and water pollution. Apart from that, now industrial waste is also responsible for atmospheric pollution directly or indirectly. There are different types of industries are responsible for producing different types of waste material that cannot be easily recycled. For example, automobile, medical, electrical, electronics, agriculture, and petrochemical industries are responsible for the production of various types of waste materials [1].

However, waste management is the only way to overcome this problem. This is the time to identify the different types of waste material from the different industries that we can easily recycle by the different recycling approaches. There are different types of recycling techniques are available for recycling waste material. The types of recycling techniques depend upon the behavior of the waste material. Now, we are discussing the biomass converted materials. There are different types of biomass converted materials that can be easily converted into useful material. So, we need to identify these types of materials from the different types of waste material that can be easily recycled by the different recycling technique

2. LITERATURE SURVEY

Here, we are discussing the different types of biomass converted material and different types of recycling processes. Zhou et al., have reported that microwave-assisted pyrolysis of biomass by using ZSM-5 composite catalyst supported by SiC foam. After facing a bunch of issues pertaining to the application of the catalyst in a catalytic fast pyrolysis system, SiC foam supports coated with the composite catalyst of ZSM-5 was developed and further, it is tested for up-gradation of pyrolytic vapors in ex-situ catalytic reaction. During the study it has been observed that using the composite catalyst over the various configuration of catalysts placement, there is a significant improvement in the quality of bio-oil and this is achieved without compromising the yield. Further, they studied the bio-oil composition and product yields for various catalysts to biomass ratio values and concluded that there is a direct relationship between the catalyst to biomass ratio and bio-oil quality at the rate of its yield. This means, with the increase in catalyst to biomass ratio there is a significant increase in the quality of bio-oil and also its yield. But after the value of catalyst to biomass ratio reaches 1/10, the catalyst activity of composite catalyst starts declining. Anyhow the composite catalyst can be reused after regeneration and even after seven reaction-regeneration cycles, its catalytic activity was well preserved. This suggests that ZSM-5/SiC could be a promising composite catalyst for ex-situ catalytic fast pyrolysis operations on a large scale [2].

However, the production process of carbon nanotubes from mixture carbon gas containing CO and CH_4 which were primarily obtained from biomass gasification gas or biomass pyrolysis, and is carried out in a fluidized bed and thermogravimetric reactor. Since the gas product i.e., the mixture gas has a very high carbon source, it can be used to synthesize CNTs. During this process, the decomposition of mixture gas results was compared with the results of the CO disproportionation process and pure CH_4 decomposition process by considering reaction temperature and partial pressure as parameters. After a series of experimentation it has been noted that with the increase in partial pressure ratio of CO/CH_4 mixture, there is a significant increase in the quality and yield of carbon nanotubes. Till the partial pressure ratio of CO/CH_4 value is 2. The carbon weight gain yield was 8 folds more than that of CH_4 decomposition under the same partial pressure. Further, the thermogravimetric results showed that there is an increase in carbon weight gain under the low temperature of CO/CH_4 gas mixture than the decomposition of pure CH_4 and also with the increase in CO partial pressure, the final carbon weight gain was enhanced. Specifically, at 550°C the carbon weight gain was enhanced by 5 times in mixture than in the pure gas. It was clear that in CO/CH_4 decomposition process, both CH_4 and CO with molecular hydrogen (H_2) as the medium played a coadjutant role in promoting the decomposition reaction for each other. As a result, the carbon nanotubes (CNTs) produced were well graphitized, smoother and longer compared to that of pure CO and CH_4 gas decomposition. So, it is feasible to synthesize carbon nanotubes (CNTs) using CO/CH_4 mixture gas [3].

Moreover, Lofty et al., applied a novel pyrolysis system to prepare carbon nanotubes (CNTs) using hydro chars of rice straw (RS), where it is unwanted biomass. In this regard, various RS pulps were synthesized namely sulfite pulp (RS-SP), the alkaline pulp (RS-AP) and neutral pulp (RS-NP). These pulps which are a source of carbon in carbon nanotubes were evaluated in comparison with raw rice straw. The so obtained hydro char precursors from the pulping process were used in the preparation of carbon nanotubes and these were affected by thermal degradation and chemical constituents of rice straw. Also, the results of transmission electron microscopy and scanning electron microscopy revealed that the rice straw precursor type plays a key role and has a notable effect on the size and shape of carbon nanotubes produced. From the pyrolysis of neutral pulp (RS-NP) hydro chars and hydro chars of raw RS, multi-walled carbon nanotube bundles were produced and their diameter ranging from 14.6 nm to 47.9 nm and 15 nm to 40 nm respectively. While both, sulfite pulps (RS-SP) hydro chars and alkaline pulp (RS-AP) hydro chars can produce needles like carbon nanotubes which were grown in between Nanosheets of graphene with the diameter ranging from 4 to 8 nm and 2.5 to 6.8 nm respectively. However, when compared with the treated rice straw, raw rice straw produces carbon nanotubes with relatively large surface area (188 m^2/g). Whereas greater methylene blue (MB) adsorption capacity (283.3 mg/g) was found in carbon nanotubes obtained from hydro chars of sulfite pulp (RS-SP-CNTs) [4].

Thereafter, Abbas et al., have discussed graphene quantum dots (GQDs) synthesis techniques from biomass-waste and also its immense applications in various fields. Due to the humongous increase in population, there is an increase in customer needs which in turn led to environmental deterioration and energy crisis. As a result, scientists started researching green routes to produce advance materials using renewable resources. Apart from various materials that were explored, nanomaterials that were based on carbon especially graphene quantum dots (GQDs) attracted enormous attention due to its special properties like long life, high conductivity, good biocompatibility, extensive surface area, and mainly low toxicity. Graphene quantum dots can be fabricated using various techniques. These techniques are basically categorized into the bottom down approach and a top-down approach based on the reaction mechanism. In a bottom-up approach, small organic building blocks were assembled chemically and this helps in synthesizing graphene quantum dots through various reaction mechanisms involving intermolecular coupling or carbonization process. But these approaches are complicated and time-consuming. In the top-down approach, large sp^2 carbon domains are converted into smaller ones by adopting mechanical shearing or chemical oxidation (etching or cutting), which results in high yield Graphene quantum dots. These methods are time saving and simple. After synthesizing, numerous methods were used to obtain desired and defined properties of Graphene quantum dots. GQDs so formed has a variety of applications in the biomedical field, light-emitting diodes and, PL sensors. It also has various energy applications like in supercapacitors, batteries, photovoltaics, etc. and due to its tunable properties, GQDs have very vast application area.

Moreover, Wang et al., have demonstrated the synthesis of supercapacitors containing solid-state asymmetric carbon by using the concept of all-round utilization of biomass. The concept, all-round utilization of resources is mainly used to increase economic and environmental benefits. All the subassemblies which form a supercapacitor are derived from orange peel using various techniques. With the help of a step carbonization process of natural orange peel, a monolithic porous carbon (OPHPC) is prepared. The OPHPC obtained has a large surface area of 860 m^2g^{-1} and it is naturally doped with 1.2% of nitrogen. Further, by using a simple hydrothermal process, the composite electrode is synthesized and it contains MnO_2 in major proportion. Due to its high porosity (74.6%) and natural porous channel structure, orange peel serves as a good material for a separator. And the electrolyte needed for a supercapacitor is obtained from orange peel juice since it has the best retention property in the natural separator. Due to the flexibility/plasticity of orange peel, the electrode shape can be customized. The ultimate supercapacitor obtained by combining all these components has a high areal capacitance. This method of synthesis not only uses the natural structure and inherent chemical components of biomass but also utilized the biomass inherent properties like flexibility or plasticity. This gives a positive hope in utilizing the biomass resources to its fullest [5].

However, the fabrication process of silver nanoparticles ($AgNP/SiO_2$) by using silica rice husk, uneconomical biomass as support via a template-free, novel, and facile sol-gel route. By introducing the acid metal solution into sodium silicate, silver nanoparticles were formed with a diameter of 25 nm on average and this is carried out at room temperature. The synthesized nanoscale silver particles were analyzed using various spectroscopic techniques and the following observations were evident. Silver inclusion into the framework of silica was found successful from EDX and FTIR analysis. From X-ray powder diffraction analysis, it was confirmed that in $AgNP/SiO_2$ amorphous nature of silica is preserved. From N_2 adsorption-desorption studies, it was evident that $AgNP/SiO_2$ exhibits high mesoporosity. From TEM analysis, it was observed that spherical silver nanoparticles were well dispersed in the silica framework. Also, the synthesized $AgNP/SiO_2$ has a Brunauer-Emmett-Teller surface area of 514 m^2g^{-1} and it has a uniform pore size of pore volume 0.50 CCG^{-1}. This green technology used, is a promising technique in synthesizing uniform silver nanoparticles from biomass particularly agro waste [6].

Moreover, Noushad et al., have demonstrated the synthesis technique of Nanohybrid silica composite using inexpensive rice husk. These spherical Nanohybrid particles were used as fillers in the dental composite. It has a low specific surface area of about 30 m^2/g and was extracted via a facile precipitation method by using chemicals that are basically non-toxic in nature. Two filler/matrix ratios, Composite A: 40/60 and Composite B: 50/50 were fabricated and tests were conducted between these two experimental dental composites. The result of the tests showed that dental composite B exhibits a flexural modulus of 6.2 GPa, Vickers' hardness of 39 HV1, flexure strength of 107 MPa, the surface roughness of 0.057 Ra and compressive strength of 191 MPa. The above promising results direct us to use silica (which was basically extracted from rice husk) as a filler material in Nanohybrid dental composites [7].

WenLi Zhang et al., have proposed a salient method to produce hierarchical porous carbon with angstrom-sized high-volume pores by utilizing biomass. This porous carbon with angstrom-sized pores has potential applications in supercapacitors because of their characteristics like high specific capacitance along with high pack density. Here onion is considered as the typical biomass which acts as the carbon precursor. Cellulose, lignin, and hemicellulose are the typical components in onion. Lignin-potassium-salts/cellulose composite is formed when the chemical bond between cellulose molecules breaks and this reaction takes place when the onion is etched by KOH solution. The formation of onion derived porous carbon (OPC) occurs when the composite as-prepared undergoes pyrolysis in an N2 atmosphere. During the carbonization process potassium sites of lignin, potassium salts act as activation agents known as porogen. N2 adsorption/desorption and scanning electron microscope were used to study morphology as well as the porous structure of the onion derived porous carbon. By conducting electrochemical impedance spectroscopy, galvanostatic charge-discharge and cyclic voltammetry performance characteristics, if onion derived porous carbon for supercapacitor applications, can be noted down. Good cycling stability and high specific capacitance are the dominant characteristics seen when OPC is used as electrode material in supercapacitors [8].

Samane Maroufi et al., have demonstrated the means of utilizing waste materials to produce silicon fibers/particles economically in a more efficient way. The carbothermal reduction process was adopted for the synthesis of high-grade silicon carbide using waste materials. The two waste materials used here are e-waste glass (EWG) and waste tyre rubber (WTR) where both of them are referred to as high volume global waste streams and problematic. EWG is the source of SiO2. After the processing of these two materials for the synthesis of SiC by carbothermal reduction technique for a duration of 150 minutes at 1550oC, the resultant composite material consisted of a mixture of nanofibres as well as distributed nanoparticles. Nanofibres of length up to 5 microns and diameter varying from 10-150 nm were present. Also, nanoparticles with higher particle surface area (85.1-67.3 m2/g) and size of particle ranging from 30-40 nm along with micropores of dominant pore size lesser than 2 nm. The reaction in the gas phase between SiO and CO resulted in the nucleation of Silicon carbide nanofibre and later formation of Silicon carbide nanofibre. Due to the deposition of SiO on the nanoparticles of carbon derived from waste tyre rubber, the formation of Silicon carbide nanoparticles was achieved. Usually, the production of SiC nanomaterials is costlier and this can be replaced by synthesizing SiC nanomaterials cost-effectively by utilizing waste materials as mentioned above [9].

Yunyi Wang et al., have studied the synthesis of Biomass-derived novel carbons (BDNCs) along with its function. Also, parameters like the electrochemical effect of structural diversity, porosity and surface heteroatom doping of the carbons in lithium-sulfur batteries have been discussed. Biomass, which is one of the major sources of carbon is the prime factor for the genesis of fossil fuels. Contribution of nature to biomass is viewed as biomass with different microstructures including a set of pores namely micropores, hierarchical pores or mesopores. Biomass-derived novel carbons (BDNCs) exhibit salient characteristics such as eco-friendly nature, changeable physical/chemical properties and economical. Biomass finds its typical applications in energy storage devices, particularly in lithium-sulfur (Li-S) batteries. Analysis of market demands of BDNCs for lithium-sulfur batteries is carried out by comparing the lithium-sulfur batteries with lithium-ion batteries [10–12].

However, in biomass conversion, there is a wide usage of acid groups like SO_3H which acts as a solid acid catalyst. The mesoporous carbon which is sulfonated has a high amount of these acid groups which can be accessed. Transportation of substrates and other products are helped by mesoporous. The stability of these materials is taken care of by its hydrophobic nature because conversion usually occurs in the water medium. There may be chances of helping in adsorption of reactants by few weak acid groups or exertion of catalytic function which is synergistic in nature. The maintenance of mesoporous in harsh conditions becomes much easier when silica phase co-exists. The hybrid surface provides specific polarity in addition to the potentiality for the purpose of multi-functionalization. Producing this mesoporous carbon from the initial stage is discussed in detail along with the application it provides in esterification and furan-derivative conversion. The problems related to stability have been taken care of and finally, an outlook into the future prospects of these materials is presented [13].

By using camphor leaves that are dead, graphene can be obtained which does not affect the environment and can be obtained by a simple and economical method called one-step pyrolysis. Atmospheric nitrogen which is in under flowing condition along with dead leaves of camphor at a high temperature of about 1200^0C and that too at the rate of 10^0C/min and then once the heating is done its temperature was brought down to the temperature of a room by making sure that there are no external hindrances. Some of the layers of graphene and end pyrolytic components were separated with the help of centrifugation and also by bringing in contact with P-Tyrosine using pi-pi interaction. The gradual degradation of camphor leaves especially at a temperature less than 800^0C is observed by using TGA. Different tests like BET, TEM, etc among others were conducted for understanding the properties of graphene. The result shows about seven layers of graphene on an average [14].

It is essential to understand the spectral characteristics of electronic devices like sensors and also of opting electrical components in order to improve its usage in various fields. The sum-over-state method, as well as the tight-binding model, is employed. The absorption spectra which has linear polarizability along with twisted angle is first calculated. Initially, based on PPP, the carbon nanotubes model is established. The position of carbon nanotubes is changed to a particular angle by positioning carbon atoms. As bond length changes calculation is done for the torsional structure model of nanotubes. The analysis is done by tabulating the absorption spectrum obtained for various twisted angles, the method of analysis is SOS. In the carbon nanotubes, significant changes are observed in spectra due to twisted angles. There is variation in a number of peaks of absorption as well as the position due to the torsional angle, but its effect on spectra is mostly at low frequency. The effect on absorption spectra is even more important when exact changes are made to twisted angles. This forms a blueprint on how to go about with the fabrication of carbon nanotubes which is used particularly in photonic devices. Photonic devices are used in various fields off late [15].

There are various problems in the use of catalysts in pyrolytic systems which have to be very fast. So taking this into account, coatings of a catalyst ZSM-5, which is a novel composite, on foam supports of SiC was developed and for improving the pyrolytic vapors tests were carried out. The arrangement of catalyst placement was varied each time for the purpose of comparison and it showed that the composite catalyst helped in enhancing the quality bio-oil significantly without much reduction in yields. By compensating on yield, the quality of bio-oil could be enhanced by adding more catalyst to biomass ratio. Till the biomass ratio reached of 1/10, the composite catalyst was able to maintain its activation energy. One more important aspect of this catalyst is that it could be reproduced and used multiple times, also its properties and catalytic activity could be kept intact up to seven cycles [2].

Rice husk ashes are used in order to extract amorphous silica with two different varieties named as cateto and agulhinha. A thorough study was done on various parameters like optimum time, reaction, etc, for producing SiO2 which has good sustainability and should be clean when obtained from the rice husk ashes. Analysis of structures was done with the help of nitrogen adsorbing-desorbing isotherms. There was an increase of 11 and 12% in the amount of silica for the two varieties respectively. The rice husks under treatment in RHAs, silica was mostly or amorphous or powder form and the temperature revealed the same.2h at 7000C was the ideal or perfect condition for calculating hysteresis of type H1 along with IV type of isotherm was observed. It also showed an excellent specific surface area. In the case of cateto, the percentage of extraction yield was found to be between 80 and 99 and for agulhinha, it varied from more than 83 and less than 97 percent [16].

Carbon materials, which are activated by nanostructures of biomass which are porous in nature, have garnered importance in areas that deal with energy storage. This is because not much is to be spent on these materials and they have good conducting properties and are also environment-friendly. By activation and carbonization, hierarchical carbon materials (porous in nature) are doped with nitrogen by using celery which acts as a biomass carbon precursor. Even polymerization is employed. Polyaniline composite material is synthesized by using the prepared activated carbon which is obtained from celery. This process is done by adopting wet-chemistry. Supercapacitors use the porous carbon material as the active electrode. A specific capacitance 402 F g-1 is shown by this material. Capacitance retention of about 97% is possible even after 1000 cycles. The excellent performance of the carbon material boosts its usage in supercapacitors [17].

3. RESULTS AND DISCUSSION

In this work we have discussed the different types of biomass converted materials like pyrolysis of biomass, biomass-waste, gasification of harmful gases, carbon-based supercapacitor, silica, macadamia shell waste, lithium-sulphur battery, and rice husk, etc. from the different industries and try to convert that material in to a useful materials by using different types of conversion approaches. There are different types of conversion techniques are available for biomass conversion like CVD, hydrothermal process, thermolysis, pyrolysis, combustion, and chemical treatments. Nowadays, waste material from different sources is the biggest challenge for the country and worldwide. These waste materials responsible for many dangerous diseases that cannot be cured easily. So to reduce the harmful effects from the waste material has to reduce, reuse and recycle the waste material [13,14].

4. CONCLUSIONS

In this work, we have discussed the different types of biomass converted materials and recycling approaches. Accordingly, we need to explore new recyclable materials from different types of industrial waste. Still, we need to explore a better way to recycle the different grades of industrial and domestic plastic waste because plastic waste is one of the dangerous waste for human beings and the atmosphere. Thereafter, rubber waste from different industries is also a major area to explore a better way of recycling. Recently, electronics technology produced different types of gadgets and human-friendly equipment that makes human life easy and simple. Apart from that, these equipment industries produced a huge amount of Electronics waste that is very harmful to human beings and atmospheric conditions. So, we need to explore the new and better way to recycle this kind of new and advanced kind of E-waste material.

Key finding and future scope

The key finding of this work to explore and identify the new biomass material that can be easily recycled into useful material. Because existing materials are responsible for the various types of pollutions and also harmful to human life and atmosphere. To save the environment and reduce the harmful effects of non-bio converted material we need to increase the use of biomass convertible materials in the industries and daily uses.

Acknowledgment

The manuscript is prepared by taking assistance from Accendere Knowledge Management Services Pvt. Ltd. We are thankful to them. We also express our gratitude to our teachers and mentor for guiding us throughout the work.

REFERENCES

[1] Nivea Raghavan, Sakthivel Thangavel G V 2017 A short review on preparation of graphene from waste and bioprecursors *Applied Energy* **7** 246–54

[2] Zhou N, Liu S, Zhang Y, Fan L, Cheng Y, Wang Y, Liu Y, Chen P and Ruan R 2018 Silicon carbide foam supported ZSM-5 composite catalyst for microwave-assisted pyrolysis of biomass *Bioresource technology* **267** 257–64

[3] Zhang B, Piao G, Zhang J, Bu C, Xie H, Wu B and Kobayashi N 2018 Synthesis of carbon nanotubes from conventional biomass-based gasification gas *Fuel processing technology* **180** 105–13

[4] Lotfy V F, Fathy N A and Basta A H 2018 Novel approach for synthesizing different shapes of carbon nanotubes from rice straw residue *Journal of Environmental Chemical Engineering* **6** 6263–74

[5] Wang C, Xiong Y, Wang H and Sun Q 2018 All-round utilization of biomass derived all-solid-state asymmetric carbon-based supercapacitor *Journal of colloid and interface science* **528** 349–59

[6] Andas J and Adam F 2016 One-pot synthesis of nanoscale silver supported biomass-derived silica *Materials Today: Proceedings* **3** 1345–50

[7] Noushad M, Ab Rahman I, Husein A and Mohamad D 2016 Nanohybrid dental composite using silica from biomass waste *Powder technology* **299** 19–25

[8] Zhang W, Xu J, Hou D, Yin J, Liu D, He Y and Lin H 2018 Hierarchical porous carbon prepared from biomass through a facile method for supercapacitor applications *Journal of colloid and interface science* **530** 338–44

[9] Maroufi S, Mayyas M and Sahajwalla V 2017 Waste materials conversion into mesoporous silicon carbide nanocermics: Nanofibre/particle mixture *Journal of Cleaner Production* **157** 213–21

[10] Imtiaz S, Zhang J, Zafar Z A, Ji S, Huang T, Anderson J A, Zhang Z and Huang Y 2016 Biomass-derived nanostructured porous carbons for lithium-sulfur batteries *Science China Materials* **59** 389–407

[11] Yang K, Gao Q, Tan Y, Tian W, Qian W, Zhu L and Yang C 2016 Biomass-Derived Porous Carbon with Micropores and Small Mesopores for High-Performance Lithium–Sulfur Batteries *Chemistry–A European Journal* **22** 3239–44

[12] Chen M, Jiang S, Huang C, Wang X, Cai S, Xiang K, Zhang Y and Xue J 2017 Honeycomb-like Nitrogen and Sulfur Dual-Doped Hierarchical Porous Biomass-Derived Carbon for Lithium–Sulfur Batteries *ChemSusChem* **10** 1803–12

[13] Zhong R and Sels B F 2018 Sulfonated mesoporous carbon and silica-carbon nanocomposites for biomass conversion *Applied Catalysis B: Environmental* **236** 518–45

[14] Shams S S, Zhang L S, Hu R, Zhang R and Zhu J 2015 Synthesis of graphene from biomass: a green chemistry approach *Materials Letters* **161** 476–9

[15] Wang W, Tian L, Song W, Lv L and Tu Z 2018 Twisted angle effects in the absorption spectra of carbon nanotube *Optik* **171** 845–9

[16] Costa J A S and Paranhos C M 2018 Systematic evaluation of amorphous silica production from rice husk ashes *Journal of Cleaner Production* **192** 688–97

[17] Du W, Wang X, Sun X, Zhan J, Zhang H and Zhao X 2018 Nitrogen-doped hierarchical porous carbon using biomass-derived activated carbon/carbonized polyaniline composites for supercapacitor electrodes *Journal of Electroanalytical Chemistry* **827** 213–20

IOP Conf. Series: Materials Science and Engineering **748** (2020) 012004 doi:10.1088/1757-899X/748/1/012004

Two Finger Gesture Imitating Animatronic Hand

Y Verma and V Rastogi

Mechanical, Production and Industrial and Automobile Engineering Department, Delhi Technological University, Shahbad Daulatpur, Bawana Road, Delhi-110042, India

E-mail: yashasviverma22@gmail.com

Abstract. This paper poses the discussion regarding animatronics that has been aided with modern trends in technological innovations. The ease of availability and procurement of present day items incorporate economical factors into research. To accomplish the aforementioned, a simple two fingered hand model has been developed that can be remotely controlled using a glove. The glove can be manipulated via a hand. It has been electronically modulated using Arduino Uno and servo motors. The discussion has been furthered to the amalgamation of future developments that can pave the way for commercialization of such devices. Hardships and challenges faced during building of the model and experimentation of the finalized model are detailed.

1. Introduction

The word "Animatronics" entomologically merges "Animate" and "Electronics". Thus, as suggested it is the technology behind animating objects to imitate humanoid gestures and actions. It can be commonly observed in motion pictures beginning with Disney's Mary Popins. The concept dates back to 3rd Century BC where it finds mention in the script of Liezi (Chinese Scripture). Its first modern day usage was for novelty clocks that were adorned with automated birds in 12th century and with the rapid advances in robotics, anatomy and mechantronics has expanded its scope to a multitude of areas such as prosthetics, industrial robotics and anthropomorphism. Robotics is advancing from remote location usage to being integrated in technology that we interact with on a daily basis. A way to humanize these robot-human interactions is to make robots imitate human gestures. Animatronics has been a step in this direction. There are many prevailing industry situations where human contact can prove to be dangerous or even lethal such as handling of hazardous chemicals, contacting Super-hot or Super-cold fluids or even entering narrow enclosures. A robotic industrial hand can help to improvise this situation to perform human-like actions of inspection or handling of injurious materials. Mishra et. al. [1] discuss the exploitation of Bluetooth feature in smartphones to transmit gesture information for remote control of industrial robots. They cancel the processing time lag enhancing the advantages of distant controllers. For specially-abled persons, animatronics is the basic step towards the development of prosthetic limbs. A hand that is capable of delivering humanoid movement to an artificial limb using electrical sensor can be a boon to many. This is achieved by anthropomorphizing the advanced animatronic limbs and extending the expanse of sensors. Kumar and Kumari [2] present a review of the control features of animatronically powered prosthetic hands. They elaborate upon the current complications in prosthesis and develop on the various technologies in utilization at present. The utilization of animatronics is expansive, as explained by Oo et. al [3] in their innovative approach to reinforce the aid provided to the deaf and dumb by incorporating a gesture governed robotic arm. A major consideration during such a development is also keeping the financial aspect in check as technology has to be affordable to even small scale industrial houses and economically weaker section of physically challenged people.

There has been present day sophisticated research on such a concept. Pradeep and Paul [4] illustrate the gesture based control of a robotic arm. They achieve this by using accelerometer to sense the gestures that are then processed and transmitted to the robotic arm using an RF transceiver module. They generate the user friendly and accurate model to aid human intervention in industrial processes. Crisman et. al [5] present their innovative technique regarding the design of a robot hand. For this they incorporate the functioning of the tendon structure of a human hand into their mechanical model. They optimize the work freedom available to fingers by developing their model with pulleys and algorithm for control using switches. Ramaswamy and Deborah [6] lay out a comprehensive research on the contemporary robotic hands that are commercially accessible. They include in the descriptive survey and the advancements along with the multitude of problems encountered. Jog et. al [7] worked on a wireless animatronic had that coalesces the modern technology of wireless communication using a XBEE module and the programming ease of Arduino Uno as a microcontroller. The result was a robotic hand with possible distant interaction but with scope of future effort to make movements more precise. Similarly, Gupta et. al [8] talk of the scope of extending the wireless robotic hand technology to a prototype arm. They dwell on the use of XBEE as an interface for wireless communication and involve the sophisticated signal processing of Arduino Uno. Utture et. Al [9] dwell on the concept of animatronics and its timeline. They discuss the amalgamation of robotics and animatronics as well as the prescribed methods for using these rapidly developing innovations. Bilgin et. at [10] has written about the purposeful usage of flexible sensors on finger tips of controller gloves to increase the precision of the animatronic movement. They discuss the use of RF 435 MHz module for wireless functions and embedding of the system with a digital interface for the proper implementation of the same. Chowdhury [11] has worked up the basic model for an animatronically controlled hand. She employs Arduino Uno and servo motors for developing communication between the controller and receiver. She goes on to discuss the economic limitations due to usage of flex sensors, the unnecessary employment of excessive sensors and the loud servo motor sounds.

In this paper, a basic model of an animatronic hand is developed. The basic outline of the logic employed is depicted in figure 1. The controlling glove has been provided with flex sensors to transmit signals corresponding to motion in the fingers of the person who dorns the glove. Syed et. al. [12] have explored in detail the mapping of a hand motion to subsequent robot model. The wireless communication basis has been concluded that enhances the employment of the mentioned robots. The signal communicated via the sensors is converted by an Arduino Uno and sent to the servo motors which control the sample tendon structure mounted on the robotic hand. This paper serves to discuss the socio-economic impact factor with the improvement in present day technology. The challenges faced with the use of flex sensors have been addressed and the ease of making the basic animatronic research available worldwide has been developed.

Figure 1. Flow chart of the methodology employed

2. Methodology, Design and Experimental Setup

2.1. Design Specifications
The gesture imitating hand was divided into two sections, one being the controller part or the part consisting of the glove to be worn by the controller and the Arduino Uno microcontroller to convert the flex sensor signals to signals compatible with servo motor movement, the other part being the mimicking part or the model hand that mimics the movements as gestured by the controller and the servo motors that cause the motion in the model fingers. The design of entire system began from the controller part. An everyday usage glove was procured. The glove should be a tight enough fit on the hand of the controller so that the flex sensors can precisely detect any movement of the fingers. For the basic model, two fingers ie. the thumb and the ringer finger were taken into consideration. The flex sensors were attached to the abovementioned fingers. These flex sensors were electrically connected to a bread board and then an Arduino Uno. The microcontroller was fed information from a laptop and was connected to two servo motors through another bread board. The servos were fed two strings which were attached to the model hand. In this case, the two-fingered hand was modeled using cardboard. The fingers were capable of bending in three stages like an anthropoid and the strings were fastened across the finger length.

2.1.1. Flex Sensors. Flex sensors are devices that convert the angle at which they are bent to a change in resistance, with the increase in the bending angle, the resistance of the device increases. It is usually connected across a voltage divider circuit that converts this increase in resistance to a proportionally linear increase in voltage. This voltage signal can be easily manipulated for further usage. For this project, two 2.2 inch flex sensors were procured. The two terminals apart from being present in a voltage divider circuit were connected to the ground terminal and the terminal of a power source present on the Arduino board.

2.1.2. Resistors. Two resistors were employed to successfully convert the resistance signal from the flex sensor to a voltage signal for downstream ease of manipulation. For this, two resistors were acquired with the Red Red Brown Gold colour code that translates to 220 Ohm.

2.1.3. Arduino Uno. This is a microcontroller board that employs 6 input analog pins and 14 digital I/O pins and is easily programmable using the Arduino IDE which is an application that is cross platform and can be worked upon with the basic knowledge of C or C++ languages with just a few changes in the structuring. The board is USB base. It was used to convert the analog voltage signal input to digital signal output compatible with the servo motors.

2.1.4. Bread Boards. They are convenient and cost effective devices to produce circuits using jumper wires without having to solder. This effectively reduces the connection time and also professes safety. The financial aspect is another advantage as they are inexpensive and widely available. They come in a variety of sizes. For this model, a full (6.5cm X 17cm) and half size (5.5cm X 8.5cm) breadboard were utilized.

2.1.5. Servo Motors. Since the model employs two fingers to configure a basic circuit. Two SG90 servo motors were used. A servo motor has 3 terminal wires which are brown, red and yellow in colour. The brown wire is connected to ground terminal and yellow to the respective output terminals of the digital section of the Arduino Board. The red wire in connected to an external battery source, which in this case was taken as a 9V battery. Two strings that acted as tendons were mounted on the Servos. The servos convert the signal to the desired angle of the blade. This exerts a pull on the strings which subsequently move the fingers of the model hand.

2.2. Methodology
There are two major sectors that required focus, one being the electrical circuit and the other being the program fed into the Arduino.

2.2.1. Electric Circuit. While developing the circuit, the main aim was to manipulate the signals received from the flex sensors into angles of movement of the Servo motors that would invariable pull on the tendonic strings to imitate the movement in the model hand. Starting from the controller section, the basis of the electrical circuit was to join a terminal of both the flex sensors to the ground terminal of the Arduino via the negative terminal of the breadboard. This has been depicted by the black coloured circuit lines emerging from the Flex Sensors in figure 2. The other terminal of the flex sensors were supposed to be connected to a fixed voltage source via the positive strip of the breadboard as shown as the red coloured lines originating from the flex sensors in figure 2. On the same connection, resistance drop had to be added to develop a voltage divider circuit which was then connected to the 2 analog input terminals of the Arduino board. This can be evidently followed along the brown coloured circuit lines concluding in the T1 and T2 input terminals of Arduino Board in figure 2. Moving to the controlled part of the prototype, the digital I/O terminals of the Arduino board were then connected to one terminal of the Servo. These brown connection lines (figure 2) serve the purpose of transmitting the manipulated signals to the Servo motors to move the hand extension. The other two terminals were connected to negative terminal of the battery as well as the ground terminal of the Arduino board and positive terminal of the battery. These can be respectively traced as the black and red coloured circuit lines traversing towards the battery respectively in figure 2. The servos were connected using a tendon functionality thread. A constant power source was required to locomote the modeled two finger hand. This was provided through the Arduino Uno being connected to a power source, abetting the red power lines in figure 2.

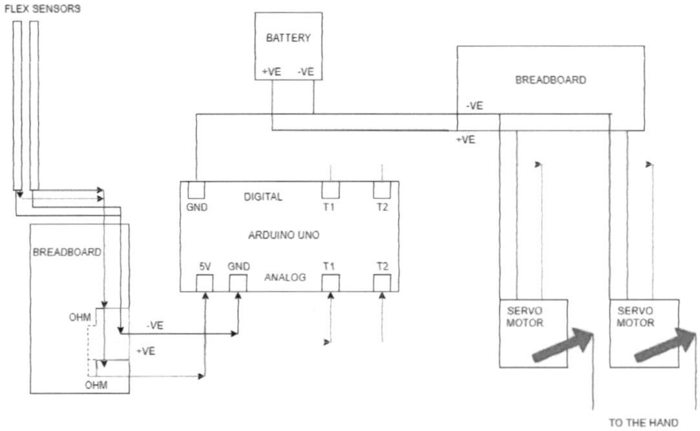

Figure 2. *Electronic circuit used for development of design*

2.2.2. Arduino Program. Figure 3 shows the code which was fed into the Arduino IDE and uploaded into the microcontroller. The code helped change the voltage analog signal received into a change in angle of the attachment rod mounted onto the servo. As the angle of the rod changed it exerted a pull onto the thread which further caused bending or flexing of the fingers of the model arm. Badamadsi [13] elaborates the functionality principles of Arduino Uno as a microcontroller by expanding on not only the hardware but the software as well. The method of compiling codes for further project development have been thoroughly examined as well.

```
#include <Servo.h>
Servo servo_motor1;
Servo servo_motor2;
int flex1=0;
int flex2=1;
void setup()
{    servo_motor1.attach(9);
  servo_motor2.attach(10);
}
void loop()
{
    int flexpos1, flexpos2, servopos1, servopos2;
    flexpos1 = analogRead(flex1);
    Serial.println(flexpos1);
    servopos1 = map(flexpos1, 1020, 1023, 0, 180);
    servopos1 = constrain(servopos1, 0, 180);
    flexpos1 = analogRead(flex1);
    servopos1 = map(flexpos1, 1020, 1023, 0, 180);
    servopos1 = constrain(servopos1, 0, 180);
    flexpos2 = analogRead(flex2);
Serial.println(flexpos2);

 servopos2=map(flexpos2, 1020, 1023, 0, 180);
    servopos2 = constrain(servopos1, 0, 180);
     servo_motor1.write(servopos1);
    servo_motor2.write(servopos2);
}
```

Figure 3. The Arduino code fed to Arduino IDE

3. Experiments and Results

The controller having access to the glove flexed his thumb and ring finger, first one at a time and then together. The model hand was seen performing the same activities. On checking the angle of the finger flex in case of the controller and the model hand, the difference between the angles was more than 15%. The error in precision of movement could be attributed to the fingers made out of proportion. If the length of the fingers of the animatronic hand are made proportional to real life fingers and the length of each section of the finger too made more anthropomorphic, this could reduce the error down to the scientifically acceptable level. These levels are set by each industry or phase of utility that demand the employment of such technology. A possible solution to such an issue can be the amalgamation of cost effectiveness and anthropomorphism. Bae et. al. [14] embellish such a concept to ensue upon a four fingered hand equipped with actuation. They dwell on the viability of controlling objects and the versatility of the applied science. At present, the resistors had to be replaced with ones having ohmic capacity of 22k Ohm. The reason for replacement being that the low level resistance caused early drainage of the 9V battery. If a long lasting source is available at the same level then 220 ohm resistors can be used. But, for optimized results replacement is advised. Chowdhary [11] expatiated over the probable use of Neoprene bend sensors for cost effective applications. With rapid globalization accompanied with torrid advances in technology, flex sensors are now easily obtainable at much lower costs. They can also be developed at home for more cost saving. Their easily availability denoted ease of procurement and hence time saving. Flex sensors are more sensitive than bend sensors as the later work on sensing pressure. As we are aiming towards a more precise model for industrial applications and enhancing gesture sensitivity, flex sensors should be the way to go. They no longer pose a cost barrier. Using a single flex sensor for more than one finger may not be the proper path because each finger has its own characteristic bend and flex. This bend has to be mimed accurately for furthering the project applications. The Servos were mounted on acoustic foam. This foam is capable of absorbing noise created by the Servos. This effectively solves the issue of lowering noise levels on the industrial floor when the animatronic hand advances to such employment. The project was developed at a low cost and hence suggested that with newer technology, the impact of such a device can be re-evaluated for the better.

4. Discussion

Technology is the science that can be made practically available to its beneficiaries. The gesture imitating basic animatronic hand was developed to check its socio economic standing with the advancements in present day technology. Be it the ease of component availability or the cost of their procurement, the hand was developed cost-effectively. This feature can be exploited in later stages where the hand can be anthropomorphized and made more precise. The time to develop such a project is also optimized with the worldwide obtain-ability of instruments. Any noise in the system is also efficiently reduced by deploying acoustic sponges which are again not a very substantial investment. The breadboards are definitely a superior substitution to soldering in terms of safety and material cost. These implications of the technical project talk bounds about its future evolution into more sophisticated models. The problem of precision in movement persist which remain to be addressed. In conclusion, the desired model and its predicted financial importance were achieved. The subsequent research can deal with the rectification of precision errors bearing in mind the reduction of possible inflation in costs that comes when using futuristic technology. The animatronic hand holds the potential for betterment of human-robot interaction.

5. Conclusion

The paper has developed a basic model for the gesture enabling remote control in an animatronic two-fingered hand. The problems faced during the concentrated effort on socio-economic steer as furnished by modern technology were elaborated upon. Certain possible ameliorations were suggested bearing in mind the basic aim of the research. Furtherance of the model may include the introduction of mechanisms for flat object picking and dynamics of humanoid object transfer.

6. References

[1] Mishra S, Tamboli S M, Yamgar M M and Prakash R 2017 Smart device based gesture controller for industrial applications *International Conference on Nextgen Electronic Technologies: Silicon to Software* 463-7

[2] Kumar V V N and Kumari P N 2015 A review on embedded control of animatronic prosthetic hand *International Journal for Technological Research in Engineering* **2** 614-7

[3] Oo H M, Tun K T and Thant M L 2019 Deaf sign language using automatic hand gesture robot based on microcontroller system *International Journal of Trend in Scientific Research and Development* **3** 2132-6

[4] Pradeep J and Paul P V 2016 Design and implementation of gesture controlled robotic arm for industrial applications *International Journal of Advanced Scientific Research and Development* **3** 202-9

[5] Crisman J D, Kanojia C and Zeid I 1996 Graspar: a flexible, easily controllable robotic hand *IEEE Robotics and Automation Magazine* **3** 32-8
[6] Ramaswamy C V V and Deborah S A 2015 A survey of robotic hand-arm systems *International Journal of Computer Applications* **109** 26-31
[7] Jog S, Dwivedi A, Ashtankar S and Gautam G 2016 Animatronic hand using wireless module *International Research Journal of Engineering and Technology* **3** 1529-31
[8] Gupta A, Jain R, Bang D and Kori J 2016 Wireless animatronic arm *International Conference and Workshop on Electronics and Telecommunication Engineering* 105-11
[9] Utture O, Mane N S, Varute S S and Waingade A B 2017 Multidisciplinary technology of animatronics : a review *Asian Review of Mechanical Engineering* **6** 4-7
[10] Bilgin S, User Y and Mercan M 2016 Robotic hand controlling based on flexible sensor *International Journal of Engineering and Applied Science* **8** 49-58
[11] Chowdhury S 2013 Animatronic hand controller 10.13140/RG.2.2.25148.74889
[12] Syed A, Agasbal Z T H, Melligeri T and Gudur B 2012 Flex senor based robotic arm controller using micro controller *Journal of Software Engineering and Applications* **5** 364-6
[13] Badamasi Y A 2014 The working principle of an arduino *11th International Conference on Electronics, Computer and Computation* 1-4
[14] Bae J, Park S, Park J, Baeg M, Kim D and Oh S 2012 Development of a low cost anthropomorphic robot hand with high capability *International Conference on Intelligent Robots and Systems* 4776-82

IOP Conf. Series: Materials Science and Engineering **748** (2020) 012005 doi:10.1088/1757-899X/748/1/012005

CREATING EMPLOYMENT IN SOME INDIAN INDUSTRIES BY REDUCING THE WORKING SHIFT TIMING

Mr. Anil Kumar[1], Ms. Renu[2] and Dr. Sanjeev Bansal[3]

[1]Assistant Professor (MAE Deptt.), GBPEC, New Delhi,

[2]Phd Scholar, (MBA Deptt.), (SLIET, Sangrur)

[3] Professor, (MBA Deptt.), SLIET, Sangrur

E-mail: anilapmae@gmail.com

ABSTRACT-

A new programme, called 'Skill India', was launched in March 2015. The main goal of "Skill India" is to create opportunities and scope for the creation of the new talents of the Indian youth and also to identify new sectors for skill development. This programme targets of providing training and skill development to 500 million youth of our country by 2020, covering entire India. Now as the time is passing and we are getting new skilled labour we have a new problem of providing job to this skilled youth. The Start Up programme is set to meet such requirements but it is not enough to fulfil the job requirements of the skilled human power. Also, this is not very easy to set a new industry in a short span of time. So in this paper, I have tried to provide a new solution for increasing job opportunities in the existing industry. I tried to increase one more shift i.e. from existing 3 shifts to 4 shifts per day by reducing the working hours from 8 to 6. I got very exciting results which can be recommended for other industries also.

1. INTRODUCTION-

In India a new programme, 'Skill India', is being launched in 2015 with an objective as a multi-skill programme [1]. The objective of this programme is to create new jobs and to identify new sectors for skill development [2]. The different objectives of the **'Skill India' programme are as under-**

- To train the young guns of India so that they can get employment.
- To provide support and guidance for all jobs.
- To generate new areas as scope of occupation
- To arrange training programmes of international standard.

'Skill India', Programme of Indian Government observed the urgent need of time that not a single person must have to be unemployed. So the government is looking to boost this project with full support. With this advance approach the government can definitely move ahead towards its targeted results.

1.1 PROBLEM FORMULATION

But now the question arrives that ***"How we can be able to utilize the strength of such a great number of skilled people"?*** This can be done by number of ways-

1. By opening new industries or organizations (Start Up) - This option seems to be very good but for that purpose number of barriers exists-

- Land ,Huge Capital and Market for Finished Products
- Specialized Training and Need of highly skilled people

Also the results will be achieved after a long span of time. We also have a large history of Bankruptcy of Firms. So this could not be an immediate solution for providing the instant solution of unemployment problem.

2. By sending skilled people to other countries where demand of such people is more-

This is also not very easy as the family conditions and migration rules are not so easy to grasp easily.

3. By adjusting them into the exiting industries or organizations-

This option seems to be very good as this can help in increasing the employment and production at immediate time.

So we have to find some suitable option for adjusting these skilled people in the existing industries. This can be done by –

1. Expansion of existing plants
2. Increasing the Working shifts

So the Problem is how to create jobs for these people?

For this problem we can propose one solution---***"REDUCING THE WORKING SHIFT FROM 8 HOURS TO 6 HOURS".***

1.2 OBJECTIVES OF THE STUDY

- ➢ To create job opportunities for the increasing number of skilled persons through Skill India Programme.
- ➢ Finding feasibility of increasing the number of shifts in the existing industries.

1.3 LITERATURE REVIEW

A six-hour workday will increase the personal time.

The deputy mayor of Gothenburg tested 30-hour week for staff at a nursing home with 68 employees. The results showed that the employees felt healthier, that there was less absenteeism and patients were better cared for, the Dutch daily newspaper Trouw reported Wednesday. This change, however, did not add to various types of extra costs. The nursing home had to employ an additional 17 extra staff. Daniel Bernmar, who is managing the municipality's elderly care, told Bloomberg, "It's far too expensive to carry out a general shortening of working hours within a reasonable time frame."

Marion Collewet at el. [3] in their research studied the effects of working hours per day on productivity in the Netherlands .He studied the sample of 332 people and collected the data regarding –

a. information on the number of daily working hours, and

b. Workers individual performance.

He observed the decreasing returns to hours for workers with shorter tenure.

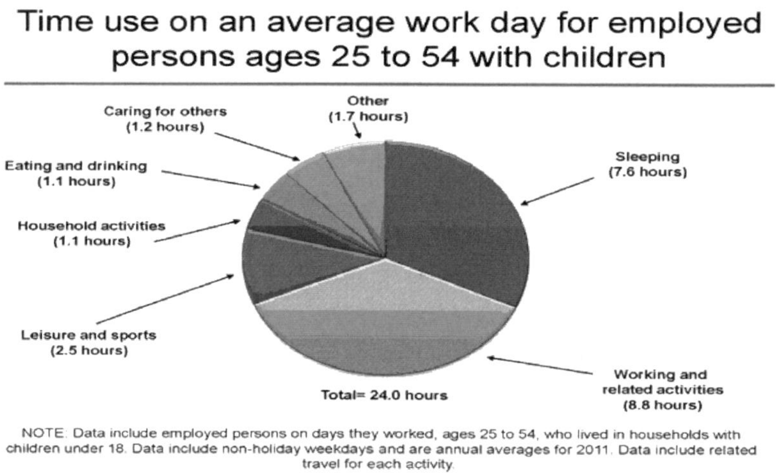

Figure 1- Use of Time by employee in 24 hours

Filimundus introduced a six-hour workday last year by reducing the number of staff meetings and asking employees not to spend office time on personal tasks such as social media. Feldt said this led to higher energy levels among employees, better teamwork and fewer workplace conflicts, as staffers were less fatigued and grumpy about being at work.

According to Roland (Paulsen University of Lund), ***"Productivity has doubled since the 1970s, so technically we even have the potential for a four-hour working day".***

After all, Greeks work some of the longest hours of any EU country, but their productivity, or output per hour, is particularly low – and that hasn't also worked out terribly well for them.

At Toyota's service centre on the country's west coast of Sweden, shifts were reduced for mechanics and with this implementation the company reported a swift increase in profit, and this profit rise, motivated the company to permanently adopt the idea of reduced working hours.

Anttila T. et al. [5] found in their research that 6-hour shifts, in comparison to other forms of working time reductions, had the strongest impact on the decrease in conflict arising from work and affecting family.

Steve Langerud, the workplace consultant based in Grinnell, Iowa [8] also agreed to the points that shorter working hour leads to "Quality of life, stress reduction, and engagement in our work". Large number of people working from home may help prevent burnout, but it also shows that workers are always connected to their jobs. According to John Pencavel an economist at Stanford University (A 2014 study), longer working hours always do not necessarily yield more output.

Smith L. et al. [10] in their study found that, 12 hour shifts leads to less stress levels, better physical well-being, improved durations and improvements in family relations. The disadvantages included the fatigue and safety.

2. METHODOLOGY

2.1. Selection of proper place for implementation of reduced working hour plan -

This study has been performed to find the effects of reducing the working shift hours from 8 to 6 hour. For this purpose selection of an organization is required where we can implement our project rules.

So we selected the - **Kalaam Powder Coating (Delhi).** The industry has been chosen as it is situated in the Delhi NCR region. The industry is involved in electroplating of the different parts. This is a company which is working on 8 hours shift with fixed production rates in 8 hours. So we have one limitation with this research that we have collected data for fixed production rate in 8 hours.

2.2 Defining the current performance and Data collection -

The first task in this section was to define the current performance of the industry Reports from the production and operation department would be gathered. The second task was to identify a process map of the production process. The purpose of a process map was to understand the sequence of activities from the beginning through the end of the process.

Then the production related complete Data will be gathered. The following Production factors were considered-

1. **Cost of production** (Transportation, Maintenance, Power and water utilization cost).
2. **Employee's ability-** (Health Working level Family relations and Employment).

2.3 Data analysis

The method used for different calculations has two components -

- The interview with workers and management for change in workplace environment.
- Analytical approach to calculation of overall labour effectiveness.

3 EXPERIMENTAL WORK

3.1 Place of Experimental work -- We have tried a number of firms and after a lot of struggle the owner of the Kalaam Powder Coating allows us to make study for this project. So we chose Kalaam Powder Coating (Delhi). The industry has been chosen as it is situated in the Delhi NCR region.

3.2. Interview with Workers and Management –

Table 1- Questionnaire for Different level people of Organization

A. For Labourers--	
Questions	**Answers**
• Are you happy with this change?	Yes, I am so glad with this change.
• How other co-workers are reacting after this?	Now all workers just focusing on their individual goal of production & no unnecessarily talk happening among us which was usual after lunch period due to which we wasted our time.
• How has this change affected your work life balance, is it improved?	Yes, it's improved & now I am giving more time to my family.
• Do you feel more productive?	Yes, I am feeling more productive nowadays.
• Would you appreciate more changes like this?	Definitely, I will appreciate this type of changes which provide benefit like this.
• Would you still work for fewer hours if salary is reduced according to it?	Sorry, not at all.
• What is negative about this change?	From labour prospective, I don't think there is any negativity in this change.
• How upper level employees are behaving with you?	They all are happy from us because after this change we are not wasting our time in useless talks & activity.
• Is it stressful to do same amount of work in fewer hours?	No, I don't think so because whenever I thought that if I complete my task of a day within six hour then I got 2 hours for myself. So ultimately this thought makes me more enthusiastic towards my task.
• What is the best thing about this change according to you?	Getting two hours each day for myself within same salary.
b. For Managers--	
• Do you appreciate this change?	Yes, I appreciate this change.
• Is it easy or hard to keep up with production rate under less available hours?	For me, it is difficult to manage the no of workers & the production rate details due to increase of one shift in a day.
• How work environment has affected?	No of workers increases & the only working environment u can see everywhere.

• How productivity increased of an individual?	Productivity is same of an individual worker but saving of 2 hours is appreciable which is used to make one more shift timing.
• Which part of the management has positively affected?	Getting no time to waste due to work load.
• Which part of management has negatively affected?	Work load increases enormously for a management team.
• How workers are behaving after this change?	More energetic after the change.
• How work life balance of an individual has improved?	More works, more tiredness. Otherwise everything is fine.
• How did this change affect you?	Made me busier in office hours.
• What is the best thing about this change according to you?	Productivity has increased due to more shifts.
c. For Owners	
• How these changes affect the productivity of the company?	In working session it makes profit for me otherwise in off-session this methodology is not suitable.
• How work environment can be improved?	The determination towards the production goals of the company from all individual will improve the environment.
• Is there any change in the rate of production?	Yes, rate of production has increased but the individual rate production of worker is same as before.
• Is it okay to pay same amount of money for less working hours?	Yes, the worker is completing the same task in less hours. So that's their rights.
• How economy of the company has affected by this change?	In working session it has proved marvellous economy change but in off-session this method provides me loss due to salary of workers.
• What do you say is it positive or negative change?	It's not totally positive or negative change, it just depends on the demand of work. According to that we have to take decision of change.
• Would you appreciate any change like this in future?	Yes, I would appreciate this type of change until it makes profit for me.
• Does this change having affects as you expected?	According to my experience I was known what changes will occur by this methodology.
• How management level of the company has changed?	Work load has been increased on management department.

• What is best thing about this change according to you?	The consistency of workers which is maintained for totally 6 hours towards their work is outstanding.

3.3 Calculation of Overall Labour Effectiveness (OLE)

A. Measuring availability Calculation:

Availability = Time operators are working productively / Time scheduled

B. Measuring performance Calculation:

Performance = Actual output of the operators / the expected output (or labour standard)

C. Measuring quality Calculation:

Quality = Saleable parts / Total parts produced

Effective use of* Overall Labour Effectiveness *discloses the information that determines root-cause analysis, the points for corrective actions and the trends that can be used to diagnose more subtle problems.

Calculation: OLE = Availability x Performance x Quality

4. RESULT AND DISCUSSION-

Overall labour effectiveness for 8 vs. 6 hours at Kalaam Powder Coating (Delhi)-

Measuring availability

Table 2- Measuring availability

Shift Hours	Manpower	Unscheduled downtime (mints.)	Availability (%)
8	5	60	87.23
6	5	40	91.48

Measuring performance

Table 3- Measuring performance

Shift Hours	Manpower	Unscheduled downtime (mints)	cycle time (mint)	Total Part	Performance (%)
8	5	60	30	13	90
6	5	40	30	10.5	93.75

Measuring quality

IOP Conf. Series: Materials Science and Engineering **748** (2020) 012005 doi:10.1088/1757-899X/748/1/012005

Table 4- Measuring quality

Shift Hours	Manpower	Target Production	Actual Production	Quality (%)
8	5	20	18	96.55
6	5	16	15	98.98

Overall labour effectiveness for 8 hours

OLE = Availability x Performance x Quality

=87.23 % x 90 % x 96.55 % = 75.79 %

Overall labour effectiveness for 6 hours

OLE = Availability x Performance x Quality

=91.48 % x 93.75 % x 98.98 % = 84.88 %

Figure 2- Comparison between 8 hours shift and 6 hours shift

5. CONCLUSION

Due to availability of skilled manpower, the question arrives that ***"How we can be able to utilize the strength of such a great number of skilled people"?***

For the solution of the aforesaid problem we suggested the following solution- "**By adjusting them into**

IOP Conf. Series: Materials Science and Engineering **748** (2020) 012005 doi:10.1088/1757-899X/748/1/012005

the exiting industries or organizations". This can be done by –

1. Expansion of existing plants
2. Increasing the Working shifts.

Bur in such short span of time, the expansion of existing industries is not possible. So we opted for the second solution i.e. "***REDUCING THE WORKING SHIFT FROM 8 HOURS TO 6 HOURS".***

After implementing the solution in "Kalam Powder Coating" and thorough analysis of the data collected, we found the following effects on different factors of production-

A. Effects of reducing working hours for the individual's health--

1. **Reducing working hours leads to low level of stress, depression.**
2. **By reducing working hours, workers may have more free time for themselves**. They can use this time to recover themselves and to eliminate fatigue.
3. **Less Occupational injuries- After long w**orking the workers get tired & become less conscious about risk prevention. In shorter working hours, they will be more energetic and concentrated on their work.

B. Effects of reducing working hours on the Work Life Balance --

If we reduce the working hours, Individuals will be spending more time with their families and they can improve their mental health. Also Individuals will feel happy in such harmonious relationship and will be more relaxed.

C. Effects of reducing working hours on Productivity of the Industry --

The study shows that if we will optimize our resources we can produce the same quantity and quality products as that in 6 hours shift instead of 8 hours shift. So this can lead to an improvement in the production and ultimately more orders can be fulfilled within the same time. So this will proved to be an effective tool for improving the productivity of the industry.

D. Effects of reducing working hours on growth of Country –

This research can help the country also in the following aspects-

1. Improved productivity and hence improved GDP.
2. Improved employment opportunities for the skilled youth.
3. Also we can help the transport service of the country by altering the timing of duty shift which ultimately allow us to use the transport system with more flexibility.

So we can treat is as an important aspect/ solution to increase the employment and to consume the products of Skill India in the exiting industries up to certain quantity which ultimately lowers down the unemployment rate and we can achieve the real goal of Skill India initiative.

IOP Conf. Series: Materials Science and Engineering **748** (2020) 012005 doi:10.1088/1757-899X/748/1/012005

6. SCOPE

This study can be used to –

- *Bring more employment in peak time.*
- *More shifts can be devised.*
- *Can also reduce traffic problem by alternating the shift timings as per 4 shifts per day.*

REFERENCE

1. Skill Development in India, 2015, A FICCI Report.
2. Phukan R. S., December 25, 2014, "Skill India Programme - Objectives, Features, Advantages, Differences and more".
3. Marion Collewet et al., 2017, "Working Hours and Productivity", IZA – Institute of Labour Economics, Discussion paper series.
4. Bosch Gerhard et al. 2001, "Working-Time Reduction and Employment: Experiences in Europe and Economic Policy Recommendations", *Cambridge Journal of Economics*, V- 25, pp 209–243.
5. Anttila, T. Et al., 2005, "Experiments of reduced working hours in Finland: Impact on work–family interaction and the importance of the socio-cultural setting", Journal of Community, Work & Family, V- 8, Issue 2. pp 187-209.
6. Bruck C. S et al., 2002, "The relation between work–family conflict and job satisfaction: A finer-grained analysis", Journal of Vocational Behaviour, V-60, pp 336-353.
7. Fagan C. et al., 2012, "The influence of working time arrangements on work-life integration or balance: A review of the international evidence", Geneva, Switzerland: International Labour Office.
8. Margherita A., et al. 2009, "Reconciliation between work, private and family life in the European Union", Eurostat, Luxembourg: Office for Official Publications of the European Communities.
9. Voydanoff P., 2005, "Work demands and work-to-family and family-to-work conflict: Direct and indirect relationships", Journal of Family Issues, V-26, pp 707-726.
10. Smith L. et al., 1998, "Work shift duration: a review comparing eight hour and 12 hour shift systems", Occupation Environment Med; V- 55, pp 217–229.

Augmentation of VIVACE by means of Thermal Buoyancy

Mohammad Athar Khan[1], Saif Masood[2], Md. Abu Shahzer[3], Syed Fahad Anwer[4]

[1]Research Scholar Computational Aerodynamics Laboratory, Department of Mechanical Engineering, ZHCET, AMU, Aligarh, UP, India
[2]Research Scholar Computational Aerodynamics Laboratory, Department of Mechanical Engineering, ZHCET, AMU, Aligarh, UP, India
[3]Research Scholar Computational Aerodynamics Laboratory, Department of Mechanical Engineering, ZHCET, AMU, Aligarh, UP, India
[4]Associate Professor, Department of Mechanical Engineering, ZHCET, AMU, Aligarh, UP, India

*Email- atharkhanlko@gmail.com

Abstract. The present numerical simulation is carried out to evaluate the effect of Richardson Number on cylinder displacement and in turn the power extraction from the movement of cylinder. An in-house code based on arbitrary Lagrangian Euler scheme is used to tackle fluid structure interaction. The following fluid parameters is used in a present study Reynolds Number (Re) = 100, Prandtl Number (Pr) = 7.1 and Richardson number (Ri) = [-1, 0]. Structural parameters are as follows Reduced velocities (U_{red}) = 3 and 8, Reduced mass (M_{red}) = 2, damping coefficient (ξ') is set to be zero. With the aid of thermal buoyancy there is significant increase in power is obtained.

Keywords: VIVACE, vortex shedding, thermal buoyancy.

1. Introduction

Vortex-induced aquatic clean energy (VIVACE) is a new proven approach to extract energy from water currents and vortex-induced vibration. Vortex Induced vibration finds its application in many engineering fields such as heat exchanger, wind engineering tall structures and oil rigs. The life of the structure is reduced because of VIV, for suppressing the vortex induced vibration there are several passive and active methods that have been employed in the past. Suppression of VIV have been reported in research article of Rashidi et al. [1]. Vortex induced vibration of bluff bodies also helps to extract energy from fluid energy. Soti et al. [2] in their numerical study discussed the energy extraction from vortex induced vibration of circular cylinder. Lee et al. [3] proposed VIVACE, that uses the flow stream kinetic energy to generate electricity. In addition, the VIV of finned cylinder helps in thermal augmentation of cylinder, as described by Izadpanah et al. [4].The VIV of a circular cylinder with Reynolds number $\leq$ 200 has been studied by many researchers. Anagnostopoulos and Bearman [5] conducted an experiment of VIV of a circular cylinder which was mounted elastically in a water channel at Reynolds Number = 90-150. They were the first researchers to report that the maximum oscillation is observed at the beginning of the lock-in region in a low Reynolds number VIV of circular cylinder. Prasanth and Mittal [6] numerically investigated the characteristics of VIV for the flow on a circular cylinder. The cylinder was free to oscillate in the direction of the transverse flow for a range of Reynolds No. = (60-200) and with a relatively low mass ratio, m = 10. They reported that there are "initial" and "lower" branches at low Re, which correspond to vibrations of low ("initial") and high ("lower") amplitude. They investigated the effect of the blocking ratio on the hysteresis behavior of the response of cylinder In a follow-up study, Prasanth et al. [7] investigated the combined effect of blockage and mass ratio on hysteresis in the VIV of a circular cylinder for a range of Re =(60 -150). Details of physics involved in the Vortex Induced Vibration can be found in review articles from [8]–[10].

Vortex shedding of bluff body is supressed or agitated by means of thermal effect or thermal buoyancy [11,12]. We have considered a square cylinder for our numerical study to find out the energy extracted from enlarge vibration of the cylinder which has not been investigated so far to the best of author's knowledge. From the literature review we can conclude that for harnessing the energy, cylinder displacement plays the major role. So in the present study, effect of negative thermal buoyancy on square cylinder displacement and in turn the power extraction from it has been investigated for two structural parameter U_{red} = and U_{red} = 8. An in house code for capturing the fluid structure interaction (FSI) is used to simulate the motion of the cylinder submerged in the fluid medium. It was concluded from the numerical simulation, that there is substantial agitation found in the displacement of the cylinder when we change the Richardson Number from 0 to -1.

In the present study numerical simulation is conducted for the parameters given as: Richardson number (Ri) = [-1, 0], at a interval of 0.25, Prandtl number (Pr) = 7.1 ,Reynolds number (Re) = 100, reduced velocity (U_{red}) = 3 and 8 and reduced mass (m_{red}) = 2.

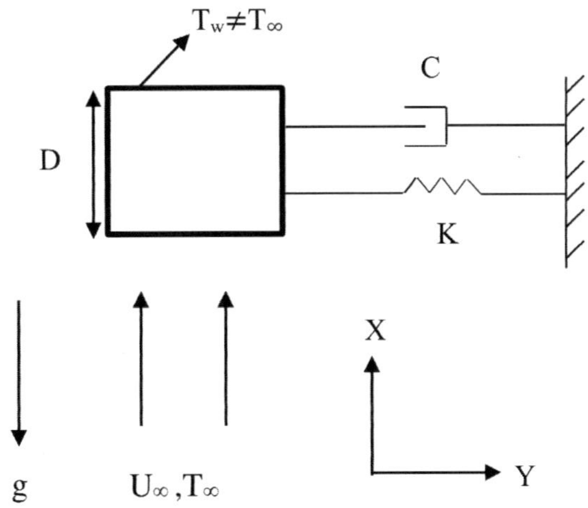

Fig. 1 Schematic diagram of an elastically mounted square cylinder

The problem definition is depicted in fig. 1. The cylinder is allowed to vibrate in transverse direction to the incoming flow. Flow is coming from negative y-direction i.e, opposite to gravity vector and cylinder is free to move in x-direction only. The cylinder is elastically mounted with a single spring and damper connected in x-direction. The spring is assumed to be linear. The structural parameters; mass, natural frequency and damping coefficient are converted in to dimensionless forms; reduced mass (m_{red}), reduced velocity (U_{red}) and damping coefficient ($\xi' = 0$). The Reynolds number of the flow depends on the cylinder diameter D.

In fig. 2 a non-orthogonal, O-type body fitted grid is shown that is wrapped around the cylinder. The outer domain of grid is set to be as 40 D. After the study of grid independence a grid of 305*375 is taken and time step of 0.0004 is taken for the present simulation.

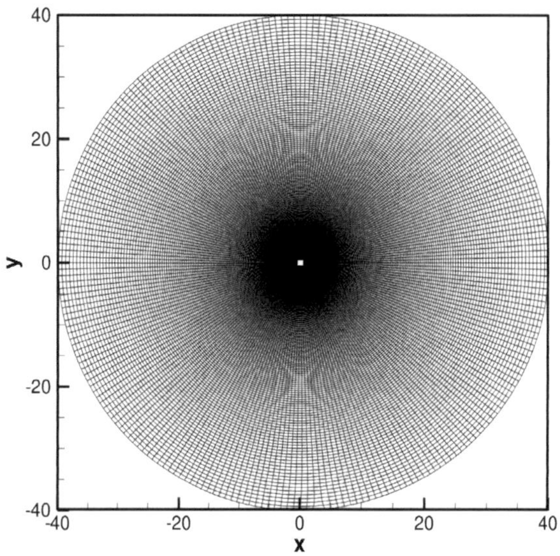

Fig. 2 Grid structure

Continuity:

$$\left(\xi_x\frac{\partial}{\partial\xi}+\eta_x\frac{\partial}{\partial\eta}\right)U+\left(\xi_y\frac{\partial}{\partial\xi}+\eta_y\frac{\partial}{\partial\eta}\right)V=0 \tag{1}$$

X - Momentum:

$$\left(\frac{\partial U}{\partial\tau}+(U^\xi-U_g^\xi)\frac{\partial U}{\partial\xi}+(U^\eta-U_g^\eta)\frac{\partial U}{\partial\eta}\right)=-\left(\xi_x\frac{\partial p}{\partial\xi}+\eta_x\frac{\partial p}{\partial\eta}\right)+\frac{1}{Re}\nabla^2U \tag{2}$$

Y - Momentum

$$\left(\frac{\partial V}{\partial\tau}+\left(U^\xi-U_g^\xi\right)\frac{\partial V}{\partial\xi}+\left(U^\eta-U_g^\eta\right)\frac{\partial V}{\partial\eta}\right)=-\left(\xi_y\frac{\partial p}{\partial\xi}+\eta_y\frac{\partial p}{\partial\eta}\right)+\frac{1}{Re}\nabla^2V+Ri\theta \tag{3}$$

Energy:

$$\frac{\partial\theta}{\partial\tau}+\left(U^\xi-U_g^\xi\right)\frac{\partial\theta}{\partial\xi}+\left(U^\xi-U_g^\xi\right)\frac{\partial\theta}{\partial\eta}=\frac{1}{Re.Pr}\nabla^2\theta \tag{4}$$

Vibration equation:

$$\frac{d^2x}{dt^2}+\frac{4\pi\xi'}{U_{red}}\frac{dx}{dt}+\frac{4\pi^2x}{U_{red}^2}=\frac{1}{2M_{red}}C_x \tag{5}$$

The non-dimensionalized form of rigid body motion equation is shown in equation (5). The reduced mass of the cylinder is defined as $M_{red}=\frac{m}{\rho_f D^2}$; where m is mass of the cylinder per unit length and ρ_f is the fluid density. C_x is the instantaneous transverse force coefficients of the body. U_{red} is defined as: $U_{red}=\frac{U_\infty}{f n D}$ where, f_n is a natural frequency of cylinder.

IOP Conf. Series: Materials Science and Engineering **748** (2020) 012006 doi:10.1088/1757-899X/748/1/012006

Validation:

The present numerical simulation has been carried out with help of in house code based on arbitrary Lagrangian-Euler scheme. The results are validated against published research papers and presented in Table 1 and found good agreement with the available research.

Table 1. Comparison of mean Nusselt number and mean drag coefficient of a square cylinder at Re=100, Pr = 0.71 and Ri = 0

Literature	$\overline{C_D}$	Nu
Sharma and Eswaran [11]	1.559	4.070
Sohanker et al. [13]	1.460	-
Present	1.477	4.031

Results and Discussion:

The variation of Ri and cylinder displacement (x) is listed in Table 2 (a) and Table 2 (b) for two different reduced velocities, U_{red} = 8 and 3. It is clearly observed from the Tables 2 (a) and (b) that as Ri changes from 0 to -1, the displacement increases. The displacement also depends up on reduced velocity, (U_{red}). For U_{red} = 8, the cylinder displacement is 0.08 D at Ri = 0 whereas at Ri = -1 there is a significant change in cylinder displacement and it becomes 1.266 D. Similar results are obtained for U_{red} = 3 where at Ri = 0, x = 0.02 D whereas at Ri = -1, x = 0.08 D. This clearly shows that structural parameters have strong impact on the cylinder displacement for the same variation in the values of Ri.

Table 2. Comparison of Richardson Number (Ri) and Cylinder Displacement (a) U_{red} = 3 and U_{red} = 8 of a square cylinder at Re = 100

U_{red} = 8	
Richardson Number (Ri)	Cylinder Displacement (x)
Ri = -1.0	1.266 D
Ri = -0.75	0.75D
Ri = -0.50	0.18 D
Ri = -0.25	0.11 D
Ri = 0	0.08 D

(a)

U_{red} = 3	
Richardson Number (Ri)	Cylinder Displacement (x)
Ri = -1.0	0.08 D
Ri = -0.75	0.07 D
Ri = -0.50	0.04 D
Ri = -0.25	0.024 D
Ri = 0	0.02 D

(b)

The power extraction from the cylinder movement due to the incoming flow is tabulated in Table 3. Table 3 (a) shows the power extracted for U_{red} = 3 at Ri = 0 and -1. It can be seen from the Table 3 (a) that with the aid of thermal buoyancy (Ri = -1), the power available is more as compared to no thermal buoyancy (Ri = 0). At Ri = -1 the power extracted is 0.24 watt whereas

Table 3. Power obtained for different U_{red} (a) $U_{red} = 3$ (b) $U_{red} = 8$

U_{red} = 3			
Ri = 0			
Displacement	Velocity	Energy	Power
0.04 m	0.016m/s	0.256 J	0.1 watt
Ri = -1			
0.18m	0.0352m/s	1.245 J	0.24 watt

(a)

U_{red} = 8			
Ri = 0			
Displacement	Velocity	Energy	Power
0.16 m	0.062m/s	3.93 J	1.54 watt
Ri = -1			
2.52 m	0.56m/s	319.25 J	71.5 watt

(b)

at Ri = 0 power extracted 0.1 watt for $U_{red} = 3$. Similar results are obtained in case of $U_{red} = 8$ as shown in Table 3 (b). The striking feature in case of Ured = 8 is that power extracted in case of Ri = -1 is much larger than that of $U_{red} = 3$. However, displacement of cylinder at Ri = 0 for $U_{red} = 3$ and $U_{red} = 8$, do not have significant difference. This clearly shows the dependency of thermal buoyancy on structural parameter reduced velocity (U_{red}).

Fig. 3 (a) and (b) shows the variation of x direction cylinder displacement with time for U_{red} = 3 and 8 at Re = 100. As we can see from the figure, cylinder displacement is increased as we change the value of Richardson number from 0 to -1. Therefore it is clearly observed that the effect of buoyancy helps the cylinder to move with larger displacement.

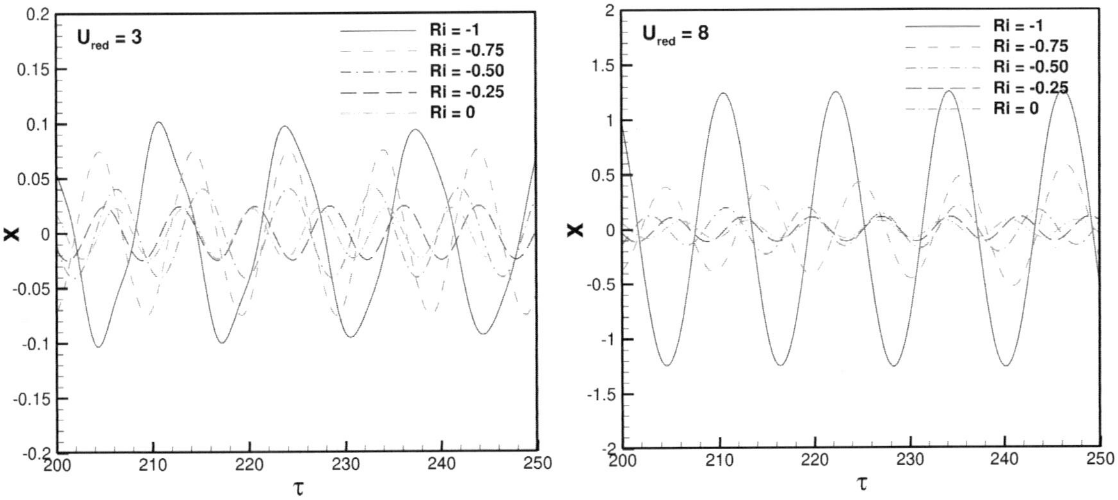

(a) **(b)**

Fig. 3 shows the variation of x-direction cylinder displacement with time for Ured = 3 and Ured =8 at Re = 100

Fig. 4 shows the instantaneous vorticity contours for different Richardson number for $U_{red} = 8$. As we can clearly observe from the fig.4, wake region in the downstream of the cylinder is wider for Ri = -1 in comparison to Ri = 0. The vortex shedding mode for all Richardson number is found to be "2S"(two vortices are formed in one vibration period of the cylinder). At Ri = 0 vortices occur alternatively and in case of Ri = -1 vortices are formed wider. The vortex pattern for $U_{red} = 3$ at different Ri was observed similar as shown in fig. 4, therefore its is not presented here. The stationary cylinder wake also becomes wider in case of Ri = -1 as reported by Sharma and Eswaran [11].

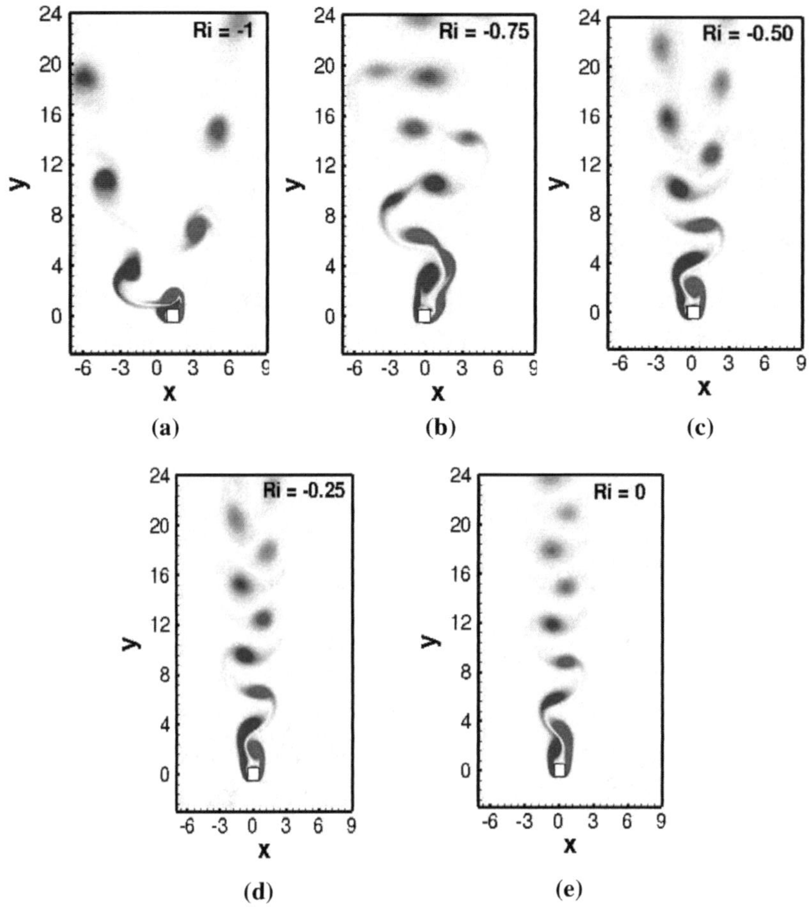

Fig. 4. Instantaneous vorticity contours for different Richardson number (Ri)

Conclusions:

It is concluded that the higher negative value of Richardson number moves the cylinder with larger displacement. Hence power extracted is more with the aid of negative thermal buoyancy. The vortex shedding pattern for Ured= 8 at Ri = [-1, 0] is obtained as 2S mode of vortex shedding.

References:

[1] Rashidi S, Hayatdavoodi M, and Esfahani J A 2016, Vortex shedding suppression and wake control: A review, *Ocean Eng.* **126**, 57–80

[2] Soti A K, Thompson M C, Sheridan J, and Bhardwaj R 2017, Harnessing electrical power from vortex-induced vibration of a circular cylinder, *J. Fluids Struct.* **70**, 360–373

[3] Lee J, Xiros N, and Bernitsas M 2011, Virtual damper–spring system for VIV experiments and hydrokinetic energy conversion, *Ocean Eng.* **38**(5), 732–747

[4] Izadpanah E, Ashouri A, Liravi M and Amini Y 2019, Effect of vortex-induced vibration of finned cylinders on heat transfer enhancement, *Phys. Fluids* **31**(7), 073604

[5] Anagnostopoulos P and Bearman P 1992, Response characteristics of a vortex excited cylinder at low Reynolds numbers, *J. Fluids Struct.* **6**(1), 39–50

[6] Prasanth T and Mittal S 2008 Vortex-induced vibrations of a circular cylinder at low Reynolds numbers, *J. Fluid Mech.* **594**, 463–491

[7] Prasanth T, Premchandran V, and Mittal S 2011 Hysteresis in vortex-induced vibrations: Critical blockage and effect of m^*, *J. Fluid Mech.* **671**, 207–225

[8] Sarpkaya T, 2004 A critical review of the intrinsic nature of vortex-induced vibrations,*J. Fluids Struct.* **19**(4), 389–447

[9] Williamson C H K and Govardhan R 2004 Vortex-induced vibrations *Annu. Rev. Fluid Mech.* **36**, 413–455

[10] Bearman P 2011 Circular cylinder wakes and vortex-induced vibrations *J. Fluids Struct.* **27**(5-6), 648–658

[11] Sharma A and Eswaran V 2004 Effect of aiding and opposing buoyancy on the heat and fluid flow across a square cylinder at RE = 100 *Numer. Heat Transf. Part A Appl.* **45**(6) 601-624

[12] Arif M.R, Hasan N 2019 Vortex shedding suppression in mixed convective flow past a square cylinder subjected to large-scale heating using a non-Boussinesq model, *Phys. Fluids* **31**(2) 023602.

[13] Sohankar A, Norberg C and Davidson L 1998 Low-Reynolds-number flow around a square cylinder at incidence: Study of blockage, onset of vortex shedding and outlet boundary condition *Int. J. Numer. Meth. Fl.* **26**(1)

IOP Conf. Series: Materials Science and Engineering **748** (2020) 012007 doi:10.1088/1757-899X/748/1/012007

Optimization of significant factors for improving compressive strength of ABS in Fused Deposition Modeling by using GA & RSM

D Chhabra[1], S Deswal[2]
1 Assistant Professor, UIET, MDU, Rohtak, India
2 Ph.D. Research Scholar, UIET, MDU, Rohtak, India
E-mail address: sandeepdeswal86@gmail.com

Abstract

The current research work is focused on optimization of FDM 3D printer input factors for compressive strength of ABS fabricated parts. It is a carbon chain copolymer thermoplastic possessing high impact strength and rigidity. Response Surface Methodology (RSM) and Genetic algorithm (GA) integrated with each other to optimize the input factors. Five factors viz. Layer thickness (LT), Build Orientation (BO), Raster angle (RA), Raster width (RW) and Air gap (AG) have been studied and modeled for compressive strength utilizing RSM. In present experimentation, quadratic model has been suggested for compressive strength. Build Orientation is most significant factor that imparts considerable effect on compressive strength as compared to other factors. Finally, GA-RSM, a popular hybrid evolutionary approach is used to optimize the factors.

Keywords: Fused deposition modeling, Response surface methodology, Genetic algorithm, Compressive strength.

1 Introduction:

A category of technology that fabricates part by depositing material in the form of layer one above other sequentially is called rapid manufacturing (RP) or additive manufacturing. These technologies automatically fabricate the physical model or prototype from computer designed replica in a shortest available time without considering any geometrical intricacy. This is the reason that discriminate these technologies from other traditional manufacturing technologies. Since the introduction of additive manufacturing technique in 1987 various additive fabrication techniques such as stereo-lithography, fused deposition modeling (FDM), selective laser melting (SLM) etc. have come in to notice. Among these FDM make a distinction in terms of fabrication nature, material used, cost effective, clean and degree of accuracy with minimum consumption of material. In FDM, 3D designed part fabrication process starts by depositing the heated and extruded filament material in form of layers in additive way on a heated bed (1). The basic principle of FDM is extrusion process and the whole FDM process is controlled by computer i.e computer numerical controlled. Therefore after depositing one layer the heated bed is moved lower and the progression is keep continual till the whole manufacturing procedure is completed. The application areas of FDM are in automobile, medical, prototype making industries, space, rapid tooling manufacturing and many more (2). The main obstacles for FDM development is the strength of fabricated parts. Diffusion welding phenomenon during fabrication process leads to bonding between the adjacent deposited layers (3). Sun Q. et.al. have studied that quality of bonding between layers depends upon the surrounding temperature or temperature conditions around the fabricated parts as convective heat transfer takes place during fabrication process. Es Said OS et. al. found that due to volumetric shrinkage of extruded material from heated nozzle, the bonding between the interlayer decreases. Therefore magnitude of internal resisting force decreases and lead to failure of part under application of force. Wang TM. et. al. studied that temperature gradient exists as part fabrication process progress from bottom to top. Due to uneven temperature conditions, inner stresses to be developed which lead to distortion of layers and finally failure of parts. Residual stresses developed during fabrication process also lead to the deformation of deposited layers (7). All these observable facts jointly affect the strength of parts. Many other researchers studied the affect different FDM input factors on mechanical properties of FDM fabricated components. Lee et. al. studied that built direction and raster angle have considerable affect on mechanical properties. Lee et. al. used ANOVA analysis and pointed out that layer thickness, air gap, raster angle and width are the significance FDM input parameters that considerably influence the performance of fabricated parts. Sahu et. al. used Taguchi technique to study the effect of different input factors like layer thickness, orientation, raster angle/width and air gap on measurement preciseness. Laeng et. al. pointed out using Taguchi and ANOVA procedure that air gap, layer thickness, raster angle and width are the significant factors that have an effect on the performance of FDM fabricated ABS material parts. Using these techniques optimum parameters cannot be obtained as studied in this paper. Therefore brittle and anisotropic nature of FDM fabricated parts makes it necessary to study the effect of unlike input factors on mechanical properties with their optimization. In this study based on above studied literature layer thickness (LT), part built (BO) orientation, raster angle (RA), air gap (AG) and rater width (RW) are chosen significant process parameters at different levels. Hence, the intend of this study is groping the effect of above stated parameters. Therefore, in this research work, response surface methodology (RSM) and genetic algorithm (GA) are used jointly to set up input factors optimization model for Acrylonitrile Butadiene Styrene (ABS) material FDM fabricated parts. The hybrid evolutionary algorithms combine with RSM can be used efficiently for process parameters optimization (12,13,14). To show a relationship for input factors to compressive strength of fabricated parts RSM model has been used. To get optimum input factors setting GA has been used for optimization.

2 Methodology:

All test pieces to be fabricated were premeditated and mocked-up using computer aided design software and exported as Standard Tessellation Language (STL) file. The hot end pathway was created for all test pieces having decided factors value in the FDM machine. A total of 32 specimens were made-up on the R*P2200i FDM technique 3D printer using ABS thermoplastic as a filament material having test pieces size according to the ASTM standard D-695. ABS is a thermoplastic copolymer having long carbon chain structure with good rigidity and impact strength. Further, the compressive strength of these test pieces was tested on UNITEK-94100 UTM machine. The tests were performed according to the ASTM D-695 standard test specifications. The machine testing speed was used to as 5mm/min during testing. Therefore, according to the above specified standards, experimental design data was processed on the UTM to calculate the maximum compressive strength. Based on studied literature (9,10,11) five process parameters as shown in table 1 are considered with their range and levels. The experimental design matrix having a total number of 32 experiments with their output response value has been described in Table 1. These parameters are Layer thickness (LT), Build Orientation (BO), Raster angle (RA), Raster width (RW) and air gap (AG). The other FDM machine parameters are maintained at their fixed values. The output value of each trial was an average of five runs. The high and low value of parameters is decided according to the machine specifications and studied literature. Response Surface Methodology (RSM) a assortment of numerical and geometrical technique is used for examination and figuring of input factors in this research work. In RSM the effect of input factors on output value is studied by considering the linear or square polynomial function value. To develop an experimental design pattern and learning the effect of quadratic, linear and cubic models of five input parameters, three level RSM based Face central composite design (FCCD) is used. In this study second order quadratic model is used to build up the numerical model and establishing the relationship among factors and compressive strength (CS) by means of equation (i). The compressive strength expected and observed values have been stated in Table 3.

$$Z = \gamma_0 + \sum \gamma_i y_i + \sum \gamma_{ii} {y_i}^2 + \sum \gamma_{ij} y_i y_j \qquad \text{i}$$

Z= predicted outcomes, y_i & y_j = coded variables, γ_o = constant tenure of the regression equation, γ_i & γ_{ii} = linear regression coefficient & each factor square term, γ_{ij} = interaction terms coefficient

Table 1: Input factors and their levels

Sr. No.	Input Factors	Unit	Symbol	Control Factors		
				Low(-1)	Centre(0)	High(1)
1	Layer Thickness	mm	LT	0.15	0.2	0.3
2	Build Orientation	Degree	BO	0	20	40
3	Raster Angle	Degree	RA	0	35	70
4	Raster Width	mm	RW	0.5	0.55	0.6
5	Air Gap	mm	AG	0	0.003	0.006

Table 2: Five factors experimental run order in face centre central composite design

Sr. No.	LT (mm)	BO (Degree)	RA (Degree)	RW (mm)	AG (mm)	CS (MPa)
1	0.15	0	0	0.5	0.006	18.12
2	0.3	0	0	0.5	0	15.41

3	0.15	40	0	0.5	0	13.61
4	0.3	40	0	0.5	0.006	13.87
5	0.15	0	70	0.5	0	17.86
6	0.3	0	70	0.5	0.006	16.1
7	0.15	40	70	0.5	0.006	8.448
8	0.3	40	70	0.5	0	17.98
9	0.15	0	0	0.6	0	14.89
10	0.3	0	0	0.6	0.006	17.18
11	0.15	40	0	0.6	0.006	13.13
12	0.3	40	0	0.6	0	11.64
13	0.15	0	70	0.6	0.006	14.58
14	0.3	0	70	0.6	0	18.49
15	0.15	40	70	0.6	0	13.83
16	0.3	40	70	0.6	0.006	11.87
17	0.15	20	35	0.55	0.003	13.49
18	0.3	20	35	0.55	0.003	14.43
19	0.225	0	35	0.55	0.003	18.28
20	0.225	40	35	0.55	0.003	14.82
21	0.225	20	0	0.55	0.003	13.59
22	0.225	20	70	0.55	0.003	13.3
23	0.225	20	35	0.5	0.003	13.65
24	0.225	20	35	0.6	0.003	13.25
25	0.225	20	35	0.55	0	14.26
26	0.225	20	35	0.55	0.006	12.8
27	0.225	20	35	0.55	0.003	13.72
28	0.225	20	35	0.55	0.003	14.45
29	0.225	20	35	0.55	0.003	13.57
30	0.225	20	35	0.55	0.003	13.31
31	0.225	20	35	0.55	0.003	13.84
32	0.225	20	35	0.55	0.003	13.77

IOP Conf. Series: Materials Science and Engineering **748** (2020) 012007 doi:10.1088/1757-899X/748/1/012007

Table 3: Compressive strength experimental actual and RSM predicted value with their percentage error

Sr. No.	Actual Outcomes(MPa)	Predicted by RSM(MPa)	% error
1	18.12	18.13	0.49
2	15.41	15.43	0.93
3	13.61	13.64	0.26
4	13.87	13.88	0.28
5	17.86	17.83	0.23
6	16.1	16.05	0.16
7	8.448	8.415	1.05
8	17.98	17.95	2.33
9	14.89	14.95	0.2
10	17.18	17.21	0.2
11	13.13	13.18	0.08
12	11.64	11.70	0.13
13	14.58	14.57	0.35
14	18.49	18.49	0.08
15	13.83	13.85	0.46
16	11.87	11.86	0.36
17	13.49	13.36	2.13
18	14.43	14.36	0.39
19	18.28	18.21	1.75
20	14.82	14.69	3.43
21	13.59	13.29	2.04
22	13.3	13.40	3.24
23	13.65	13.69	0.29
24	13.25	13.00	0.98
25	14.26	14.09	0.32
26	12.8	12.77	2.54

27	13.72	13.90	0.96
28	14.45	13.90	0.64
29	13.57	13.90	00.00
30	13.31	13.90	0.32
31	13.84	13.90	00.00
32	13.77	13.90	10.36

2.1 GA-RSM optimization:

In this study to obtain the maximum compressive strength at optimum values of input factors Genetic Algorithm (GA) is used. On human genetics principle based, GA is non-traditional tool which is consists of basic four steps (15). These steps are reiterated till the maximum desires output value is achieved. In first step from given set of solutions initial population is generated. Second on the basis of fitness value of objective function, evaluation of each individual population is performed. Next to consider the one entity that has utmost fitness value for mutation and crossover. In Final step after following selection, mutation and crossover new individual parent is selected from the generated population.

The steps followed to create GA-RSM model are as follows:

a. A population is generated having set of plausible solutions.

b. Inside the predefined arrangement, the values of factors have been created.

c. Equation engendered by RSM is stacked as fitness function value for GA.

d. Creation of new-fangled population using mutation, crossover function.

e. Calculation is finished after getting the optimum factors value for maximum compressive strength at the position of meeting fitness function and the chromosomes.

The course chart of GA evolutionary practice is shown in figure 1:

Figure 1: The general course of action of GA evolutionary practice

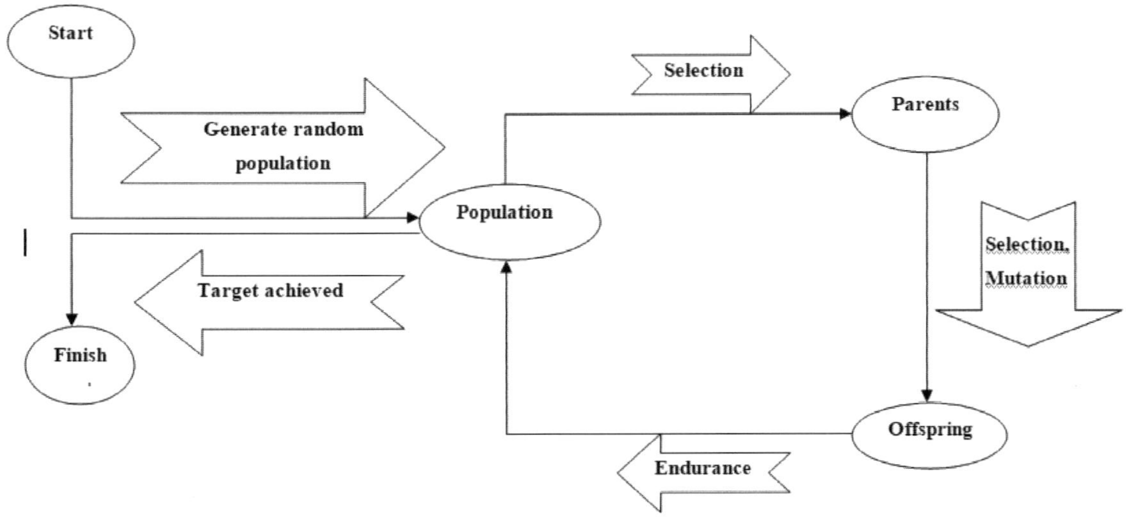

3 Outcomes and Conversation:

The paying attention in this learning is on increasing compressive strength by optimizing input factors using GA-RSM hybrid techniques. The experimental and predicted outcomes obtained by RSM are represented in Table 3 for a developed design matrix under a different set of input factors. The analysis of outcomes is done by developing numerical models by means of Design expert (6.0.8) & GA tool box of Mat Lab (16.0). In current learning the quadratic model was analyzed and considered on the basis of sum of squares, lack-of-fit value. The obtained highest value of R^2, R^2 (adj.) and R^2 (pre.) are in agreement with every one for the developed quadratic model of compressive strength as represented in Table 4. Therefore insignificant lack-of-fit and smaller *p* value for quadratic models in contrast to other model gives commendable explanation amid FDM input factors and compressive strength.

Table 4: RSM details of quadratic model developed for CS

Output Value	**Model**	**Sequential P-value**	**Lack of fit-value**	**R^2**	**R^2_{Adj}**	**R^2_{Pred}**	**Precision**	**Remarks**
Compressive Strength	Quadratic	<0.0001	0.8490	0.9926	0.9792	0.8803	sufficient	Suggested
	2FI	0.0056	0.0059	0.8573	0.7235	-3.0305	In-sufficient	

	Cubic	0.9507	0.2950	0.9937	0.9673	-28.4166	In-sufficient	Aliased
	Linear	0.0027	0.0011	0.4856	0.3866	0.0382	In-sufficient	

3.1 RSM model details:

Using RSM technique second order developed quadratic model for compressive strength is derived, analyzed and validated. The table 4 represents the RSM details for the developed model. The regression equation derived using DOE 6.0.8 to analyze the model in terms of actual factors is represented by equation (ii). Models have been validated using various plots values such as normal plot of residual and actual vs. predicted. The significant F value of 73.85 and P value of <.0001 shows that developed model was significant. The normal probability plot of residuals for compressive strength is represented by figure 2. Residuals of normal plot in a straight line indicate no abnormalities. Figure 3 present actual vs. predicted model values plots for compressive strength. Points are clustered around a straight line thereby implying that predicted values are in close adherence with actual values.

CS=-46.17002+19.21117*LT-0.3154*BO-0.032066*RA+236.55534*RW-733.63837*AG
-7.67915 * LT^2+6.36701E-003*BO^2-4.55670E-004*RA^2-221.27809*RW^2-52577.24657
*AG^2+0.19217*LT*BO+0.27076*LT*RA-42.86667*LT*RW+392.22222*LT*AG
-1.38571E-004*BO*RA-0.068000*BO*RW-9.45000*BO*AG+0.091143*RA*RW
-14.23333 *RA*AG+2596.66667*RW*AG

(ii)

Figure 2: compressive strength residual plot

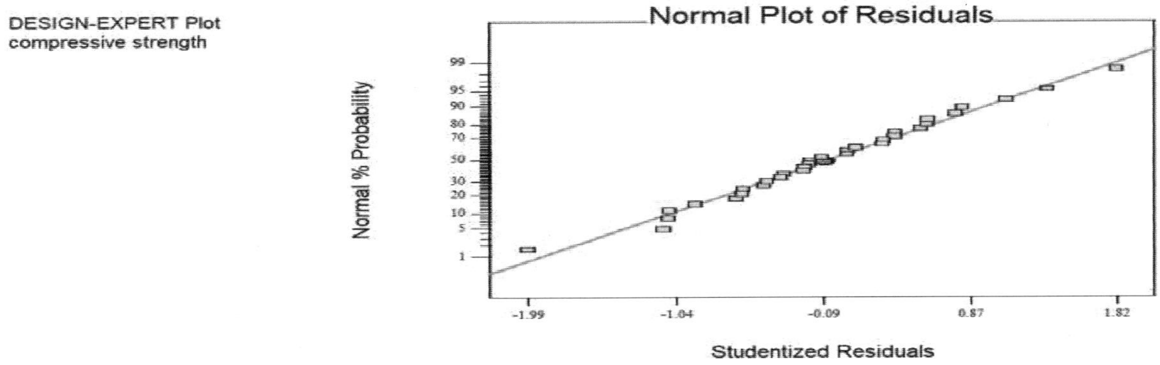

Figure 3: compressive strength predicted vs. actual plot

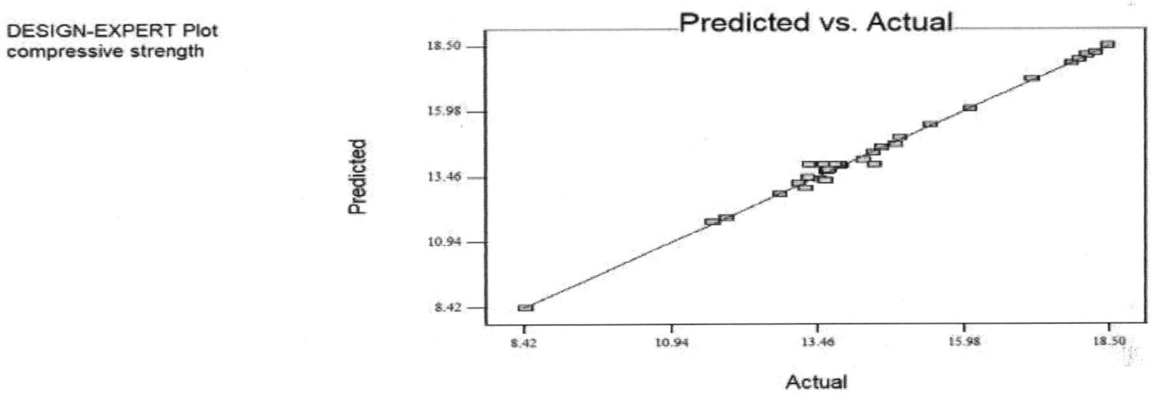

3.2 ANOVA model:

The input factors selected for optimizing the compressive strength were layer thickness, Build orientation, Raster angle, Raster width and Air gap. The CS was calculated by performing trials as indicated by Table 1. For output values, numerical model is generated according to the second order quadratic equations (i) given above. ANOVA analysis has been used to understand the statistical significance of above stated parameters. The each factor percent contribution tells about the total variation observed for each significant input factor. That contribution depends upon the obtained value of sums of square of each factor as shown in table 5. It is clearly illustrated that parameters LT, BO, RA, RW and AG are significantly inflate the compressive strength of FDM fabricated parts with varying percent contribution. ANOVA statistics showing

IOP Conf. Series: Materials Science and Engineering **748** (2020) 012007 doi:10.1088/1757-899X/748/1/012007

that build orientation is the most sensitive parameter with 38.54 % contribution with compressive strength of parts. However, LT, RA, RW and AG are also identified as influencing parameters with percent contribution 3.11%, 0.04%, 1.47% and 5.40% respectively as shown in Table 5. This study shows that BO is the only one influencing parameter that greatly affects the compressive strength of FDM parts.

Table 5: CS ANNOVA table

Source	Sum of Squares	Degree of Freedom	Mean Square	F-value	P-value		% contribution
Model	143.91	20	7.20	73.85	< 0.0001	Significant	
LT	4.51	1	4.51	46.31	< 0.0001		3.11
BO	55.87	1	55.87	573.42	< 0.0001		38.54
RA	0.06	1	0.06	0.59	0.4583		.04
RW	2.13	1	2.13	21.83	0.0007		1.47
AG	7.83	1	7.83	80.37	< 0.0001		5.40
LT^2	0.00	1	0.00	0.05	0.8321		
BO^2	15.96	1	15.96	163.82	< 0.0001		
RA^2	0.77	1	0.77	7.87	0.0171		
RW^2	0.75	1	0.75	7.73	0.0179		
AG^2	0.55	1	0.55	5.66	0.0366		
LT×BO	1.33	1	1.33	13.64	0.0035		
LT×RA	8.08	1	8.08	82.96	< 0.0001		
LT×RW	0.41	1	0.41	4.24	0.0639		
LT×AG	0.12	1	0.12	1.28	0.2822		
BO×RA	0.15	1	0.15	1.55	0.2397		
BO×RW	0.07	1	0.07	0.76	0.4022		
BO×AG	5.14	1	5.14	52.79	< 0.0001		
RA×RW	0.41	1	0.41	4.18	0.0657		
RA×AG	35.74	1	35.74	366.78	< 0.0001		
RW×AG	2.43	1	2.43	24.91	0.0004		

Residual	1.07	11	0.09				
Lack of fit	0.35	6	0.05	0.41	0.849	not significant	
Pure error	0.72	5	0.14				
Cor Total	144.98	31					

Figures 4 shows the perceptible persuade of build orientation on compressive strength. It decreases with increasing the value of orientation and then increases with further increase in build orientation at fixed value of layer thickness, raster angle , raster width and air gap. The maximum compressive strength at low value of orientation is obtained as can be seen from figure 4(a,b,c,d). Here, build orientation is momentous factor that can be seen by the spiky curve of Figure 4(a,b,c,d) and highest F-value in Table 5. The regularity of deposited layers decreases as build orientation value increases. At high value of orientation, magnitude of distortion between layers decreases due to regular deposit pattern and homogeneous cross section of deposited layers (3,16). The same pattern can be seen from surface plot as shown in figure 4. The residual stresses induced during part fabrication in additive manufacturing produce deformation in parts. As the number of layers increases the magnitude of residuals stresses increases and vice versa. Therefore compressive strength of parts at high layer thickness is more (17). The same can be seen from figure 5 that with increase in value of layer thickness the compressive strength of printed parts increases as number layers decreases. For compressive strength the interaction between raster width and air gap can be seen from figure 6(a) that it is increasing with increase in value of both factors. The reason is that with increase in raster width spacing between layers decreases or less void spacing as oppose to increase in air gap. The parts with less void spacing posses more bonding and strength. The high value of void spacing between layers leads to less bonding between layers and decreases the CS of fabricated parts (18). The maximum value of CS is in between minimum and maximum value of both these two factors. The high value of air gap also increases void spacing but more prominent effect of raster width can be seen from figure 6(a). At high value of raster width & air gap value of CS is low. This is due to improper alignment of extruded material molecules and denser deposited layers (19). The CS is increasing with increasing raster angle as shown in figure 6(b). This is in agreement with the past studied literature (3).

Figure 4: Factors interaction surface plot of (a) BO & LT (b) RA & BO (c) RW & BO (d) AG & BO for compressive strength

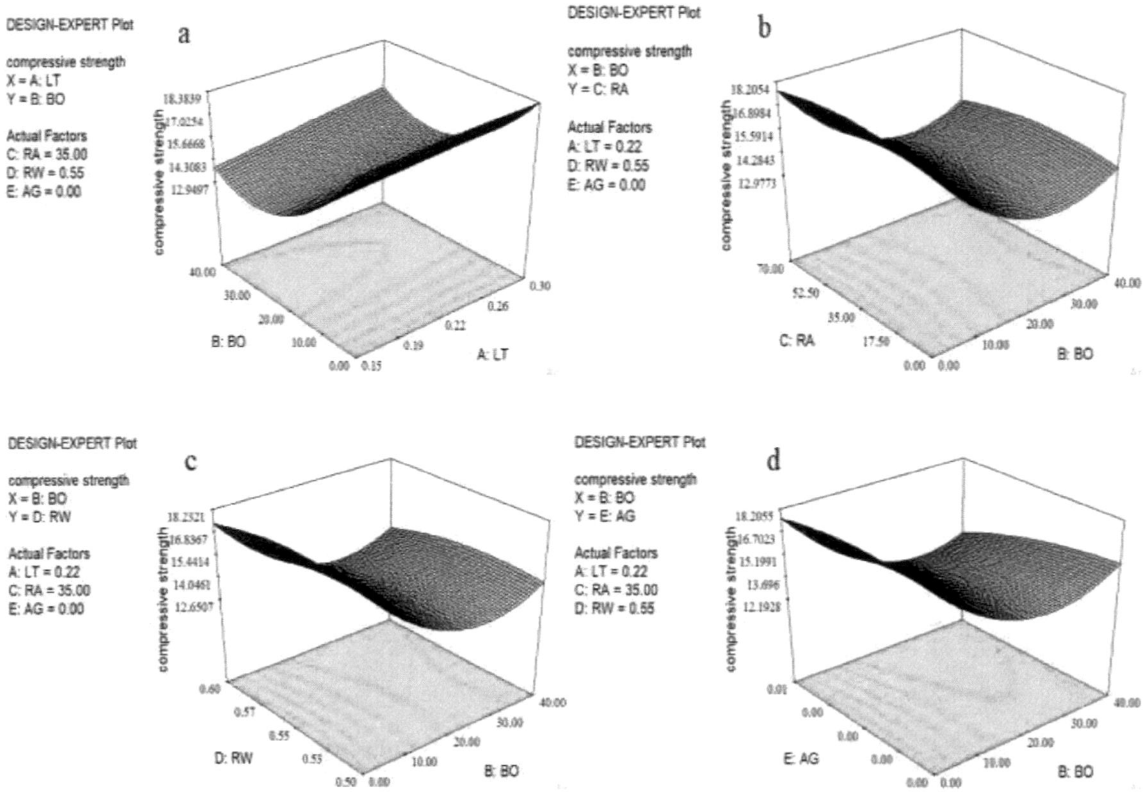

Figure 5: Factors interaction surface plot of LT with RW & AG for compressive strength

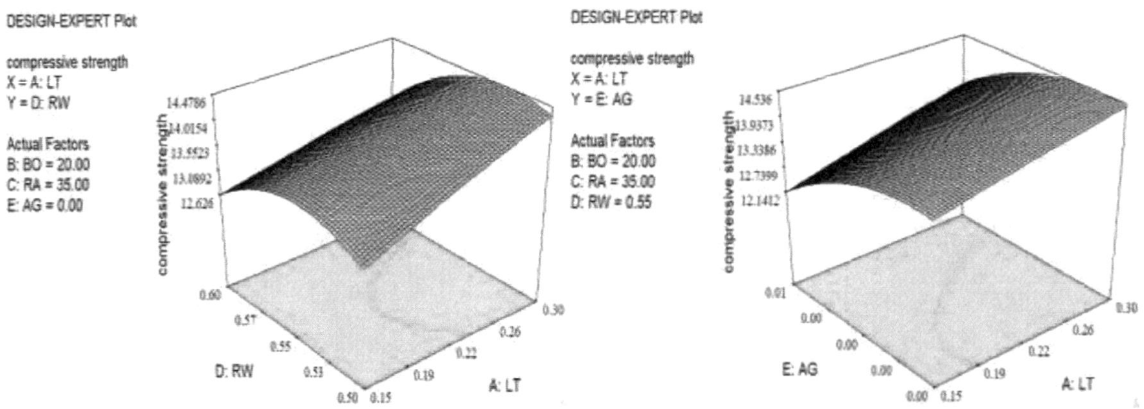

IOP Conf. Series: Materials Science and Engineering **748** (2020) 012007 doi:10.1088/1757-899X/748/1/012007

Figure 6: Factors interaction surface plot of (a) AG & RW (b) AG & RA

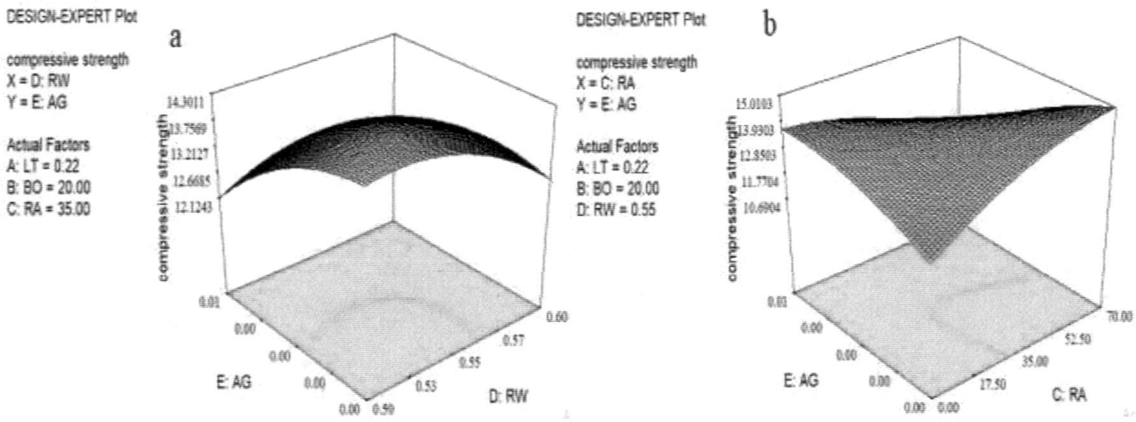

3.3 Parameter optimization by GA-RSM:

To locate the optimum factors settings some evolutionary algorithm are required as ANOVA only gives the significant contribution of each individual factors. In this study GA is used to optimize the given problem within limit constraint. For various numbers of generations in GA epochs are used to determine the performance and number of cycles. It is observed that the optimized setting of GA tool is that 50 generations, 250 population size, crossover fraction value 0.8, elite count 0.05 of population, crossover function use- constraint dependent, mutation use-adaptive feasible. Best fitness value with number of generation is shown in figure 7. The process is continued until to get optimum factors settings for maximum compressive strength in GA tool by setting up higher and lesser bound factors limits. The maximum compressive strength obtained with GA-RSM is 19.92 MPa at process parameters (LT 0.3 mm, BO 0^0, RA 70^0, RW .52 mm and AG 0.0 mm) and its contrive as shown in figure 7.

Figure 7: compressive strength best and mean fitness value plot with number of generations

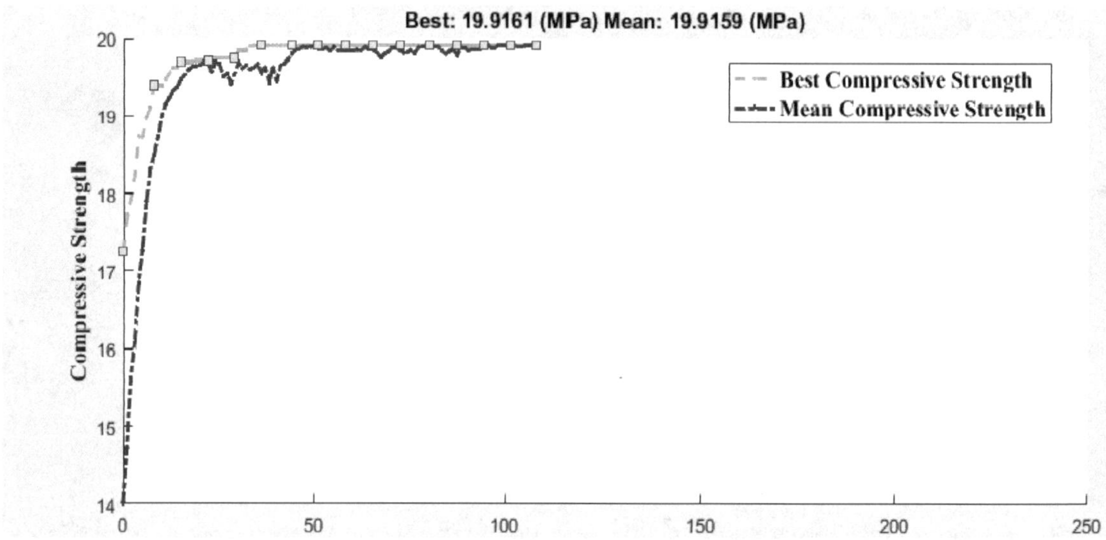

3.4 Validation of GA-RSM:

The optimum factors blends sets acquired from the GA-RSM created models were utilized to approve the model. From table 5 three experiments having different combination of input parameters have been performed for confirmation of output values. The percentage error has been estimated between optimized GA-RSM and experimental obtained values as represented in table 6. The maximum error obtained is 3.62%. The experimental and optimized outcomes values are very close to each other. Therefore, it can be concluded that the percentage of prediction errors is much less and hence the optimized performance of the GA-RSM is quite satisfactory. This validates the numerical optimized outcomes obtained from GA-RSM.

Table 6: GA-RSM validation result

Number	LT	BO	RA	RW	AG	Optimized Value	Experimental value	% error
1	0.3	0	70	.52	0	19.92	19.44	2.41
2	0.3	0	56.86	.52	0	19.41	18.85	2.88
3	0.3	0	68.85	.52	0	19.87	19.15	3.62

4 Conclusion:

In this study, Response Surface Methodology and Genetic Algorithm have been used to optimize the input factors for compressive strength of ABS material fabricated parts by FDM 3D printer. The experimental study has been carried out by developing experimentation order using RSM based FCCD methodology. The effect of individual factors on compressive strength has been explained by drawing response surface graphs. ANOVA shows that build orientation (BO) is most significant factor affecting the compressive strength. The strength of FDM fabricated parts depends on the bonding between layers and deposition pattern. A high build orientation value leads to uniform cross section pattern of deposited layers. Therefore, increases the strength of parts by decreasing the magnitude of distortion between layers. The quadratic RSM build up numerical model was connected with GA to ideal the optimum settings leading to the maximum compressive strength value 19.92 MPa corresponding to input factors values (LT 0.3 mm, BO 0^0, RA 70^0, RW .52 mm and AG 0.0 mm). The interaction study between factors by ANNOVA is more helpful to elaborate the effect of different factors in comparison to traditional technique like Taguchi. Therefore fabrication of ABS parts with FDM at optimum parameters setting can increase the use of FDM technique 3D printer in industrial applications. For multi-objective optimization problem GA based hybrid optimization technique such as GA-RSM can be used effectively. Effect of different process parameters at optimum setting on compressive strength can be analyzed using multi-material part fabrication on FDM machine.

References:

[1] Peng A, Xiao X and Yue R 2014 Process parameter optimization for fused deposition modeling using response surface methodology combined with fuzzy inference system *Int. J. Adv. Manuf. Technol* **73** 87-100

[2] Mohamed O A, Masood S H and Bhowmik J L 2015 Optimization of fused deposition modeling process parameters: a review of current research and future prospects *Adv. Manuf* **3** 42-53

[3] Sood A K, Ohdar R K and Mahapatra S S 2012 Experimental investigation and empirical modelling of FDM process for compressive strength improvement *J. Adv. Res* **3** 81-90

[4] Sun Q, Rizvi G M, Bellehumeur C T and Gu P 2008 Effect of processing conditions on the bonding quality of FDM polymer filaments *Rapid Prototyp. J* **14(2)** 72-80

[5] Es-Said O S, Foyos J, Noorani R, Mendelson M, Marloth R and Pregger B A 2000 Effect of layer orientation on mechanical properties of rapid prototyped samples *Mater. Manuf. Process* **15(1)**107-22

[6] Wang T M, Xi J T and Jin Y 2007 A model research for prototype warp deformation in the FDM process *Int. J. Adv. Manuf. Technol* **33(11-12)** 1087-96

[7] Zhang Y and Chou K 2008 A parametric study of part distortions in fused deposition modelling using three-dimensional finite element analysis *Proc. Inst. Mech. Eng. Part B J. Eng. Manuf* **222(8)** 959-67

[8] Lee C S, Kim S G, Kim H J and Ahn S H 2007 Measurement of anisotropic compressive strength of rapid prototyping parts *J. Mater. Process. Technol* **187-188** 627-30

[9] Lee B H, Abdullah J and Khan Z A 2005 Optimization of rapid prototyping parameters for production of flexible ABS object *J. Mater. Process. Technol* **169** 54-61

[10] Sahu R K, Mahapatra S S and Sood A K 2014 A Study on Dimensional Accuracy of Fused Deposition Modeling (FDM) Processed Parts using Fuzzy Logic *J. Manuf. Sci. Prod* **13(3)** 183-97

[11] Laeng J, Khan Z A and Khu S Y 2006 Optimizing flexible behaviour of bow prototype using Taguchi approach *J. Appl. Sci* **6** 622-30

[12] Deswal S, Narang R and Chhabra D 2019 Modeling and parametric optimization of FDM 3D printing process using hybrid techniques for enhancing dimensional preciseness *Int. J. Interact. Des. Manuf* **13** 1197-14

[13] Chhabra D, Bhushan G and Chandna P 2014 Optimization of Collocated/Noncollocated Sensors and Actuators along with Feedback Gain Using Hybrid Multiobjective Genetic Algorithm-Artificial Neural Network *Chin. J. Eng* **2014** 12

[14] Kumar V, Chhabra D and Shukla P 2017 Xylanase production from Thermomyces

lanuginosus VAPS-24 using low cost agro-industrial residues via hybrid optimization tools and its potential use for saccharification *Bioresour. Techno* **243** 1009-19

[15] Kaushik V S, Subramanian M and Sakthivel M 2018 Optimization of Processes Parameters on Temperature Rise in CNC End Milling of Al 7068 using Hybrid Techniques *Materials Today: Proceedings* **5(2)** 7037-42

[16] Magar S, Khedkar N K and Kumar S 2018 Review of the effect of built orientation on mechanical Properties of metal-plastic composite parts fabricated by Additive Manufacturing Technique *Materials Today: Proceedings* **5** 3926-35

[17] Sood A K, Ohdar R K and Mahapatra S S 2010 Parametric appraisal of mechanical property of fused deposition modelling processed parts *Mater. Des.**31(1)** 287-95*

[18] Percoco G, Lavecchia F and Galantucci L M 2012 Compressive properties of FDM rapid prototypes treated with a low cost chemical finishing *Res. J. Appl. Sci. Eng. Technol* **4(19)** 3838-42

[19] Montero M, Roundy S and Odell D 2001 Material characterization of fused deposition modeling (FDM) ABS by designed experiments *Proc. Rapid Prototyp. Manuf. Conf* 15-17

IOP Conf. Series: Materials Science and Engineering **748** (2020) 012008 doi:10.1088/1757-899X/748/1/012008

Beneficiation of goethite iron ore through magnetizing roasting followed by magnetic separation

Raj Kumar Jaiswal, Dipak kumar Acharya, Shatrughan Soren

Department of Fuel, Mineral and Metallurgical Engineering IIT(ISM) Dhanbad

Email: raj080791@gmail.com

Abstract. The present study is carried out to recover iron value from low grade goethite iron ore. The method is based on coal based magnetization roasting followed by magnetic separation. The effect of temperature, residence time and magnetic field intensity on grade and recovery of iron ore have been studied. The magnetizing roasting was carried out at 700°C, 800°C and 900°C for different residual time viz., 10, 20, 30, 40, 50 and 60 minutes. The production obtained from magnetizing roasting was then separated in wet high intensity magnetic separator (WHIMS) at different magnetic field of 2000 gauss, 4000 gauss and 6000 gauss. The result obtained at 900°C roasting temperature for 50 minute followed by magnetic separation at 2000 gauss shows optimum Fe grade of 59.20% and recovery is 68.23%, this grade and recovery of iron ore increases with increase of temperature and time. However, the grade and recovery of iron ore decreases during prolonged roasting at 900°C. The recovery increases with increasing in magnetic field but the grade is decreasing. In the present study, the as received goethite iron ore and concentrate ore have been examined using optical microscopy to know the phases present and change of phase during processing.
Keywords: Goethite ore, magnetized roasting, WHIMS, microscopic

1. Introduction

Goethite (a-FeOOH) ore is formed in terrestrial environment, as a result of oxidative weathering and due to formation of acid soil. It is one of the most common and abundant iron oxide and oxyhydroxide minerals present in the soil [4]. Moreover, goethite constitutes the main component of the reject streams in many iron ore plant. This makes the processing of goethite ores almost inevitable [3]. The overwhelming part of these ores is represented by low-grade varieties which cannot be used in iron and steel industry without preliminary beneficiation. Although various techniques are available to improve lower grade iron ores in various part of the world, the objective of this research work focused on the mineral transformation and chemical of goethite ores from Langalota region of Odisha India by reduction roasting and magnetic separation.

The goethite iron ore which comprises of hematite, goethite, kaolinite and quartz also contains high levels of silicon and aluminum. Vitreous goethite which is rich in iron is hard and crystalline whereas Ochreous goethite is mixed with aluminum through clayed and grained particles. Thus, the prime objective is to investigate how to reduce the impurities such as Al and Si. The existence of iron in ferromagnetic form, such as magnetite, could induce paramagnetic properties in the material, and this paramagnetic material could be attracted along the magnetic field towards the points of higher magnetic field intensity. Magnetite shows the strongest ferromagnetic properties although hematite and siderite can be transformed into magnetite using reduction roasting which can enhance their

IOP Conf. Series: Materials Science and Engineering **748** (2020) 012008 doi:10.1088/1757-899X/748/1/012008

separation trough magnetic separation method [2]. The sample fed in the magnetic separation consists of two products (i) the magnetic products (ii) the non magnetic products; consequently, the magnetic separation was used to separate both products the magnetic (iron ores) from the non magnetic (rejects).

A number of researchers have describe the techniques of magnetizing processes by roasting-reduction in the range of temperature 700-900°C but several of them unsatisfied with the recovering rate and cost balance. In this paper, an experiment was performed, to developing a technique of "reduction and magnetic separation" and showing the difference of feed sample and roasted sample of optical microscopy [2].

2. Experimental

The sample of goethite iron ore was taken from an iron ore tailing from the Langalota, Odisha in India. In the sample bigger size of particles was 12.5 mm. The sample divided into different size ranges by screening, mineralogy was characterized by optical microscopy and the chemical composition of the sample. And the coal is collected from garokhasi, Jharkhand, India the sample is characterized by proximate analysis, ultimate analysis, gross calorific value.

In this process coal and iron ore ratio is constant (1:9) and temperature varies from 700°C to 900°C and time varies from 10 minute to 60 minute. After roasting the roasted product is separated in magnetic separator at 2000 gauss, 4000 gauss and 6000 gauss.

The mixed sample of iron ore and coal is roasted at three different temperature 700°C, 800°C and 900°C for 10, 20, 30, 40, 50, 60 minutes and then separate in Wet High Intensity Magnetic Separator (WHIMS) magnetic field of 2000 gauss, 4000 gauss, 6000 gauss. The optical microscopy was utilized in order to analyze the phase transformation of goethite after the beneficiation of roasted ore samples by WHIMS (Wet High Intensity Magnetic Separation). This analysis was carried on to determine the iron ore recovery after treating it through dehydroxylation and magnetic separation.

3. Results and Discussion

3.1. Sample characterization

3.2. There is 40.212% Fe, and 19.04% SiO2 and 8.83% Al2O3 in the head sample. The chemical analysis of the main samples are shown in table 1.

Table 1. Chemical analysis of goethite (head) sample

Constituents	Fe	SiO2	Al2O3	LOI
Assay%	40.212	19.04	8.83	10.46

The sample was also screening into 4 different sizes, from +12.5 mm to 1 mm, (12.5mm, 6mm, 3mm, and 1 mm) for size by size analysis. The particle size distribution curves are presented in Fig. 1. Nearly 50% of sample contains coarser than 3 mm. And the d_{80} is calculated as 8.9 mm.

The optical microscopic analysis report the main mineral components of the samples are shown in Fig. 2. The iron ore sample was then taken for microscopic study under reflected light. The study showing iron ore and gangue are very finely interlocked. Due to its poor liberation characteristic it is very difficult to upgrade it by using physical beneficiation process.

Table 2. Size distribution of feed

mesh size(mm)	Nominal size(mm)	wt. (g)	wt.%	Cum wt.% Retain	cum wt.% Passing
+12.5	12.5	375	3.57	3.57	96.42
-12.5 +6	6	3971	37.86	41.44	58.56
--6 +3	3	2572	24.52	65.96	34.03
-3 +1	1	1434	13.67	79.64	20.36
-1 +0	0	2135	20.35	100	0.00
		10487	100		

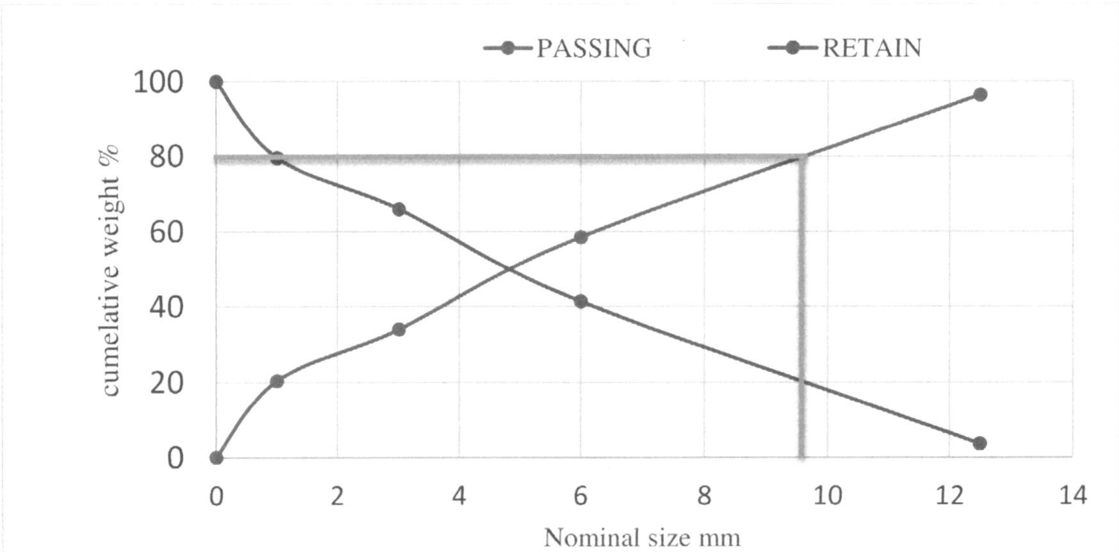

. **Figure 1.**size distribution curve

It required very fine grinding for liberation. The sample is then grinded to -75 micron by using jaw crusher, roll crusher and ball mill. The fine ore was again taken for microscopic study. The sample was then observed under both plain polarized and cross polarized light to know the different phase of iron present in it. The ore contain very less amount if Magnetite and high amount of hematite and goethite.

As the sample contain high amount of goethite and hematite reduction roasting is used to convert these phases to magnetite. So that the sample can be easily upgraded by using magnetic separation.

Table 3. Proximate analysis of coal sample (-75 micron)

Constituents	Moisture	Volatile Matter	Ash	Fixed Carbon
%	1.03	18.51	16.37	64.09

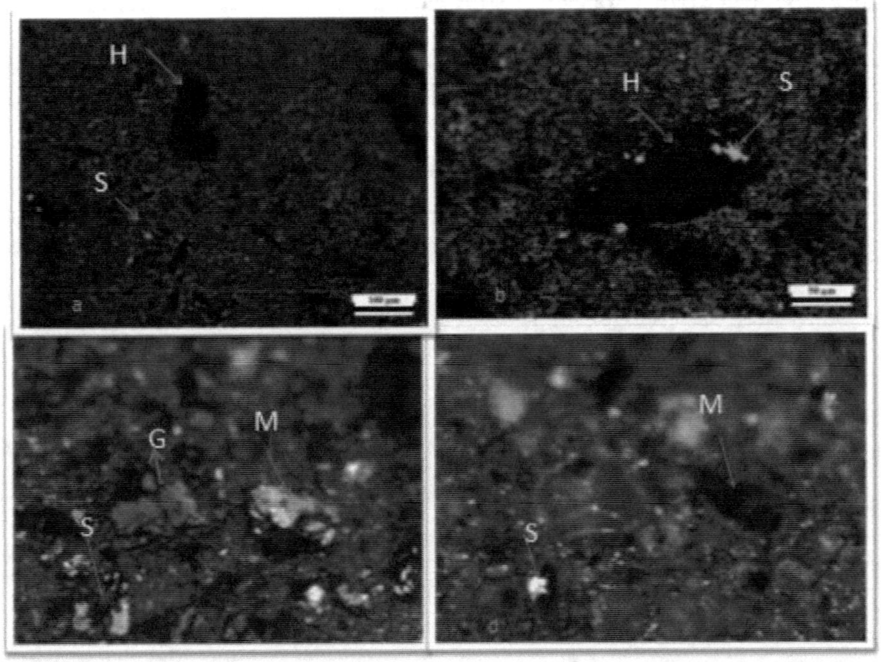

Figure 2.a Microscope image of feed in reflected light at 100 μm **Figure 2.b** Microscope image of feed in reflected light at 50 μm **Figure 2.c** Microscope image of -75 μm feed in plane polarized light G=Goethite, M=Magnetite, S=Silica **Figure 2.d** Microscope image of -75 μm feed in cross polarized light

3.2.1. Gross calorific value The grinded coal was then subjected to Bomb Calorimeter GCV = 7683 Kcal/kg.

3.2.2. Ultimate Analysis The sample was subjected for ultimate analyses to know the C, H, N, O, S %

Table 4 Ultimate analysis of coal

constituents	C	H	N	S	O
%	68.74	4.16	1.42	0.67	25.01

3.3. Magnetizing roasting and magnetic separation

The main factor affecting the quality of roasting process are coal and iron ore ratio, time temperature, and in magnetic separation main factor affecting grade and recovery is magnetic field intensity.

In this process coal and iron ore ratio is constant (1:9) and temperature varies from 700°Cto 1000°C and time varies from 10 minute to 60 minute. After roasting the roasted product is separated in magnetic separator at 2000 gauss, 4000 gauss and 6000 gauss.

IOP Conf. Series: Materials Science and Engineering **748** (2020) 012008 doi:10.1088/1757-899X/748/1/012008

3.4. Roasting at 700°C

The mixed sample of iron ore and coal is roasted at 700°C for 10, 20, 30, 40, 50, 60 minutes and then separated in magnetic field of 2000 gauss, 4000 gauss, 6000 gauss. With increase in time the grade and recovery both increases because with increase in time more iron ore phases converted to magnetite and recovery increases. With increase in magnetic field the recovery increases and grade decreases. In 700°C when sample is roasted for 60 minute and separated in 2000 gauss highest grade of 52.49 % obtained with a recovery of 70.3%.

3.5. Roasting at 800°C

The mixed sample of iron ore and coal is roasted at 800°C for 10, 20, 30, 40, 50, 60 minutes and then separated in magnetic field of 2000 gauss, 4000 gauss, 6000 gauss. With increase in time the grade and recovery both increases because with increase in time more iron ore phages converted to magnetite and recovery increases. With increase in magnetic field the recovery increases and grade decreases. In 800°C when sample is roasted for 60 minute and separated in 2000 gauss highest grade of 55.73 % obtained with a recovery of 73.97 %.

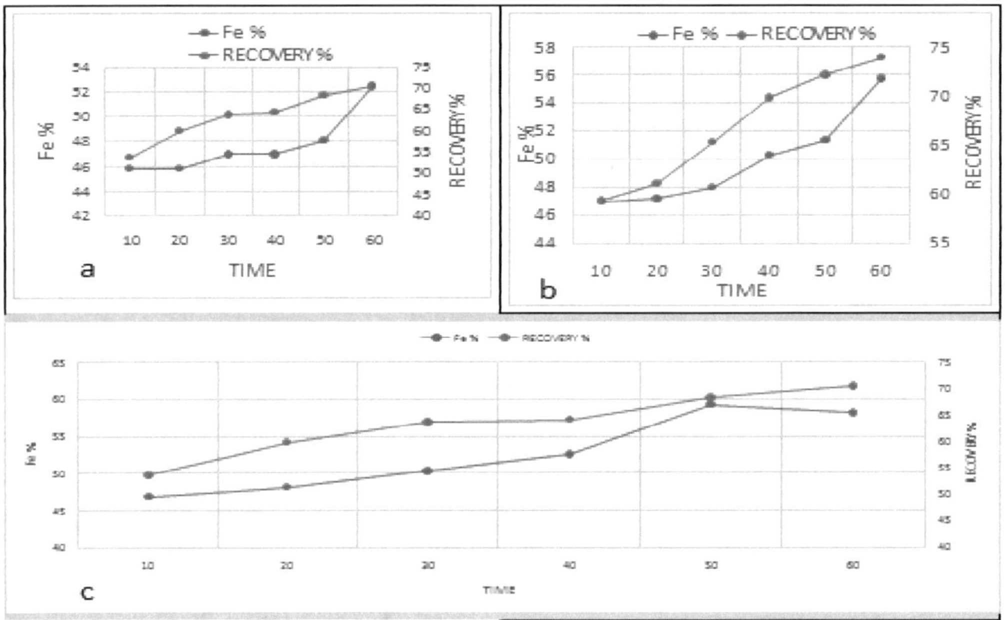

Figure 3: (a) Roasting for different time at 700°C and separation at magnetic field (2000 gauss). **(b)** Roasting for different time at 800°C and separation at magnetic field (2000 gauss) **(c)** Roasting for different time at 900°C and separation at magnetic field (2000 gauss)

3.6. Roasting at 900°C

The mixed sample of iron ore and coal was roasted at 900°C for10, 20, 30, 40, 50, 60 minutes and then separated in magnetic field of 2000 gauss, 4000 gauss, 6000 gauss in WHIMS by using ball matrix. With increase in time both grade and recovery increases up to 50 minute but in 60 minute it decreases. Here up to 50 minute when temperature increases the grade and recovery decrease because different iron ore phases like goethite hematite are converted to magnetite but in 60-minute grade decreases because here some FeO produce which is les magnetic in nature. When the separation was at 2000 gauss the grade was high because the magnetic field-attract high magnetic particles like magnetite and less hematite, goethite.

When the separation was at 4000 gauss the magnetic field attract some amount of hematite and goethite also with magnetite so the recovery increases but the grade was decreased. During separation at 6000 gauss grade was very low and recovery was very high because at this magnetic field the magnetic field attract high amount of goethite and hematite with magnetite.

After magnetic separation microscopic study for magnetic part was performed. After roasting and magnetic separation iron ore phases changes to magnetite but silica amount was not reduced significantly.

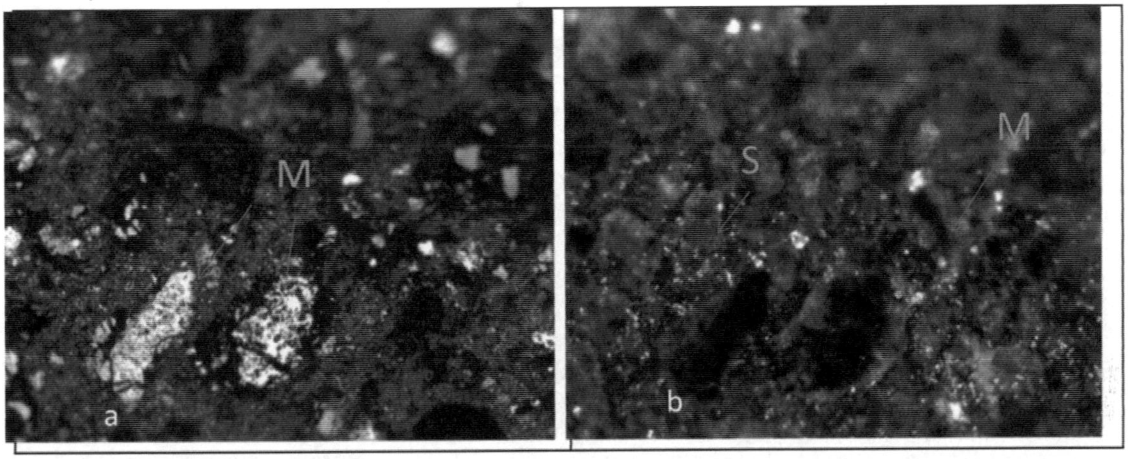

Figure 4: (a) Microscopic study of product after roasting and magnetic separation in plane polarized light. **(b)** Microscopic study of product after roasting and magnetic separation in plane polarized light.

4. Conclusions

In this article, the mineral transformation and chemical of goethite done by roasting and magnetic separation. The iron ore (goethite) sample of 12.5 mm head size was taken from an iron ore tailing from the Langalota, Odisha region in India. The size analysis carried out a essential amount of fine particles: 30 wt% < 1 mm. Mineralogical and Chemical analysis was conducted using optical microscope and volumetric analysis. The chemical analysis of the sample had 40.212% Fe, 19.04% silica and 8.83% alumina. Optical microscope analysis had the iron ore and gangue are very finely interlocked and magnetite particles shows bright colour in plane polarized light and black color in cross polarized light because it is isotropic in nature. The content of aluminium and silicon in the mineral phases was of the reverse (increasing) order. The laterite iron ore which contain 40.12 % Fe can be upgraded to 59.20 % with 68.23 % Recovery by reduction roasting at 9000C for 50 minute and magnetic separation at 2000 gauss. In reflected light microscope the changes in phase of iron ore was also studied and found that after roasting higher amount of magnetite was found in product. In this Experiment Fe % is increasing from 40.21 to 59.20 % but silica and alumina % is not changing significantly so this can't be useful for pellet making. So This process can be combined with flotation or gravity separation to decrease the silica and alumina % and its product can be used in pellet making. The product of this process can be used in pellet making blending with high grade iron ore with less alumina and silica.

References

[1] Chao Li, Henghu Suna, Jing Baic, 2010 The recovery of iron from iron ore tailings using magnetic separation after magnetizing roasting, Innovative methodology for comprehensive utilization of iron ore tailing part 1, Longtu Li *Journal of Hazardous Materials* **174** 71-77.

[2] Kyoung-oh Jang, Venkata R.M. Nunna, Sarath Hapugoda, Anh V. Nguyen, Warren J. Bruckard,

2014 Chemical and mineral transformation of a low grade goethite ore by dehydroxylation, reduction roasting and magnetic separation, *Minerals Engineering* **60** 14–22.

[3] Yan Wua, Mei Fang, Lvdeng Lan, Ping Zhang, K.V. Rao, Zhengyu Bao, 2012 Rapid and direct magnetization of goethite ore roasted by biomass fuel, *Separation and Purification Technology* **94** 34–38.

[4] LI Ke-qing, NI Wen, ZHU Ming, ZHENG Mei-juan, LI Yuan, 2011 Iron Extraction From Oolitic Iron Ore by a Deep Reduction Process, *Journal of iron and steel research, international*, **18(8)**: 09-13.

[5] Das, S.K., Das, B., Sakthivel, R., Mishra, B.K., 2010. Mineralogy, microstructure, and chemical composition of goethites in some iron ore deposits of Orissa, India. *Mineral. Process. Extr. Metall. Rev.* **31 (2)**, 97–110.

[6] Huifen Yanga, Lili Jing, 2011 Recovery of vanadium tailing with coal-based direct reduction followed by magnetic separation, *Baogang Zhangb.*

[7] Jalil Vahdati khaki, Hossein Shalchian, Ali Rafsanjani-Abbasi, 2018 Recovery of iron from a high-sulfur and low-grade iron ore, *Naghdali Alavifard Thermochimica Acta* **662** 47-54.

[8] Jianwen YU, Yuexin Han, Yanjun Li, 2017 Beneficiation of an iron ore fines by magnetization roasting and magnetic separation, *Peng Gao International Journal of Mineral Processing* **168** 102-108.

[9] Kasonde Mawejaa, Tshikele Mukongob, Ilunga Mutombo 2009 (Cleaning of a copper matte smelting slag from a water-jacket furnace by direct reduction of heavy metals).

[10] Swagat S. rath, Danda S. Rao, Barada K. Mishra, 2016A novel approach for reduction roasting of iron ore slime using cow dung, *International Journal of Mineral Processing 1***57** 216-226.

[11] Veeranjaneyulu Rayapudi, Nikhil Dhawan, 2018 Up-gradation of banded iron ores for pellet grade concentrates.

[12] Wanchao Liu, Jiakuan Yang, Bo Xiao, 2009 Application of Bayer red mud for iron recovery and building material production from alumosilicate residues.

[13] Yali Zhang, Huaimei Li, Xianjin Yu, 2012 Recovery of iron from cyanide tailings with reduction roasting followed by magnetic separation, *Journal of Hazardous Materials* **213-214** 167-17.

[14] IBM Indian Minerals Yearbook 2015.

[15] Iron and steel Vision 2011 **2020**.

Exergoeconomic and enviroeconomic analysis of semitransparent and opaque photovoltaic (PV) panels: a comparative study

R K Mishra[1*], Gagan Chaudhary[2], Rajesh Tripathi[3] and Rajendra Prasad[4]

[1, 3, 4]Department of Applied Sciences (Physics) Galgotias College of Engineering and Technology, Greater Noida
[2]Department of Mechanical Engineering, HIET, Ghaziabad, UP
E-mail:bhu.rajeev@gmail.com

Abstract. In present communication the exergoeconomic and enviroeconomic analyses of two different stand-alone photovoltaic (PV) modules viz. semitransparent and opaque have been performed. The exergoeconomic examination is very useful way to investigate the correlation between capital cost and exergy losses. The enviroeconomic analysis describes the carbon dioxide mitigation by the system. The energy (electrical) produced annually by semitransparent and opaque PV panel is also evaluated. It is reported that the semitransparent panel shows better performance as for as energy saving are concern. Environmental cost reduction is found to be 128.7 Rs. /year and 125.95 Rs. /year for semitransparent and opaque PV modules, respectively.
Keywords: Exergoeconomic, enviroeconomic, PV module, exergy.

*Corresponding author

1. Introduction

All around the world there are lots of major consequences because of using conventional fuels. One of the basic reasons for the environmental pollution is the emission of hazard gases [1, 2]. Because of the bad environmental impact and poor stability of the economics the development in the PV sector is of great interest [3]. The annual energy from panel depends on the orientation of the panel [4, 5]. Performance of PV panel using full-scale mock-up model was studied by Song et al. [6] for South Korea. Author found that the photovoltaic module inclined at 30°, facing due south, gave around 2.5 times more power as compared with the PV panel placed vertically. The electrical efficiency of photovoltaic panel is also strongly dependents temperature of the module. Park et al. [7] experimentally revealed that power was found to be decreased by 0.48–0.52% for one degree rise in temperature of the panel. The energy production from the PV module is affected by its temperature which changes as the property of glass changes. For designing and analyzing the energy systems, the thermodynamics and the cost accounting of the system should be combined. According to Rosen and Dincer [8] the unit cost based on energy is utilized in order to calculate the cost of energy conversion devices. Singh et al. [9] observed that the quantitative and qualitative losses are analyzed by exergy examination. The performance of energy systems is evaluated and improved using exergoeconomic analysis also termed as the exergetic cost analysis [10]. Environmental policies affecting the environmental cost analysis is termed as Enviroeconomic analysis. Badescu [11] analyzed that by using an electrical energy storage system connected with a photovoltaic array, the heat pump compressor can be drived. Ozgener and Hepbasli [12] observed a structured link between the total cost and loss in exergy while investigating solar operated appliances. The exergoeconomic evaluation of hybrid photovoltaic thermal module air heater was carried out by Agrawal and Tiwari [13]. Author reported that the hybrid photovoltaic thermal (PVT) system saves more energy than a PV panel. Mishra and Tiwari [14] found that a photovoltaic thermal (PVT) solar collector is a viable option not only for fulfilling the need of hot water but also for getting DC power during daytime when sun is there. Sahota and Tiwari [15] while studying the energy and exergy of a solar still, found that the carbon dioxide released annually was estimated to be 14.95 tonnes and 3.17 tonnes hybrid system (without heat exchanger) and 24.61 tonnes and 2.36 tonnes for hybrid system (with heat exchanger). Elbar et al. [16] performed exergoeconomic and environmental estimation of new solar distillation unit integrated with photovoltaic panel (PV). The result showed that solar still system, used with PV as reflector, is the most efficient technology.

In this communication, the electrical energy production, exergoeconomic and enviroeconomic analyses of two distinct types of stand-alone PV panel namely semitransparent (Fig. 1a) and opaque (Fig. 1b) are evaluated for New Delhi India [17].

Figure 1(a): Photograph of Semitransparent PV module.

Figure 1(b): Photograph of opaque PV module.

2. Thermal modeling

(i) Semitransparent PV module (Fig. 1a)

For semitransparent PV panel [18]:

$$\tau_g \alpha_c \beta_c \, I(t) bdx = \left[U_{tm,a}\left(T_m - T_a\right) + U_b\left(T_m - T_a\right)\right] bdx + \eta_c \tau_g \beta_c \, I(t)\, bdx$$

(1a)

From above Eq. (1a), the equation for Tm will be given as

$$T_m = \frac{(\tau_g \alpha_c \beta_c - \eta_m) I(t) + U_{tm,a} T_a + U_b T_a}{U_{tm,a} + U_b} \tag{1b}$$

where, $\eta_m = \eta_c \tau_g \beta_c$

The equation for electrical efficiency of a PV panel [19, 20]

$$\eta_m = \eta_{m0}\left[1 - \beta_{ref}\left(\overline{T_m} - T_{ref}\right)\right] \tag{1c}$$

where, T_{ref} = 25°C for indoor testing conditions and T_{ref} = T_a (ambient temperature) for outdoor testing conditions.

From above Eqs. (1b) and (1c),

$$\eta_m = \frac{\eta_{m0}\left[1 - \dfrac{\beta_{ref} \tau_g \alpha_c \beta_c I(t)}{U_{tm,a} + U_b}\right]}{1 - \dfrac{\beta_{ref} \eta_{m0} I(t)}{U_{tm,a} + U_b}} \tag{1d}$$

(ii) Opaque PV module (Fig. 1b)

For opaque PV panel [19]

$$\tau_g\left[\alpha_c \beta_c + \alpha_T\left(1 - \beta_c\right)\right] I(t) bdx = \left[U_{tm,a}\left(T_m - T_a\right) + U_b\left(T_m - T_a\right)\right] bdx + \eta_c \tau_g \beta_c I(t) bdx \tag{2a}$$

From Eq. (2a),

$$T_m = \frac{\left[(\alpha\tau)_{G-T,eff} - \eta_m\right] I(t) + U_{tm,a} T_a + U_b T_a}{U_{tm,a} + U_b} \tag{2b}$$

where, $\eta_m = \eta_c \tau_g \beta_c$

From Eqs. (1c) and (2b)

$$\eta_m = \frac{\eta_{m0}\left[1 - \dfrac{\beta_{ref} (\alpha\tau)_{G-T,eff} I(t)}{U_{tm,a} + U_b}\right]}{1 - \dfrac{\beta_{ref} \eta_{m0} I(t)}{U_{tm,a} + U_b}} \tag{2c}$$

The electrical power produced can be given as [21]

$$P(e) = \eta_m A_m I(t) \tag{2d}$$

3. Net present value

Let us consider P is present value, S is the salvage value, R_1 is operational costs per year and R_5's are painting and maintenance cost for the interval of five years.
NPV can be written as [22]

$$Net\ Present\ Value\,(NPV) = P + R_1 \times \left[\frac{(i+1)^n - 1}{i(i+1)^n}\right]_{i,n} + R_{5,1} \times \left[\frac{1}{(i+1)^n}\right]_{i,5} + R_{10,2} \times \left[\frac{1}{(i+1)^n}\right]_{i,10} + R_{15,3} \times \left[\frac{1}{(i+1)^n}\right]_{i,15} + \dots\dots\dots - S \times \left[\frac{1}{(i+1)^n}\right]_{i,n} \quad (3)$$

4. Exergoeconomic evaluation

Because of irreversibility, exergy is lost during a process.

Energy in- Energy out = Energy accumulation (4a)

and

Exergy in - Exergy out- Exergy consumed= Exergy accumuled (4b)

From (4a) and (4b),

Energy out = Product energy out + Waste energy out (4c)

and

Exergy out = Product exergy out + Waste exergy out (4d)

The loss rates in terms of energy and exergy can be written as:

$$\text{Energy loss rate } (L_{en}) = \sum_{\text{inputs}} \text{Energy flux rates} - \sum_{\text{products}} \text{Energy flux rates} \quad (4e)$$

and

$$\text{Exergy loss rate } (L_{ex}) = \sum_{\text{inputs}} \text{Exergy flux rates} - \sum_{\text{products}} \text{Exergy flux rates} \quad (4f)$$

From Eqs. (1a), (4e) and (4f), for semitransparent panel

$$\text{Energy loss rate } (L_{en}) = \sum \tau_g \alpha_c \beta_c I(t) b dx - \sum \eta_c \tau_g \beta_c I(t) b dx \quad (4g)$$

and

$$\text{Exergy loss rate } (L_{ex}) = \sum \tau_g \alpha_c \beta_c I_{ex}(t) b dx - \sum \eta_c \tau_g \beta_c I_{ex}(t) b dx \quad (4h)$$

where

$$I_{ex}(t) = I(t) \times \left[1 - \frac{4}{3} \times \left(\frac{T_a}{T_s}\right) + \frac{1}{3} \times \left(\frac{T_a}{T_s}\right)^4\right]$$

where, T_s is the sun temperature in K.

From Eqs. (2a), (4e) and (4f) for opaque PV panel

$$\text{Energy loss rate } (L_{en}) = \sum \tau_g \left[\alpha_c \beta_c + \tau_T (1-\beta_c)\right] I(t) b dx - \sum \eta_c \tau_g \beta_c I(t) b dx \quad (4i)$$

and

$$\text{Exergy loss rate } (L_{ex}) = \sum \tau_g \left[\alpha_c \beta_c + \tau_T (1-\beta_c)\right] I_{ex}(t) b dx - \sum \eta_c \tau_g \beta_c I_{ex}(t) b dx \quad (4j)$$

Following Rosen and Dincer [23] define,

$$R_{en} = \frac{L_{en}}{\text{NPV}} \quad (4k)$$

and

$$R_{ex} = \frac{L_{ex}}{NPV} \quad (41)$$

5. Enviroeconomic evaluation

The cost reduction parameter for environment can be given as [24]:

$$C_{CO_2} = c_{CO_2} \times x_{CO_2} \quad (5)$$

where, C_{CO_2} is the cost reduction parameter for environment, c_{CO_2} is CO_2 emission reduction price for one tCO_2 and x_{CO_2} is reduction in CO_2 mitigation per year (tCO_2/year).

(i) Glass to glass type PV module

Total electrical energy produced per annum = 79.18 kWh

Suppose that the unit power is used and reduction in energy because of bad quality of home devices is about 20%. The transferred power will be 1.25 units. Suppose there are 40% losses in the transmission and distribution channels.

The generated power in the plant is $\frac{1.25}{1-0.4} = 2.08$ units. The average CO_2 identical intensity for electrical energy from coal is approximately 0.98 kg of CO_2 per kWh at the source.

Thus, for unit power used the amount of CO_2 emission will be 2.08×0.98=2.04 kg.

The annual CO_2 emission reduction = 79.18 × 2.04 kg CO_2e

= 0.16132 tCO_2e (1 ton = 10^3 kg)

The price of international carbon was between 13 \$/tCO_2 and 16 \$/tCO_2 [24]. The average is 14.5 \$/tCO_2.

From Eq. (5), environmental cost reduction per annum = 0.16132 ×14.5 \$= 2.34 \$.

(ii) Glass to tedlar type PV module

Total electrical energy produced per annum by glass to tedlar type PV module = 77.75 kWh

The carbon dioxide emission reduction per annum = 77.75 × 2.04 kg CO_2e

= 0.1586 tCO_2e (1 ton = 10^3 kg)

Using Eq. (5) environmental cost reduction per annum = 0.1586 ×14.5 = 2.29 \$.

6. Results and discussion

The detail specifications of both types of PV panels are listed in Table 1. Figure 2 shows plot of sun intensity and environmental temperature for a clear day in June, 2019 for New Delhi, India. The changes of module surface temperature and experimental electrical efficiency on hourly basis for panel are given in Figure 3. Figure reveals that semitransparent PV panel gives more efficiency as compared with opaque PV panel. This is because of the reason that the solar energy falling on the area of the panel which is not packed by solar cell is directly pass through the glass cover of panel, but, in opaque PV panel all the radiations which falls on the area where solar cells are there is absorbed by the tedlar. So, cell temperature for opaque PV module is higher than temperature in semitransparent PV panel.

Table 1: Specifications of photovoltaic modules.

Specifications	Values
b	0.605 m
K_g	1.1W/m K
K_T	.033W/m K
L_g	.003 m
L_T	.0005 m
α_c	0.9
α_T	0.5
PF for semitransparent	0.76
PF for opaque	0.83
β_{ref}	0.0045/K
η_{m0}	0.12
τ_g	0.95
Life of PV module	30 Years

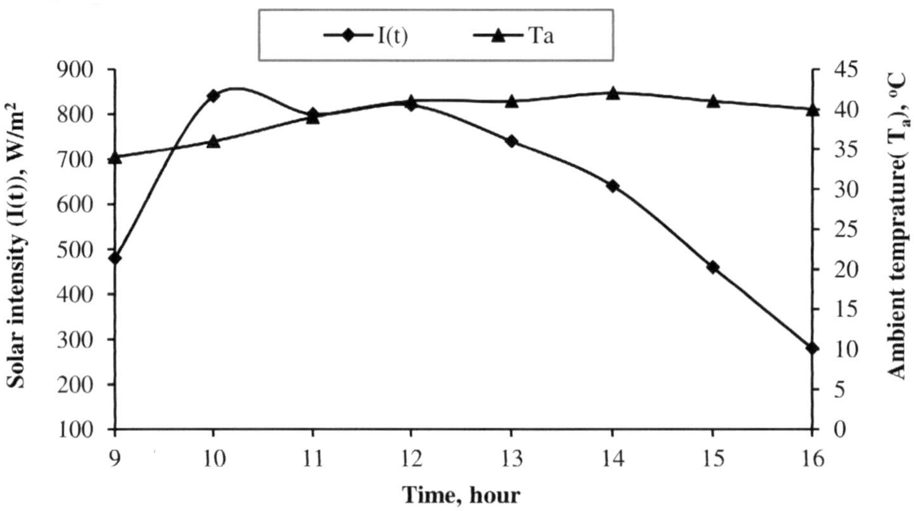

Figure 2: Plot of sun radiation and environmental air temperature for a clear day in June, 2019.

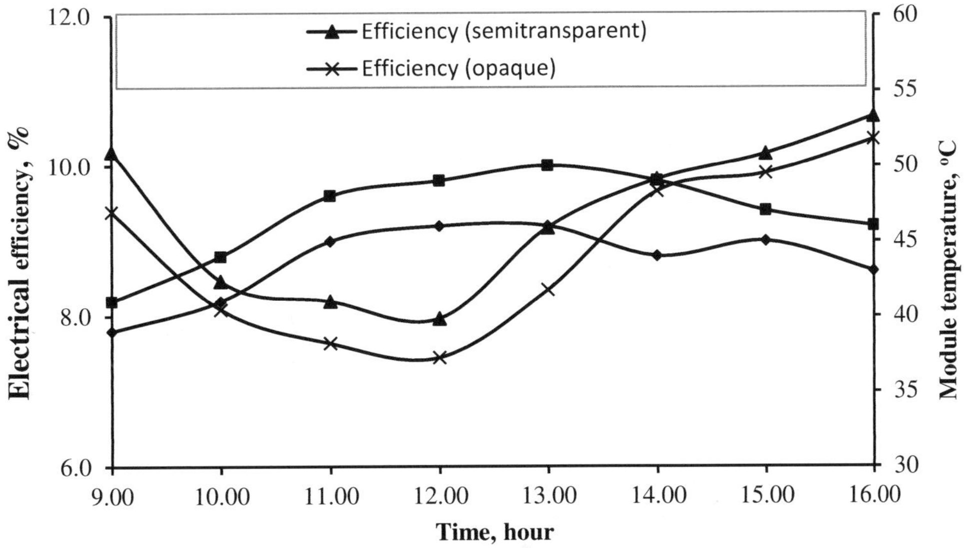

Figure 3: Plot of module surface temperature and efficiency for a clear day in June, 2019.

Figure 4 shows the hourly changes of the power obtained from semitransparent and opaque PV modules. It is observed that the power generated by semitransparent module is higher as compared to opaque module because of the higher electrical efficiency of semitransparent panel. The measured values of electrical parameters for semitransparent and opaque panels are given in Tables 2 and 3, respectively.

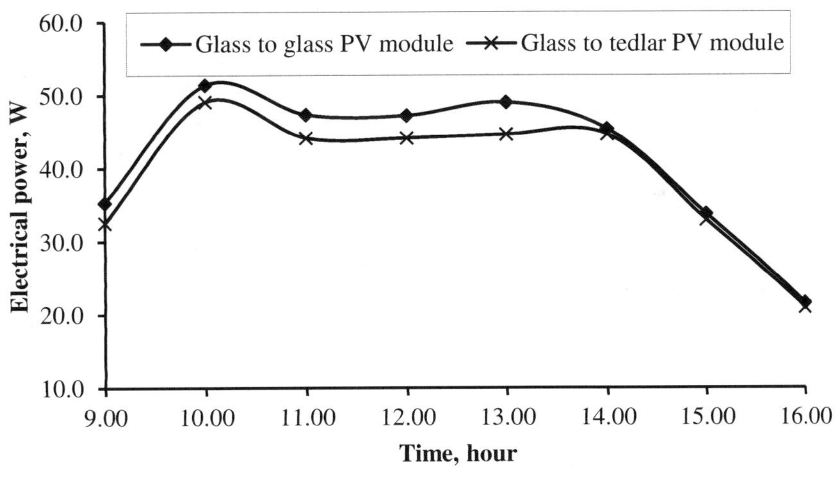

Figure 4: Plot of electrical power for a clear day in month of June, 2019.

Table 2: Parameters for semitransparent PV module June 18, 2019.

Sr no	Time (hour)	I(t) (W/m^2)	Ta (°C)	Tm (°C)	Isc (A)	Voc (V)	η_m (%)	Electrical power (W)
1	9	480	34	39	2.3	19.2	10.19	35.33
2	10	840	36	41	3.4	18.9	8.47	51.41
3	11	800	39	45	3.2	18.5	8.20	47.36
4	12	820	41	46	3.2	18.4	7.97	47.22
5	13	740	41	46	3.3	18.6	9.18	49.05
6	14	640	42	44	3	18.9	9.82	45.39
7	15	460	41	45	2.2	19.2	10.16	33.76
8	16	280	40	43	1.4	19.2	10.66	21.56

Table 3: Electrical parameters for opaque PV module for a June 18, 2019.

Sr no	Time (hour)	I(t) (W/m^2)	Ta (°C)	Tm (°C)	Isc (A)	Voc (V)	η_m (%)	Electrical power (W)
1	9	480	34	41	2.2	18.5	9.39	32.56
2	10	840	36	44	3.3	18.6	8.09	49.10
3	11	800	39	48	3.0	18.4	7.64	44.16
4	12	820	41	49	3.0	18.4	7.45	44.16
5	13	740	41	50	3.0	18.6	8.35	44.64
6	14	640	42	49	3.0	18.6	9.66	44.64
7	15	460	41	47	2.2	18.7	9.90	32.91
8	16	280	40	46	1.4	18.7	10.35	20.94

Eqs. (1d) and (2c) are used to calculate the theoretical values for PV panel. Monthly average electrical parameter of semitransparent and opaque PV modules for New Delhi, India are evaluated and the result is given in Figures 5 and 6 respectively. The annual average electrical efficiency of semitransparent and opaque PV module is found to be 10.17% and 9.92% respectively. Eq. (2d) is used to calculate the electrical energy of PV panel. Figures 7 and 8 show the monthly variation of average electrical energy produced by semitransparent and opaque PV module respectively. Figures reveal that the annual average electrical energy produced by semitransparent and opaque PV modules is 79.18 and 77.75 kWh respectively. Then the monthly average electrical efficiency and electrical power produced is calculated [15].

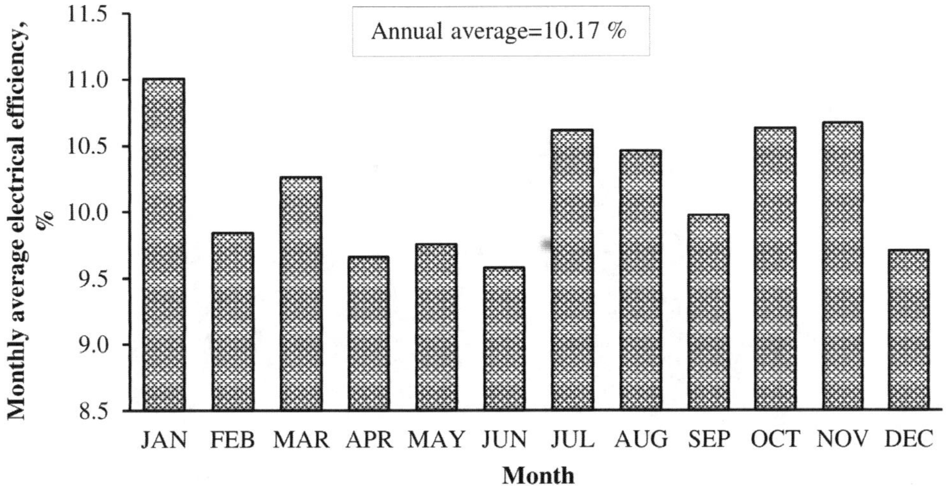

Figure 5: Electrical efficiency for semitransparent PV module.

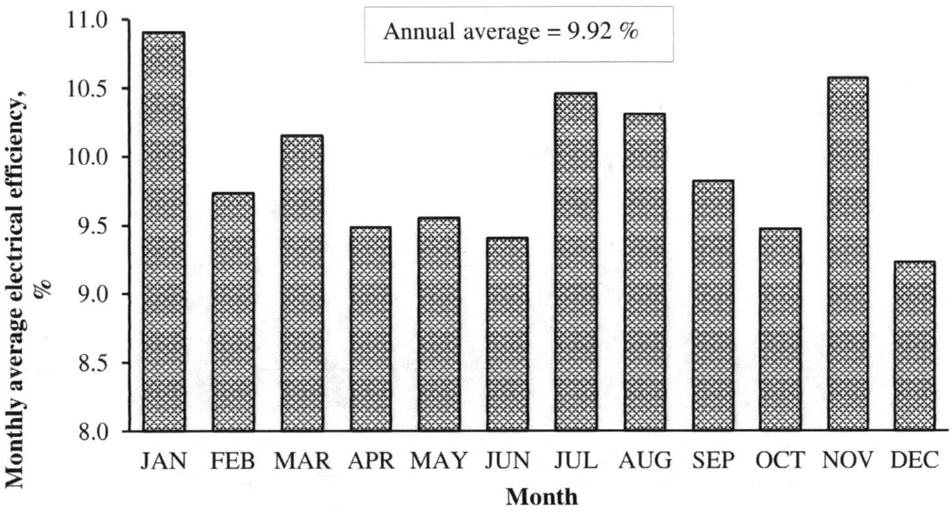

Figure 6: Month wise electrical efficiency for opaque PV module.

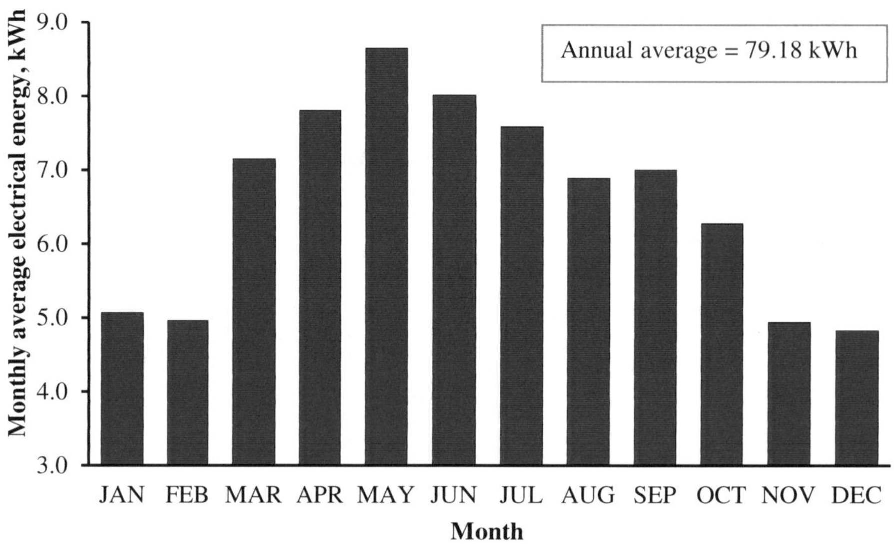

Figure 7: Month wise electrical energy obtained from semitransparent PV module.

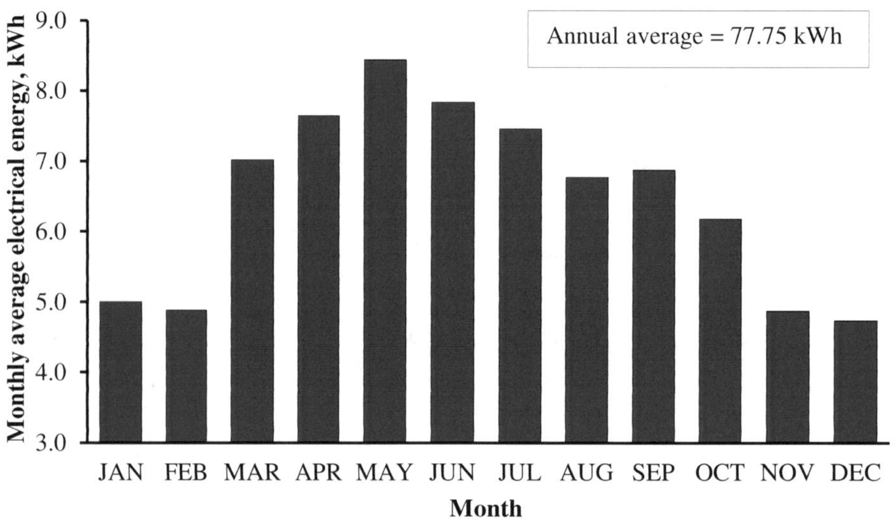

Figure 8: Month wise electrical energy obtained from opaque PV module.

Capital cost and salvages value for different component are shown in Table 4. The net present value of semitransparent and opaque PV module is calculated by using Eq. (3).

The rate of losses in energy and exergy for modules has been computed by using Eqs. (4g), (4h), (4i) and (4j). Eqs. (4k) and (4l) are used to calculate the parameters R_{en} and R_{ex}. The detail results of exergoeconomic and enviroeconomic analyses for New Delhi is shown in Tables 6. From Table it is clear that ratio of loss rate to net present value (R_{en} and R_{ex}) is lower and the cost reduction for environment per annum is higher for semitransparent PV module for both the climatic conditions. It indicates that in terms of saving of energy and environment effect, the semitransparent PV panel gives better results as compared with opaque PV panel.

IOP Conf. Series: Materials Science and Engineering **748** (2020) 012009 doi:10.1088/1757-899X/748/1/012009

Table 4: Capital cost and salvages value for different component of the systems.

Components	Qty in kg	PV module With stand ($)	Salvage value (S_S) at the inflation rate of 5% (present values of Iron scrap @$0.36/kg)
			After 30 year Iron scrap @$1.18/kg
Support structure @ $ 1.45/kg	10	14.5	11.8
PV module G-G @ $290.9/ 75Wp PV module G-T @ $218.1/ 75Wp	1 no. 1 no.	290.9 218.1	9.1 9.1
Paint @ $1.8/kg	0.5 kg	0.91	*
Fabrication charges		9.1	*
Capital cost ($)	G-G PV module G-T PV module	315.5 242.7	20.9 20.9

Table 5: The result for semitransparent and opaque PV module for New Delhi, India, climatic conditions.

PV module	NPV ($)	L_{Ten} (kWh)	R_{en} (kWh/$-year)	L_{Tex} (kWh)	R_{ex} (kWh/$-year)	C_{co2} ($/year)
Glass to glass	382.1	25239.43	2.2	25107.57	2.19	2.34
Glass to tedlar	309.5	23770.5	2.56	23238.57	2.50	2.29

7. Conclusions

Following conclusions have been drawn:

(i) Electrical power generated by semitransparent PV module is higher as compared to opaque panel.

(ii) In terms of energy saving and effects on environment, the semitransparent PV panel is better as compared with opaque panel.

References

[1] Omer A M 2008 Energy environment and sustainable development *Ren and Sus. En. Rev.* **12** 2265–300

[2] Tiwari G N and Mishra R K 2012 Advanced Renewable Energy Sources *(RSC Publishing Cambridge UK)*

[3] Chaar L E, Lamont L A and Zein N E 2011 Review of photovoltaic technologies *Ren. and Sus. En. Rev.* **15** 2165–75

[4] Mishra R K and Tiwari G N 2013 Energy and exergy analysis of hybrid photovoltaic thermal water collector for constant collection temperature mode *Int. J. Sol. En.* **91** 161-173

[5] Mishra R K, Tiwari S, Tiwari A and Tiwari G N 2017 Unit cost analysis for Sodha Bers Complex (SBC): An energy efficient building *Th. Sc. Eng. Prog.* **4** 58-70

[6] Song J H, An Y S, Kim S G, Lee S J, Yoon J H and Choung Y K 2008 Power output analysis of transparent thin-film module in building integrated photovoltaic system (BIPV) En. and build. **40** 2067–75

[7] Park K E, Kang G H, Kim H I, Yu G J and Kim J T 2010 Analysis of thermal and electrical performance of semi-transparent photovoltaic (PV) module *Energy* **35** 2681–7

[8] Rosen M A, Dincer I and Kanoglu M 2008 Role of exergy in increasing efficiency and sustainability and reducing environmental impact, *En. Pol.* **36** 128-137

[9] Singh N, Kaushik S C and Misra R D 2000 Exergetic analysis of solar thermal power system *Ren. En.* **19** 135–43

[10] Sahoo P K 2008 Exergoeconomic analysis and optimization of a cogeneration system using evolutionary programming *App. Th. Eng.* **28** 1580–8.

[11] Badescu V 2002 First and second law analysis of a solar-assisted heat pump based heating system *En. Conv. and Mang.* **43** 2539–52

[12] Ozgener O and Hepbasli A 2005 Exergoeconomic analysis of a solar assisted ground-source heat pump greenhouse heating system *App. Th. Eng.* **25** 1459–71

[13] Agrawal S and Tiwari G N 2012 Exergoeconomic analysis of glazed hybrid photovoltaic thermal module air collector *Int. J. Sol. En.* **86** 2826–2838

[14] Mishra R K and Tiwari G N 2013 Energy and exergy analysis of hybrid photovoltaic thermal water collector for constant collection temperature mode *Int. J. Sol. En.* **90** 58-67

[15] Sahota L and Tiwari G N 2017 Exergoeconomic and enviroeconomic analyses of hybrid double slope solar still loaded with nanofluids *En. Conv. and Mang.* **148** 413–430

[16] Elbar A R A, Yousef M S and Hassan H 2019 Energy, exergy, exergoeconomic and enviroeconomic (4E) evaluation of a new integration of solar still with photovoltaic panel *J. Clean. Prod.* **233** 665-680

[17] Dubey S, Sandhu G S, Tiwari G N 2009 Analytical expression for electrical efficiency of PV/T hybrid air collector *App. En.* **86** 697–705

[18] Tiwari A and Sodha M S 2006 Performance evaluation of hybrid PV/thermal water/air heating system: A parametric study *Ren. En.* **31** 2460-2474

[19] Schott T 1985 Operation temperatures of PV modules: A theoretical and experimental approach *Proc. Sixth EC PV Sol. En. Conf. (London)* p 392-396

[20] Evans D L 1981 Simplified method for predicting PV array output *Int. J. Sol. En.* **27** 555-560

[21] Skoplaki E and Palyvos J A 2009 On the temperature dependence of photovoltaic module electrical performance: A review of efficiency/power correlations *Int. J. Sol. En.* **83** 614-624

[22] Tiwari G N, Mishra R K and Solanki S C 2011 Photovoltaic modules and their applications: A review on thermal modelling *App. En.* **88** 2287-2304

[23] Rosen M A and Dincer I 2003 Exergoeconomic analysis of power plants operating on various fuels *App. Th. Eng.* **23** 643–58

[24] Den Elzen M G J, Hof A, Beltran D A M, Grassi R G, Van R M, Van V B and Van V J D P 2011 The Copenhagen accord: abatement costs and carbon prices resulting from the submissions *Env. Sci. & Pol.* **14** 28–39

Optimizing Connecting Rod Design Using Fillet Radius Modifications in ANSYS

Swet Chandan[1], Abhishek Singh[2]

[1]Galgotias University, Yamuna Expressway Greater Noida 201307, India
swetchandan@gmail.com
[2]Galgotias University, Yamuna Expressway Greater Noida 201307, India
abhisheksingh10051993@gmail.com

Abstract.
In this paper we have investigated a connecting rod design of a 2wheeler,. We have first measured the design parameters of the existing connecting rod and established the material used and cause of failure. We found that surface roughness in the big end is the main area for failure. Then a solid model of the same connecting rod was made using solidworks software. ANSYS software was employed to analyze and examine the effect of stress in different parts of the connecting rod. For the analysis we have started with the existing design parameters and then incrementally changed the fillet radius at big end with keeping the small end dimensions constant and then changing the fillet radius of the small end with keeping the fillet radius of the big end intact. Finally optimum parameters were selected for the modified design for the minimum stress conditions.

Keywords: Connecting rod, Fillet radius, stress analysis, Optimization, Finite element analysis.

1 Introduction

Connecting rod is one of the most important parts of any internal combustion engine which experiences cyclic loading. Mostly it consists of three parts: small end or pin end, the main body and the big end or the crank end. Aftzal et al (2003) has analyzed in the detail connecting rod dynamic loading and has shown how time dependent load and stress varies in a given connecting rod. Mirehei et al. (2008) has employed ANSYS software to investigate the various stress acting on the connecting rod used in the universal tractor. Webster et al. (1983) have extensively modelled a connecting rod used in the high speed diesel engine. They have employed iso-parametric 20-node, 15-node and beam element and used symmetry to reduce the computation time. Calculation of pressure distribution on the big end and small end due to force exerted by the piston is still used by most of the finite element analysis. Athavale and Sajanpawar (1991) has used modeled the connected rod using finite element and employed inertia load for the analysis. Khare et al.(2012) has done investigation on the connecting rod of a two wheeler and found that the fatigue load is the main culprit of the early failure of the given connecting rod. They have used fillet radius and contact stresses to analyze the problem. Lee et al.(2006) has used big end and small end fillet radius as a design parameters to analyze the connecting rod.

2 Model Description:

2.1 Sample Selection:

We have selected the sample of a two wheeler connecting rod after the user complains about the vibration and non smooth driving condition by the driver. After investigation we have found that connecting rod of the vehicle is damaged due to spalling. Dimensions of the connecting rod were measured and with the help of laboratory tests composition of the material was established.

Fig. 1: Actual model of sample of Yamaha Crux connecting rod

2.2 SOLIDWORKS (3D) model of sample:

After the sample selection I made the 3D model of sample in SOLIDWORKS software with the same dimension of sample. For making the model I started with following steps.

2.2.1 Design Criteria:

Connecting rod converts reciprocating motion of piston into rotary motion of the crankshaft. It consists of two parts the upper part is forked section which fits on the crosshead bearing and lower part fit on the crankpin bearing. Different dimensions of internal combustion engine connecting rod are given below. For making the model of connecting rod we have used following configuration.

Connecting rod length=12.2 cm
Piston diameter=1.8 cm
Crank radius=1.35 cm
Stroke length=2*Crank radius=3.7 cm
Mass of connecting rod=0.52894 kg
Volume of connecting rod=6738.1 mm^3

With the help of above parameters 3 D model was constructed for the connecting rod in the SOLIDWORKS 2016 software. The SOLIDWORK model of the connecting rod is shown in figure 2.

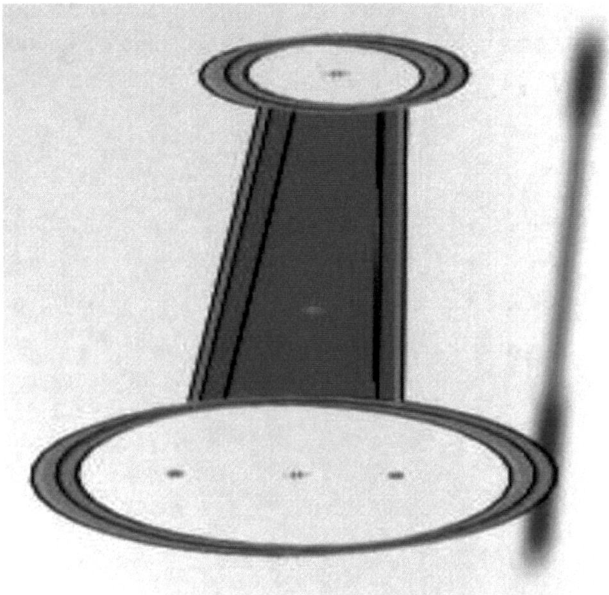

Fig. 2: SOLIDWORKS (3D) model of sample connecting rod

S No	Material Property	SAE-4140
1.	Young's modulus (MPa)	71000
2.	Density	7.85g/cm^3
3.	Poisson's ratio	0.0.27
4.	Bulk Modulus(MPa)	140e+003
5.	Shear Modulus(MPa)	80e+003

Table 1: Mechanical properties of materials(From Spectro Analysis)

3. Finite Element Analysis under the compression loading condition:

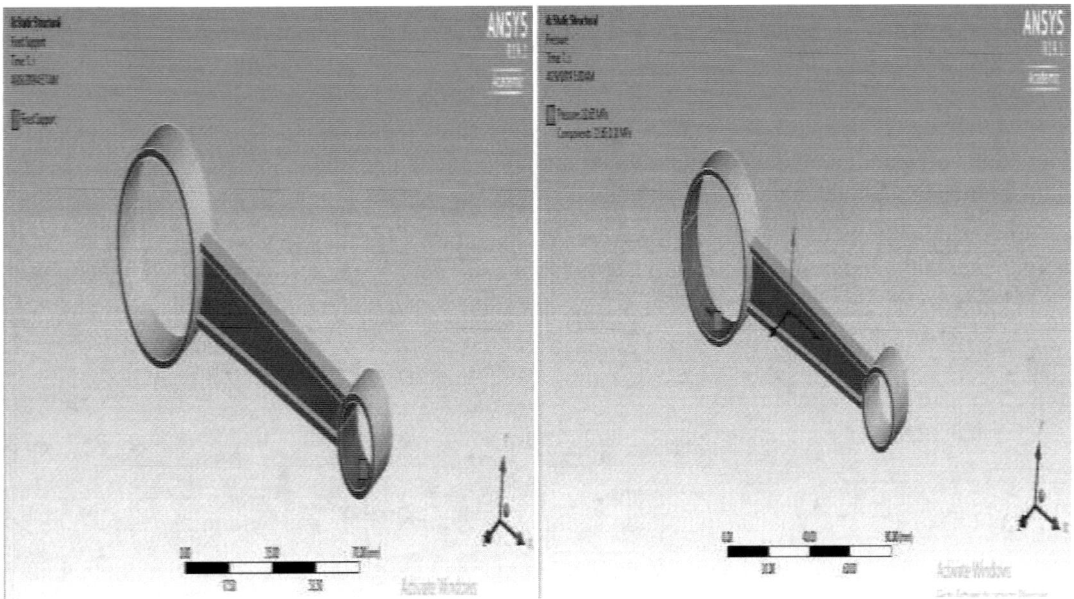

Fig. 3: Fixed support and loading condition for connecting rod in pin end

In Figure 3 indicates fixed support applied in pin end and 22.65MPa pressure applied to model of connecting rod at crank end.

3.1*Boundary conditions and Load:*

Distributed load(Webster, 1983) was applied at the small end (Pin end) internal surface while at big end(crank end) surface displacement(zero) boundary was applied.

3.2*Changing the shapes and selection of design:*

By changing the shapes like fillet radius, thickness and other parameter to find out for which design the stress is minimum. I have changed the fillet radius in CRANK SIDE (I SECTION) dimensions by using SOLIDWORKS software.

When the modification of design is completed then we have choose the best dimension which satisfied all the constrained and reduces the stress.

3.2.1 *Effect of stress in connecting rod while changing the fillet radius in CRANK SIDE (I SECTION):*

The initial analysis using ANSYS was performed with distribution of loads at big end. There were fixed displacement boundary conditions at pin end.

Finite Element Analysis with ANSYS:

The initial analysis was performed by using ANSYS software with distribution of load at big end and fixed support at small end. Figure 6 shows the modeling and analysis results of connecting rod used to determine the variation of stresses though change the fillet radius in CRANK SIDE (I SECTION)

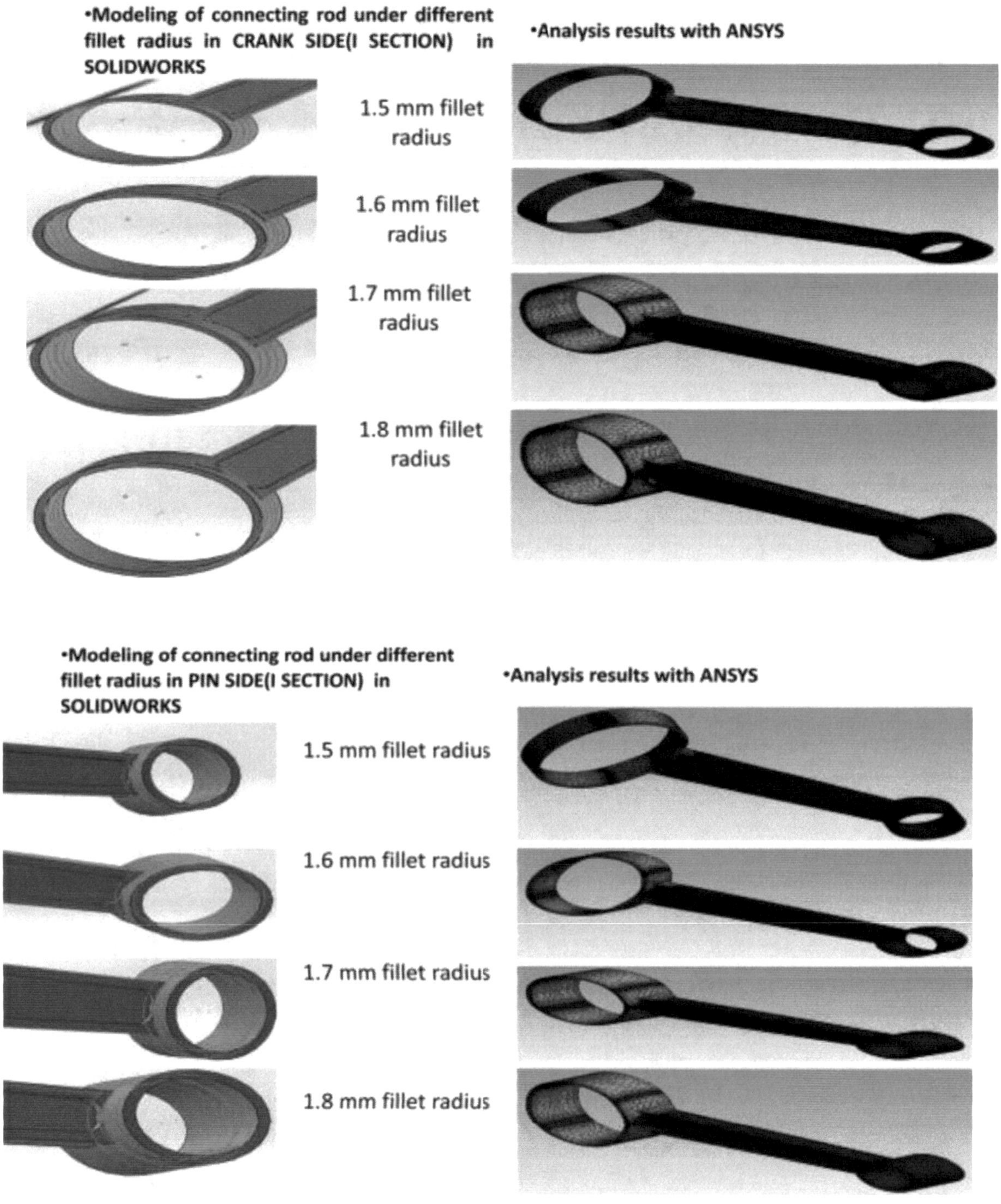

Fig. 4: Modeling and Analysis result with ANSYS

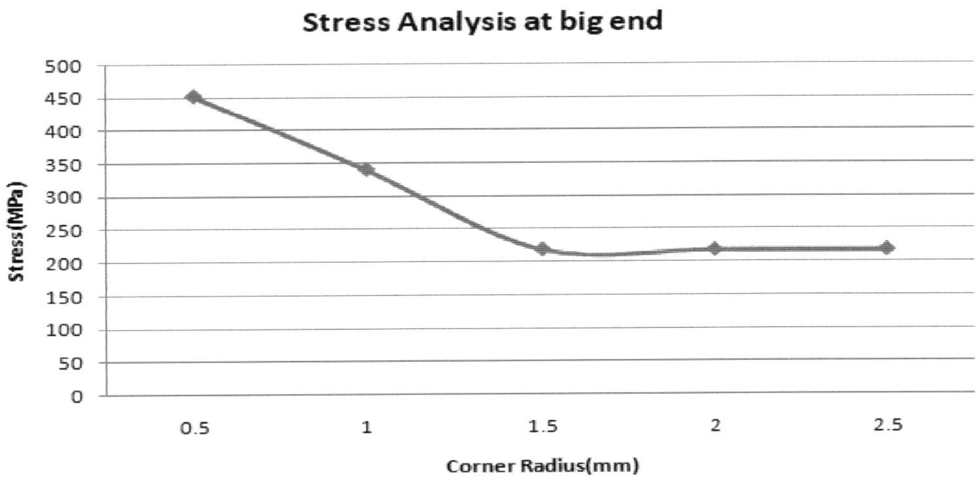

Fig 5: Stress analysis result at big end (I section)

ANSYS result examination:

After the running the simulation in the ANSYS workbench we have found that stress value peaks in the fillet area zone. Further analysis has shown that fillet radius of 1.7mm has proven to be the minimum stress(212MPa) at the big end. Further for similar investigation has shown that for the small end the minimum stress(226MPa) occurs at the 1.6mm.

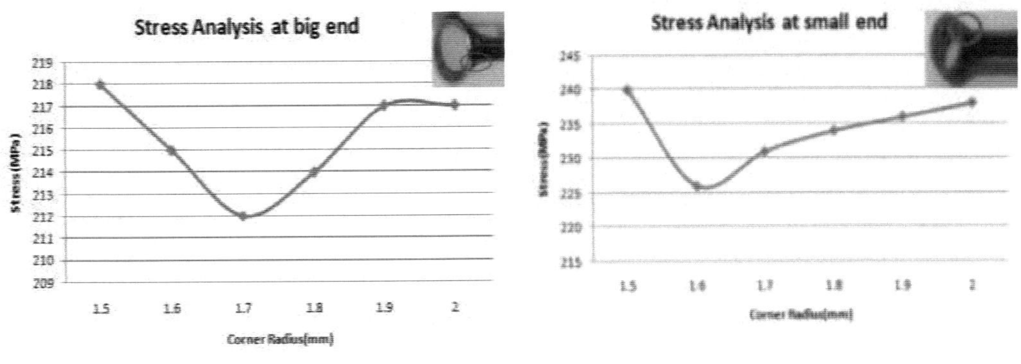

Fig 6 Analysis results

4. Conclusion:

We have change the fillet radius at crank end (I section), pin end (I section) in the model. And also we have analyzed the stresses. We have found that the value of stress is less for fillet radius 2.5 mm in small end and 3.0 mm fillet radius at crank end.

The following conclusions can be obtained from this study.

- Change in the fillet radius greatly affects the maximum stress value inside the connecting rod.

- Change in the Design parameters of the big end in the connecting rod affects more than the small end.

5. Future Work:

The effect of change in the material and use of different types of coating can be employed to study the performance of the connecting rod.

6 REFERENCES:

1. Afzal.A and Shenoy.P 2003 Dynamic Load Analysis and Fatigue Behavior of Forged Steel vs Powder Metal Connecting Rods American Iron and Steel Institute
2. Mirehei.A, Hedayati.M, Zadeh, Jafari.A, Omid A.,2008 Fatigue analysis of connecting rod of universal tractor through finite element method (ANSYS) Journal of Agricultural Technology, **vol-4(2)**, pp. 21-27
3. Webster, W. D., Coffell R., and Alfaro D., 1983 A Three Dimensional Finite Element Analysis of a High Speed Diesel Engine Connecting Rod SAE Technical Paper Series Paper No. 831322 **.**
4. Athavale, S. and Sajanpawar, P. R., 1991 Studies on Some Modelling Aspects in the Finite Element Analysis of Small Gasoline Engine Components Small Engine Technology Conference Proceedings, Society of Automotive Engineers of Japan, Tokyo PP. 379-389
5. Khare.S., Singh O.P., Bapanna K., Dora, C. Sasun, 2012 Spalling investigation of connecting rod TVS Motor Company Ltd. Research and Development, Hosur, Tamilnadu, India Engineering Failure Analysis **vol.(19)** pp. 77–86.
6. Lee. S, et.al. Failure Analysis of Connecting Rod at Big End Trans Tech Publications, Switzerland, **vol.(306-308)** pp. 345-350 (2006).

Design of Piezoresistive Sensor using FEA in ANSYS

YashwardhanSahi[1], Shivam Sharma[2], Swet Chandan[3]
[1]Galgotias University, Yamuna Expressway Greater Noida 201307, India
yashwardhansahi3@gmail.com
[2]Galgotias University, Yamuna Expressway Greater Noida 201307, India
shivam2kumarsharma@gmail.com
[3]Galgotias University, Yamuna Expressway Greater Noida 201307, India
swetchandan@gmail.com

Abstract. Piezoresistive pressure sensors are widely used in various fields such as in industries, auto- mobiles, medical applications and many others. To get proper input from the sensors we must have a good sensor having good sensitivity in every type of environmental conditions. The behavior of the sensor also depends upon its thickness, area, material properties etc. This paper is based on the piezoresistive micro pressure sensor having greater sensitivity made of silicon material. Using finite element analysis (FEA) and ANSYS the sensitivity is determined on different dimensions with different loading conditions. The piezo resistors of silicon material are arranged in different patterns and the sensitivity of each plate is determined by having all the results best design for the sensor is achieved.

1. Introduction

MEMS sensor is the short form of Micro-Electro-Mechanical-System sensor. These sensors will have very small devices with 1μm-100μm size. A MEMS sensor will have at least one mechanical element or functionality. MEMS design consideration is more important than with larger scale mechanical devices due to large surface area to volume ratio. They are generally two types of MEMS - variable capacitive and piezoresistive. Here we will choose piezoresistive sensors due to simple construction means low cost and durability. Some of the qualities of these sensors are having good resistance to shock, vibration, and dynamic pressure changes. The output is linear with pressure and the response time is typically below one millisecond. The output is also stable over time. They can be used for a wide range of pressure measurements from 21 kPa to 150 MPa. [1,2]. These sensors can be operated at higher temperatures and are more suitable for use in harsh environment. The working principle of Piezoresistive Sensor is based on change of resistance. Whenever the tilt is applied to the Piezoresistive Sensor, then a balanced mass makes a difference within the electric potential. This can be measured like a change in resistance. It contains a mass- spring system designed so that the force exerted by the spring exactly equals the force required to accelerate the mass, and the displacement of the mass (deflection) is directly proportional to the acceleration measured by the strain gauge. It contains a built-in Wheatstone bridge circuit to measure the resistance change and produce an electric output. The approach combines analysis of silicon based piezoresistive sensors with simulation and modeling in a circuit simulation environment and relies on a new class of parameterized behavioral models while giving different parameters with the use of software ANSYS [8].

2. Design with Specification

Design of Diaphragm

The diaphragm is the part where stress distribution will be taken place so for maximum sensitivity, we need different dimensions. Here we use three different case as per our approach, we use different parameters to get higher sensitivity[3, 4]. In First case we use for dimension (1200*1200*15) μm and the pressure varies from 10kPa to 160kPa. In second case we use for pressure 120kPa with thickness 15 μm and dimension varies from 700 μm to 1200 μm while in third case we take pressure at 120kPa and dimension (1200*1200) μm and thickness varies from 5μm to15μm.

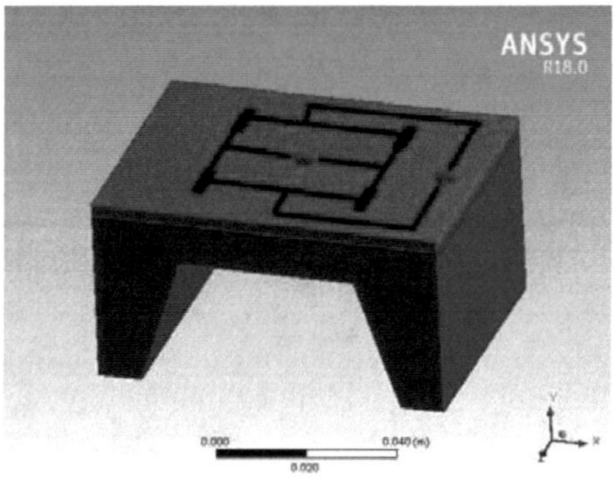

Fig. 1 Layout of diaphragm

1st case

Dimension (1200*1200*15)μm will be fixed and pressure varies from 10KPa to 160KPa.

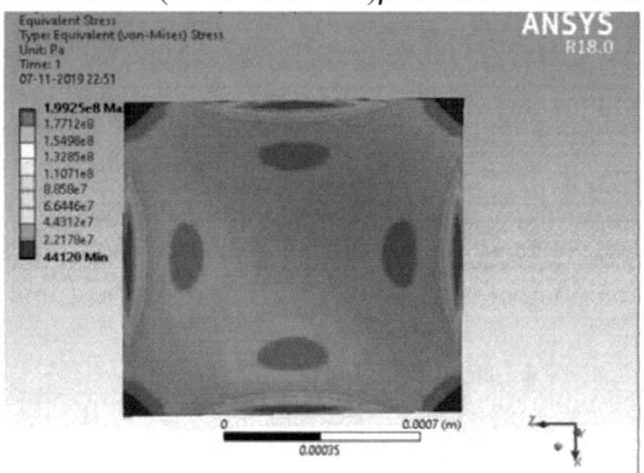

Fig.2 Maximum stress

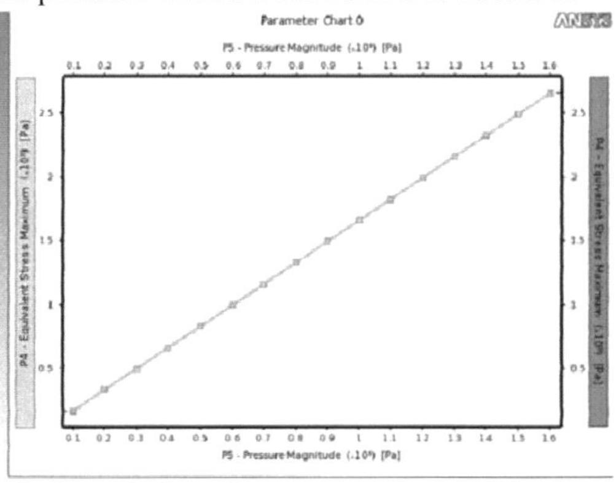

Fig.3 Variation of Pressure vs M. Stress

This indicates the equivalent (von-Mises) Stress by finite element analysis in softwareANSYS R18.0. [5]. Here the pressure varies and dimensional and thickness is fixed. In thisdeformation showing the areas where the stress is maximum and minimum.

By FEM i.e. finite element method, we analyse the data and according to it the graph is made. The result will be obtained from FEA.

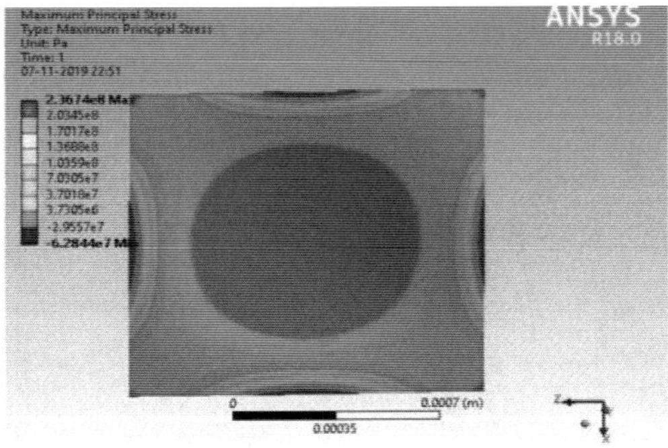

Fig.4 Maximum Principal stress

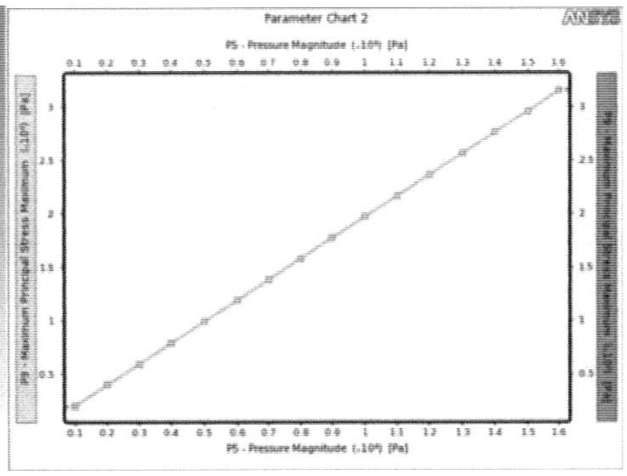

Fig.5 Variation of Pressure vs P. Stress

[5] Using FEA we will use sensor from different patterns with fixed dimension and variable thickness while some with fixed thickness and variable dimensions.

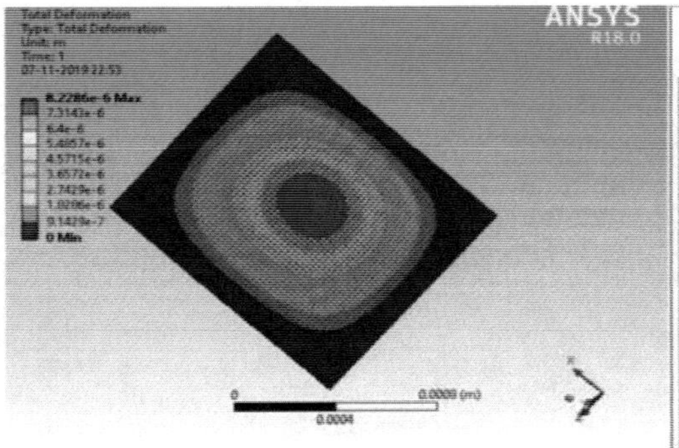

Fig.6 Total deformation

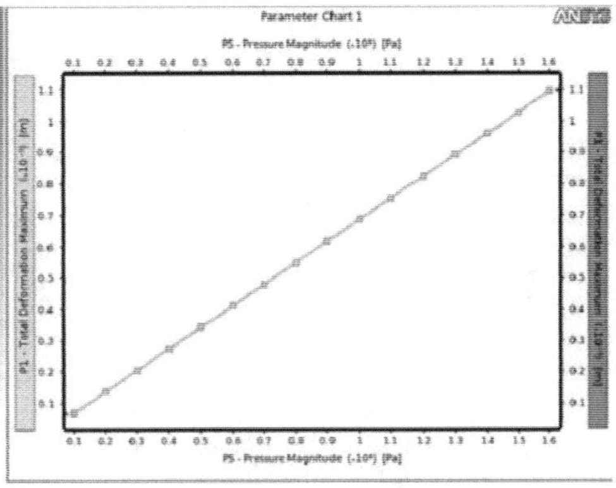

Fig.7 Variation of Pressure vs Total deformation

Here it showing the total deformation occur by putting dimension and thickness fixed and pressure varies.

*pressure is 120KPa and dimension is 1200*1200*15 μm

2^nd^case

Dimension varies from 700μm to 1200μm with fixed pressure 120KPa and thickness 15μm.

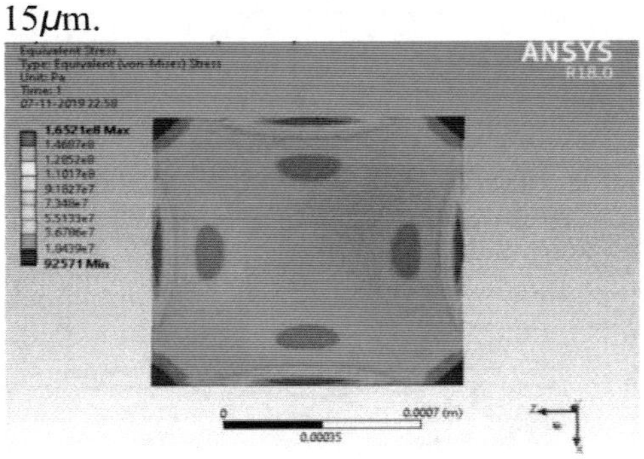

Fig.8 Equivalent stress

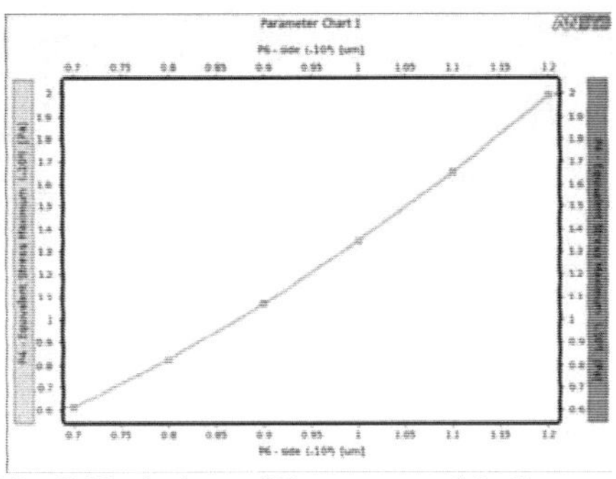

Fig.9 Variation of Pressure vs M. Stress

Again, with the given data the equivalent stress is taken out with maximum and minimum stress [5].

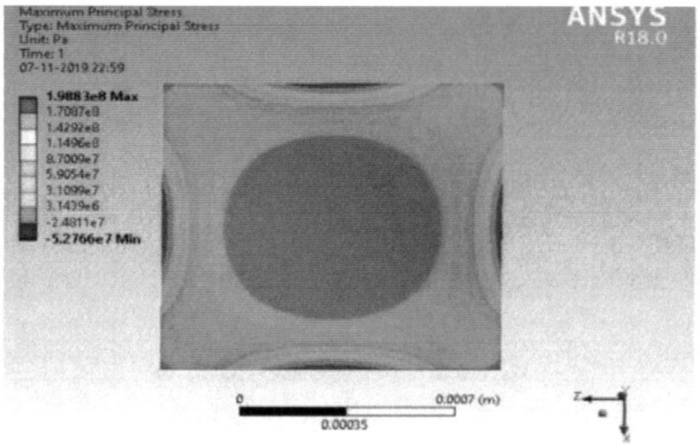

Fig.10 Maximum principal stress

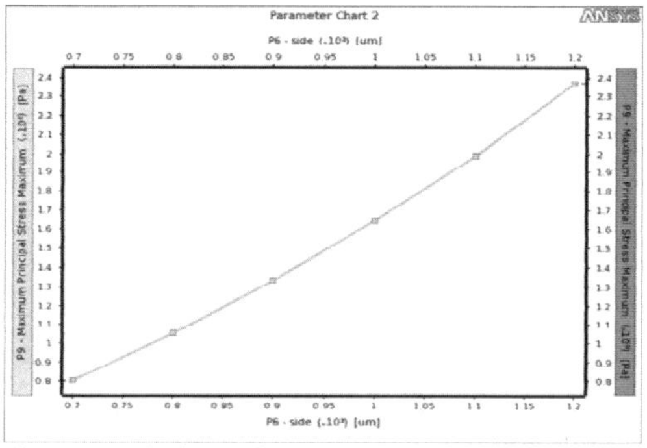

Fig.11 Variation of Pressure vs P.Stress

For the 2^{nd}case maximum principle stress is taken out as shown infig
*pressure is 120KPa thickness is 15 μm and dimension is 1100*1100 μm [5]

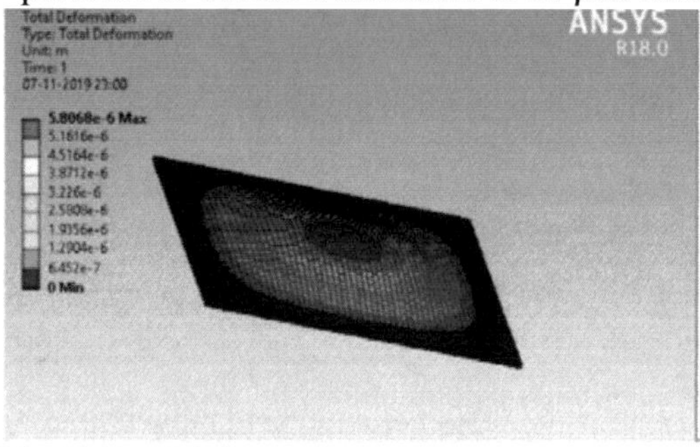

Fig.12 Total deformation

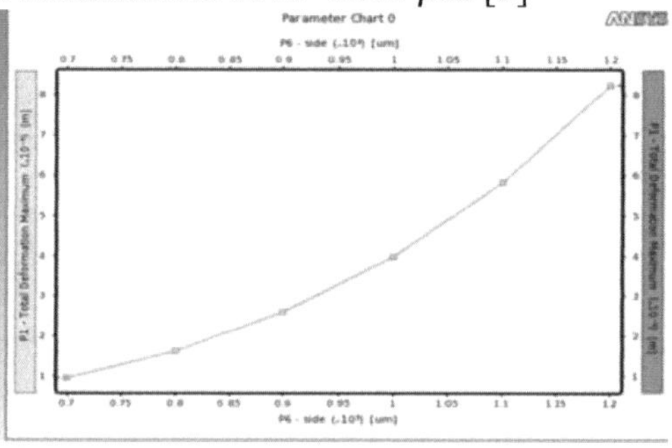

Fig.13 Variation of Pressure vs Totaldeformation

3^rd^case

Dimension (1200*1200)μm, pressure 120KPa and thickness varies 5 μm to 15μm [5].

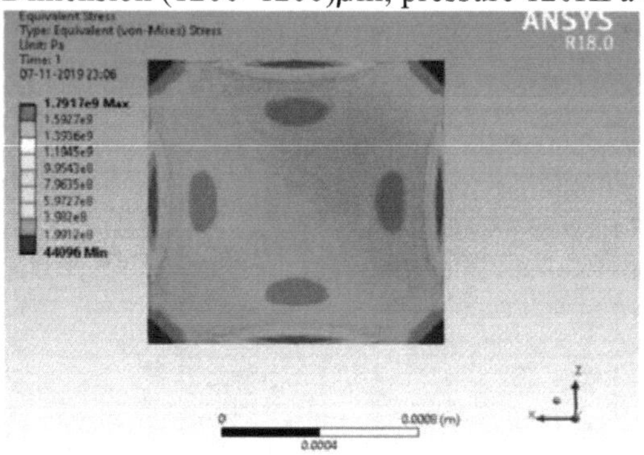

Fig.14 Maximum stress

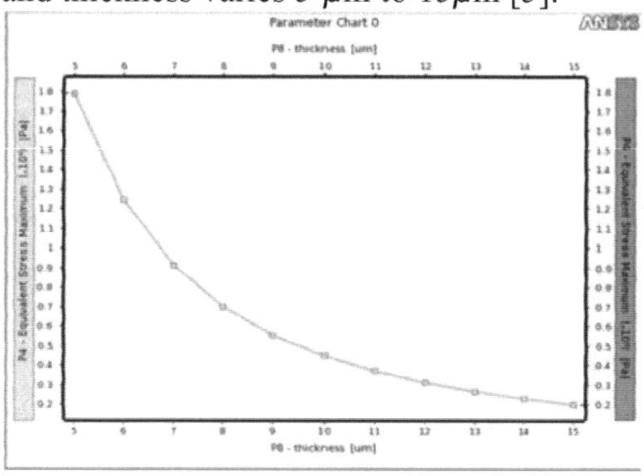

Fig.15 Variation of Pressure vs M.Stress

According to the data, here its gives total deformation and the stress is taking place at the center and minimum stress can be seen at the corners.

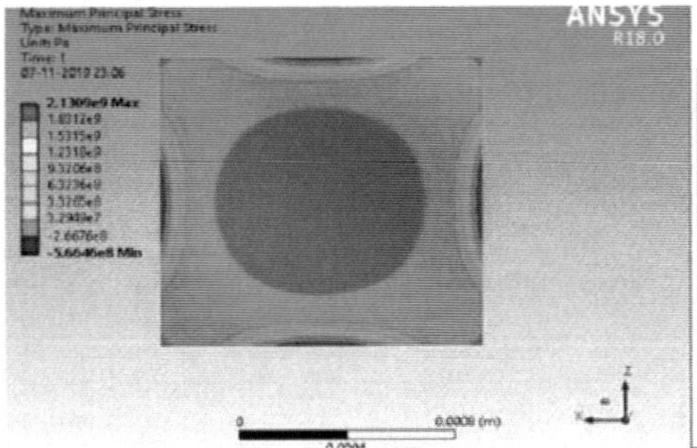

Fig.16 Maximum Principal stress

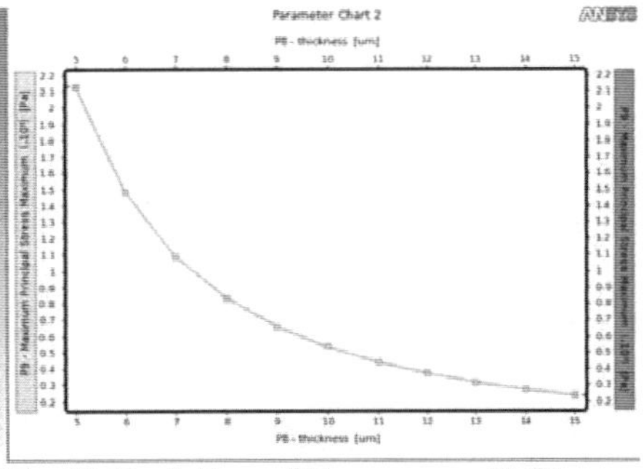

Fig.17 Variation of Pressure vs P. Stress

Here the graph for total deformation showing the thickness varies from 5μm to 15μm.

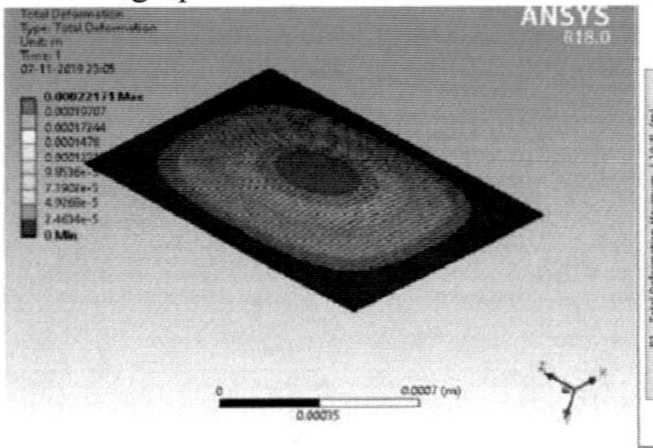

Fig.18 Total deformation

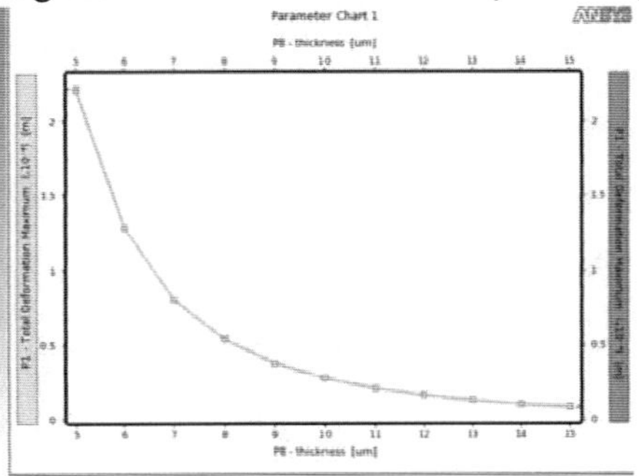

Fig.19 Variation of Pressure vs Total deformation

*pressure 120KPa, dimension 1200*1200 μm and thickness are 5 μm

3. Pattern ofdiaphragm

The design of these diaphragms is taken in 4 arrangements, considering different thickness, dimension and pressure to get higher sensitivity, quick response and accurate result [6, 7].

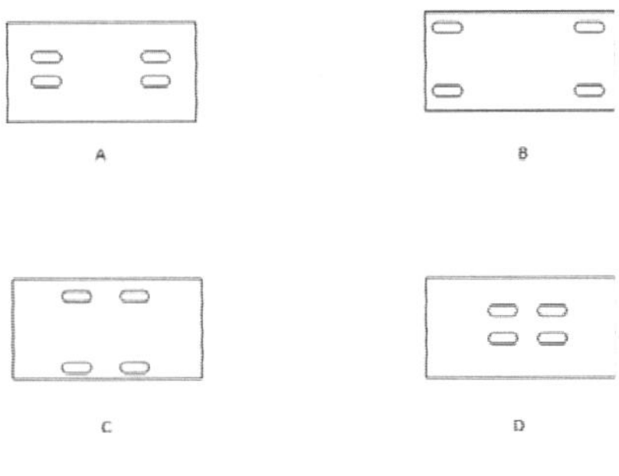

Fig. 20 Pattern of Diaphragm with different arrangement of piezoresistors

The change in piezoresistance can be calculated by using the formula:

$$\frac{\Delta R}{R} = \pi_l \sigma_l + \pi_t \sigma_t \qquad [1]$$

Where . $\sigma_t = \nu\sigma_l$, π_t and π_l are considered as transverse and longitudinal piezoresistive coefficients respectively.[9]

The value of σ_l will be taken from graph. The value of π (piezoresistive coefficient) for p-type semiconductors of silicon material is higher compared to n-type semiconductors. So the value of π_l and π_t will be taken 71.8×10^{-11} /Pa and -66.3×10^{-11} /Pa respectively.

The formation of piezoresistors are arranged in Wheatstone bridge structure as shown below:

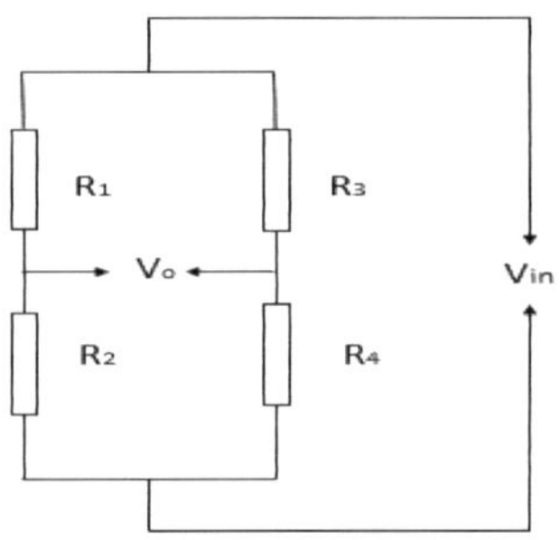

Fig. 21 Piezoresistors arrangement in Wheatstone bridge

The orientation of the piezoresistors are taken in such a way so the output should be more precise. The values of each piezoresistors are taken 1kΩ and input voltage 10V. Now to find the sensitivity following formula is used:

$$S = \frac{V_O}{V_{in}} \times \frac{1}{\Delta P} \qquad [2]$$

Where ΔP is change in pressure, V_o is voltage output and V_{in} is voltage input.[10] The result is shown in the table below by using the above expressions:

ΔP at 120kPa	Output Voltage (V_o)	Sensitivity (S)

1st case	1.098	9.15
2nd case	.91	7.63
3rd case	.98	8.36

4. Conclusion

In a sensor the main deciding factors are the dimension and thickness of the sensing element to determine the sensitivity at the maximum pressure strength of 120 kPa. In the simulation software ANSYS deformation and stresses are calculated at the different measurements of the sensing element for the varying pressure strength on keeping the safety and factors in mind. From the analysis we can conclude that sensing element having side length of 1200μm and thickness of 15μm gives the good sensitivity i.e. 9.15 which can be used in different conditions. Other sensing element having small dimensions can be used for the better results in terms of sensitivity. The piezo resistors are placed in specific locations to enhance the sensitivity of the system.

5. Reference

[1]. Mandie O 1995 a new micro-machined silicon high accuracy pressure sensor: https://doi.org/10.1016/0924-4247(94)00875-I 129-132

[2]. Mohammadi V 2009 design modeling and optimization of a piezoelectric pressure sensor based on thin film PZT diaphragm contain of Nanocrystalline powders 6th international symposium on mechatronics and its applications.

[3]. Kanda. Y 1982 a graphical representation of the peizoresistance coefficients in silicon – IEEE transaction on electron devices 29(1) 64-70

[4]. Pravinraj T, et.al Piezoresistor size effect on sensitivity of SOI piezoresistive pressure sensor- Proc. Of the international conference on science and engineering.

[5]. Suja, KJ Komaragiri, R.Photonic sens 2015 5:202 https://doi.org/10.1007/s13320-015-0239-y

[6]. Madhavi K Y et.al 2013 Design of peizoresistive micro pressure sensor using finite element analysis- International journal of computer application volume 70-No.3

[7]. Panas M, et.al 2010 a system approach to modeling piezoresistive MEMS SENSORS www.mit.edu

[8]. ANSYS guide release 18.0 ANSYS Inc. https://www.ansys.com

[9]. Petersen k 1982 silicon as a mechanical material-proceedings of the IEEE vol. 70 No.5 PP.420-457.

[10]. Pfann, W G and Thurston, RN 1962- semiconducting stress transducers utilizing the transverse and shear peizoresistance effect- journal of applied physics Vol.32 No.10

IOP Conf. Series: Materials Science and Engineering **748** (2020) 012012 doi:10.1088/1757-899X/748/1/012012

Influence of Drop Height on Impact Stresses in Aerial Bomb - A Numerical Study

Prahlad Srinivas Joshi [1*] and S.K Panigrahi[1]
[1]Department of Mechanical Engineering, DIAT, Girinagar, Pune – 411025, India

E-mail: prahjo20904@gmail.com

Abstract. A typical conventional aerial bomb is a streamlined cylinder with an outer casing, inner explosive material, a fuse to ignite the explosive filling, an arming mechanism for fuse and an optional tail unit. The outer casing is usually metallic with a pointed nose at the tip for better aerodynamic characteristics. The present work mainly focuses on Finite Element (FE) simulation of aerial bomb impact using Ansys Explicit Dynamics for imported three-dimensional model of aerial bomb using Solid Edge. FE simulation was carried out to find stresses and deformation limits for two varied impact distances between Bomb and fixed target fields. Further, the impact simulation results were validated by analytical solutions.

1. Introduction

The word bomb is derived from Latin word bombus which means 'booming'. A typical Bomb comprises of a metallic body or shell or casing with explosives filling in it. Bomb works on the principle of exothermic reaction of explosive filling to provide an extremely sudden violent release of energy. The bomb dropped from an airborne platform like aircraft is called an aerial bomb. Aerial bombs are largely deployed in offensive roles to destroy injure or kill enemy and his infrastructure. Some of the non-offensive roles include practice bombs and target indicating. Aerial bombs in offensive roles have high explosive filling which produce blast, heat & primary fragmentation (originating from casing) which are the main effects of explosion and secondary fragmentation (originating from primary fragmentation after striking to an object or structure), Ground Shock & craters which are the secondary effects of explosion. Aerial bombs in non-offensive roles produce smoke and flash [1].Publications were referred to understand the mechanics of operation of various types of bombs and the details of conventional material used in the bomb body making [2, 3, 4,5,6,7, 8 and 9]. The blast effect is characterized by the peak pressure due to explosion and so the material has to withstand this pressure. Wen- long Xu et al [10] experimentally calculated peak air and ground pressure for composition B(RDX+TNT), RDX and aluminized (RDX+TNT+ Aluminum powder) explosives weighing 2 kg each. It was observed that RDX has a peak ground pressure of 372.91 KPa at a distance of three meters and 0.030 seconds post detonation. This value was the highest among all the three explosives tested. This value is extrapolated for numerical studies in the paper.This work aims at finding impact stresses and its corresponding deformations which were induced in aerial bomb shell during drop impact using Ansys Explicit dynamics FE simulation software. The FE results presented for varying impact distance between bomb and target field. Further, the FE results were compared with analytical impact stress values for the model validation.

IOP Conf. Series: Materials Science and Engineering **748** (2020) 012012 doi:10.1088/1757-899X/748/1/012012

2. Geometric Modelling of Aerial Bomb

Approximate two-dimensional schematic aerial bomb of a fighter aircraft is shown in **Figure 1**. and it is used for making 3D model by the aid Solid Edge software.

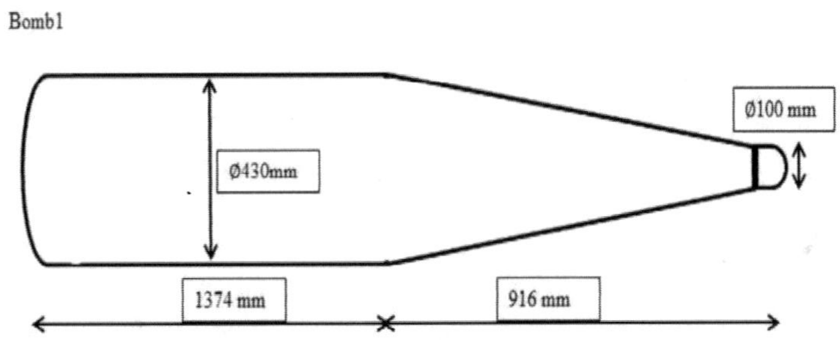

Figure 1. Schematic showing 2D graphical sketch of Aerial bomb of a fighter aircraft.

By using dimensions of the **Figure 1**, the 3D model for impact FE simulation was done using solid edge software and which is shown in **Figure 2.**

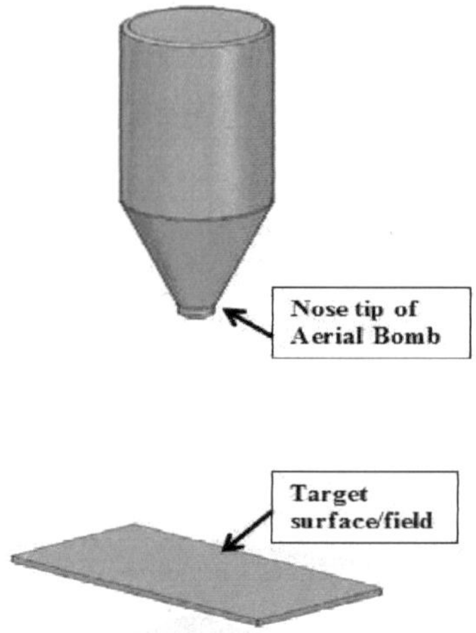

Figure 2.3D model showing of aerial bomb and target filed.

3. Finite Element Modeling/Meshing

3D model was imported form solid edge and Finite element mesh was generated with Tetrahedron elements using automatic meshing option in the Ansys software. The details of the finite element mesh for aerial bomb and target field is displayed in Figure 3.

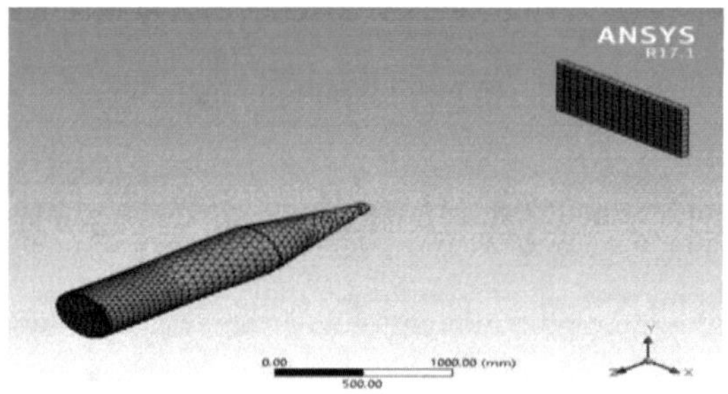

Figure 3. Finite element meshing for aerial bomb of a fighter aircraft.

4. FE Simulation of Aerial Bomb – Results and Discussions

The numerical FE simulation of aerial bomb was done using Ansys Explicit dynamics with following are the material properties and self-weight of the bomb is listed in Table 1 as inputs to the software. The various constraints used for impact simulation are illustrated in Table 2.

IOP Conf. Series: Materials Science and Engineering **748** (2020) 012012 doi:10.1088/1757-899X/748/1/012012

Table 1. Material properties and weight of the empty bomb shell for FE simulation

Bomb type	Weight of the bomb In N	Density (kg/mm^3)	Young's Modulus MPa	Poisson's Ratio	Bulk Modulus MPa	Shear Modulus MPa
Bomb1,2	4414.5	7.85×10^{-6}	2.016×10^{5}	0.26	1.4×10^{5}	80000

Table 2. Constraints used for impact simulation

Bomb type	Impact height/distance	Impact velocity	Charge pressure inside the bomb shell
Bomb1 _DHT1	2 meters	300 m/sec	34.507 MPa
Bomb2_ DHT2	10 meters	300 m/sec	34.507 MPa

4.1 Load-displacement and stresses in aerial bombs

The FE models had pre-defined drop height with respective time duration and well-defined target fields are referred to DHT1, DHT2with underscore to each Bomb, for example Bomb1 with pre-existing drop height and target field is referred to Bomb1_DHT1. **Figure 4** showing the process of penetration of Bomb 1_DHT1 in to the sand type target filed with a velocity of 3×10^{5} mm/secthrough as distance of two meters. **Figure 4 (a)** plots the correlation between directional deformations in bomb shell after impact for first case (Bomb1_ DHT1). It shows that the modeling results confirm to the general trend of the classical (analytical) results for both drop height and target field. It was seen that the large deformations both in the bomb1 shell and target field after few second of bomb impact. These large deformations are responsible for movement of metal fragments during explosions.

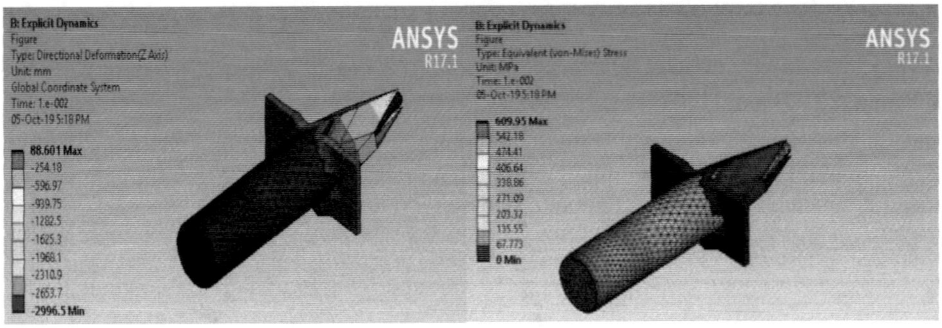

(a) (b)

Figure 4. Plot showing (a) Directional deformation and (b) Stress distribution for Bomb1_DHT1 after impacting through a drop height of 2 meter.

From **Figure 4(b)** it was seen that significant stresses (609 MPa) on the shell surface of theBomb1_DHT1 after impacting the bomb1 through a drop height of 2 meters. This stress values are well above the yield strength of the material chosen for bomb simulation. In deed as per the bomb explosive concepts the bomb has to blast/fractures just after few seconds of impact.

Similarly, the FE simulation of Bomb2 DHT2 with drop height of 10 meters impacted to same target field. The FE results of Bomb 2 DHT2 are shown in **Figure 5.**

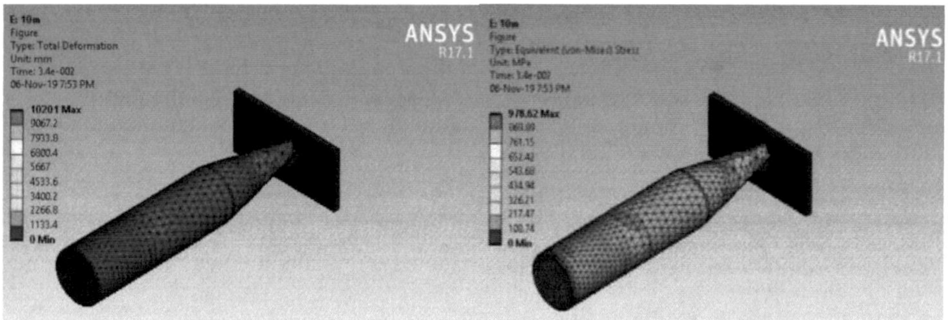

(a) (b)

Figure 5. Plots showing (a) directional deformation and (b) Stress distribution for Bomb2_DHT2 after impacting through a drop height of 10 meters

The FE results almost exhibit same trend as we seen in Bomb1 DHT1. But the higher deformations and stresses were noticed in Bomb2_ DHT2 **(Figure 5(a) and (b)),** this is due to the influence of increasing drop height from two meters to ten meters. This large deformation and stress on the shell wall of Bomb 2 is due to gained potential energy leading metal fragments moving at higher velocity resulting into collateral damages. This nature of explosion is sometimes well suited for attacks in enemy's territory but not suited in case of deployment within the land for Guerilla warfare/ Anti-Naxal operations. Further, the FE simulation results for Bomb1_ DHT1 and Bomb2_ DHT2 were validated with classical solutions and the percentage errors are listed in Table 3.

4.2 Classical solution for FE simulation results for aerial bombs

For Bomb1_ DHT1

$$\text{Impact Stress} \quad \sigma' = \frac{W}{A}\left[1 + \sqrt{1 + \frac{2hEA}{WL}}\right] \text{.....................} \textbf{(1)}$$

$$\sigma' = \frac{450 \times 9.31}{7401.59}\left[1 + \sqrt{1 + \frac{2 \times 2000 \times 206 \times 10^3 \times 7401.59}{450 \times 9.81 \times 2290}}\right]$$

$$\sigma' = 458.87MPa$$

But, in this case bomb is influenced by internal pressure due to this hoop stress will induce

$$\sigma = \frac{PD_i}{2t}$$

$$\sigma = \frac{34.507 \times 100}{2 \times 20}$$

$$\sigma = 86.26\ MPa$$

So overall impact stress

$$\sigma_{Impact} = \sigma' + \sigma \quad \text{...} \quad (2)$$

$$\sigma_{Impact} = 545.13MPa$$

For Bomb2_ DHT2

$$Impact\ Stress \quad \sigma' = \frac{W}{A}\left[1 + \sqrt{1 + \frac{2hEA}{WL}}\right]$$

$$\sigma' = \frac{450 \times 9.81}{7401.59}\left[1 + \sqrt{1 + \frac{2 \times 10000 \times 206 \times 10^3 \times 7401.59}{450 \times 9.81 \times 2290}}\right]$$

$$\sigma' = 1013.71\ MPa$$

But, in this case bomb is influenced by internal pressure due to this hoop stress will induce

$$\sigma = \frac{PD_i}{2t}$$

$$\sigma = \frac{34.507 \times 100}{2 \times 20}$$

$$\sigma = 86.26MPa$$

So overall impact stress

$$\sigma_{Impact} = \sigma' + \sigma$$

$$\sigma_{Impact} = 1099.97MPa$$

Table 3. Results comparison of FE simulation and Classical solutions

Type of Bomb	Maximum Stress value by FE Simulation	Maximum Stress values by Classical solution	Error in %
Bomb1 DHT1	609 MPa	545.13 MPa	10.48
Bomb2 DTH2	979 MPa	1099.97 MPa	10.99

5. Conclusions

The FE simulation results for two types of bombs were done successfully using Ansys Explicit Dynamics R17 by varying the drop height, the following conclusions were drawn from the above set of FE analysis. For minimal drop height (2 meter) the deformations and stresses generated are significantly low but exceed the ultimate limits of the material which is in Bomb1_ DHT1. As the drop height of the bomb increases to 10 meters as in case of Bomb2_ DHT2, it was observed that the significant enhancement in bomb shell deformations and associated stress values caused higher damage, which indicates the large metal fragments spreads at higher velocity during impact/explosions. Finally, the FE simulation results were compared with classical solutions using strength of material approach and it was seen that the error was well within the acceptable limits.

6. Acknowledgments

The authors would like to thank Dr. P.S. Shivakumar Gouda Department of Mechanical Engineering, SDM College of Engineering and Technology, Dharwad for their support and help during execution of this work.

7. References

[1] https://en.wikipedia.org>wiki>Mark 82 bomb
[2] https://en.wikipedia.org>wiki>Mark 82 bomb
[3] *Explosive Weapon effects-final report* 2017 Geneva International Center for Humanitarian Demining (GIHCD), Geneva,
[4] GIHCD Case study report charectarisationexplosiveweapons.org/studies/annex- e-mk82-aircraft-bombs.
[5] *United States Bureau of Naval Weapons British Navy publication*1960-1963 Aircraft bombs, fuses and associated components.
[6] Mott NF. 1943 Theory of fragmentation of shell and Bombs Ministry of supply report No. AC 4035
[7] *US Depatment of Army Technical manual (TM5-855-1)* 1986 Fundamentals of protective design for conventional weapons
[8] Randles PW 1996 SPH simulation of fragmentation in MK-82 bomb AIP conference proceedings
[9] Predrag elak et al 2014 Numerical simulation of fragmentation warhead mechanisms
[10] Wen- Long Xu 2019 Investigation on energy output structure of explosives near ground explosion *Journal of Defence Technology*, **online**
[11] Ansys Explicit dynamics R17.1 user manual

IOP Conf. Series: Materials Science and Engineering **748** (2020) 012013 doi:10.1088/1757-899X/748/1/012013

Development of a Sintered Porous Neck Filter for miniature Joule Thomson Cryo-cooler for cooling Infrared detectors

Mayank Singhal[1], Rajesh Kumar[2], R.S. Walia[3] and S.K. Pandey[1]
[1]Defence Research and Development Organisation, Delhi, India
[2]Delhi Technological University, Delhi, India
[3]Punjab Engineering College, Chandigarh, India

E-mail: mayanksinghal25@gmail.com

Abstract. One of the primary requirements for operation of Infrared (IR) detector systems for defence applications is the availability of a reliable miniature cryocooler. Due to a simple configuration, compact and light structure and rapid cool down characteristics, miniature Joule Thomson (JT) cryocoolers are considered ideal for IR detector systems. The need for high purity gas supply is a main concern in order to ensure long and reliable operation of Joule Thomson miniature cryocoolers. Minute solid particles are the common and problematic impurity which on being lodged in the cooler orifice hampers the cryo-cooling process of the cooler by choking which affects the operation of IR detector packaged in an Infrared detector cryochamber or dewar.

The paper focuses on realization of sintered porous neck filter, the developmental aspects of the same and its characterization. In the present work a sintered porous neck filter with 5μm pore size and a filter porosity of nearly 35% was developed. The filter developed was assembled at the warm end of the cooler and the cooler assembly was integrated in the IR cryochamber for in system evaluation and confirmation of the desired quality of the developed porous filter.

Keywords: JT cooler, sintered neck filter, porosity, pore size

1. Introduction

IR detectors in military use find application in the field of missile guidance, tracking of projectile systems and ground surveillance [1]. Though uncooled night vision technologies are available, most of IR detectors used for military applications are cooled. In terms of working ranges, resolution and ability to recognize/ track fast moving objects in dynamic infrared scenes cooled systems are still considered to be superior [2, 3].

A cooling medium must be provided, and the detector must be mounted inside a vacuum encapsulation, called IR detector cryochamber or Dewar, so that it will remain cold and still be accessible to the infrared. There are number of long life coolers available for space applications covering a wide variety of thermodynamic cycles and configuration types [4]. However in case of IR detectors thermoelectric, Stirling and JT coolers are mainly used [5].

Miniature Joule-Thomson (JT) coolers are widely used for rapid cooling of infrared detectors, cryosurgery probes, thermal cameras, and missile homing guidance systems, due to their special features of simple configuration, compact structure and rapid cool down characteristics. It is the isenthalpic expansion of gas, known as the Joule Thomson effect [6] which forms the basis of the miniature JT cooler which is the one used for cryo-cooling to achieve temperatures of ~ 77 K suitable for IR detector operation in the present study.

These JT coolers may be operated with or without filter assemblies, however the latter may be preferred on account of expected better qualitative performance by eliminating various contaminants helpful in preventing choking. However, a clear quantitative

IOP Conf. Series: Materials Science and Engineering **748** (2020) 012013 doi:10.1088/1757-899X/748/1/012013

assessment of the performance of an IR detector cooling performance by introducing a filter assembly has not yet been investigated. Hence, it is worthwhile to investigate the same and this forms the prime motivation for the present work.

2. Configuration of JT cooler

The miniature finned tube JT cooler along with its main parts including the filter assembly is shown in Fig. 1. In case when the filter is used, high pressure nitrogen gas of high purity is passed through a conical sintered porous filter, situated at the main inlet port or the neck of the JT cooler, as shown in Fig. 1(b). The importance of using a sintered porous filter of about 5μm porosity is to arrest all contaminants larger than 5μm in the incoming gas in order to avoid any choking, which normally occurs in the expansion nozzle of the JT cooler, resulting in either restricting or fully blocking the gas flow.

A conical, porous sintered filter, as shown in Fig. 1(c), is thus integrated at the inlet of the cooler assembly so that the contaminant present in the gas supply can be arrested at the beginning itself, i.e., at the main entry of the cooler. The gas after passing through the filter is cooled by the outgoing cold gases (expanded earlier), flowing between outside of the finned tube and inside of the inner Dewar tube, at constant pressure. Thus, the finned tube acts as a recuperative heat exchanger, with counter flow of incoming and outgoing gases. This heat exchanger is basically a fine tube of a bore of about 0.3mm, usually made of copper-nickel alloy with copper fins soldered to its outside surface, with a specific fin density, and wound in the form of closely of coiled helix around the tube. Thereafter, the finned tube is wound over a cylindrical mandrel made of a low thermal conductivity material, such as, Micarta or Stainless Steel (SS), to give the JT cooler its final shape. The finned tube is then soldered to the inlet port and outlet port at the two ends of the cooler. A micro-orifice cap is then integrated on the outlet port, which acts as a nozzle, where the final expansion of the gas actually occurs at constant enthalpy, resulting in cooling due to the Joule-Thomson effect.

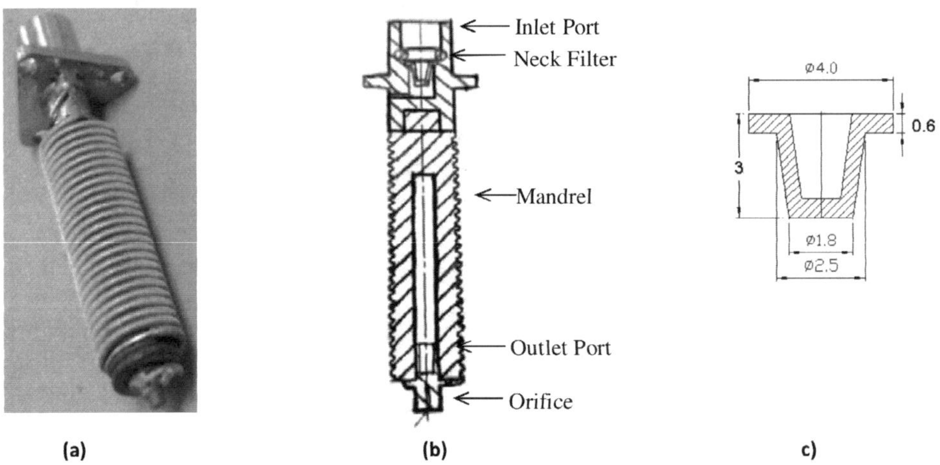

Fig. 1: a) JT Mini-cooler, b) its sectional view and c) Sintered Porous Neck Filter

3. Development of Sintered porous Bronze Filters

Technology of powder metallurgy (P/M) is used extensively nowadays as it is a highly economic method for producing porous parts. In powder metallurgy, costly machining processes are reduced or eliminated resulting in less scrap loss. [7-9].

Desired properties of metals, which would normally not alloy easily, can be achieved by combining different metal powders or mixtures of metal and non metal powders. Copper based P/M materials are widely used in industry at par with other P/M materials such as those based on iron, steel and aluminium [10]. The porous sintered parts with a porosity ranging from 20% to 90% have found many applications such as self-lubricating bearings and filters in which the pores form an interconnected three dimensional network. P/M bronze parts are frequently used for structural applications, however, low hardness and strength along with poor wear resistance limits their usage [11]. Structural parameters such as porosity and pore size depend on processing parameters such as powder size range, compacting pressure, sintering temperature and time [12-16].

Filters constitute one of the major applications of porous materials. The most commonly used powders include bronze, stainless steel, nickel, nickel based alloys, titanium and aluminium. The ability to achieve close control of porosity and pore size is the main reason that metal powders are used in filter applications.

Spherical 89/11 bronze powders are amongst the most preferred materials used to make filters [17] as these powders are capable of having closely controlled particle size. Bronze filters usually are made by gravity sintering of these spherical powders [17,18]. The effective pore sizes of filters generally ranges from 5 to 125 μm [18].

Porosity of the filter developed is ensured so that the required cooling is achieved by JT cooler which indirectly helps in trouble free operation of IR detector used for imaging. The density, pore size, pore size distribution, and permeability determine the overall performance of a porous material in an application.

The most critical challenges in developing such filters include achieving the desired porosity and its extensive characterization along with achieving the required dimensional accuracy of the components. Thus, micro-structural characterizations as well as weight, dimension and % porosity measurements are equally important prior to them being employed in a cost intensive JT cooler assembly.

The specifications of the developed Sintered Porous Bronze Filter, to be employed as neck-filter for the JT cooler are based on the orifice size of the JT cooler, the availability of space at the inlet port and integration/compatibility constraints.

Fig. 2 below describes the process flow for the development of the sintered porous neck filter and the details of each process are discussed subsequently.

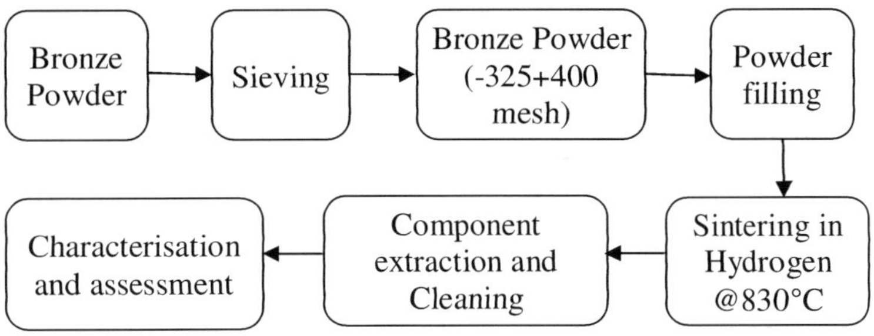

Fig. 2: Process flow for development of sintered porous neck filters

3.1 Sieving:

Sieving of the as-received bronze powder is conducted using a sieve shaker. The size fraction of -325 +400 mesh (37 to 45μm) was selected for the required filters. This narrow size fraction was obtained after repeated sieving until there was complete absence of any -400 mesh size particles. This is essential as the presence of -400 mesh size powders essentially leads to undesirable densification of the powder.

3.2 Tooling and Powder filling

Tooling which is negative shape of final part is an element of high precision and high durability. A suitable stainless steel (SS) mould was designed and fabricated prior for processing of the porous filters. A dry graphite lubricant was applied to the internal surfaces of both the halves of the mould prior to powder filling in order to avoid direct contact between the bronze powder and the mould surface. The surface smoothness was also ensured after coating as any surface roughness will leave an impression on the final developed component. Gas atomized spherical bronze powders sieved to narrow size fractions are filled in the mould and vibrated for 15 to 20 min. Necessary care was taken during powder filling, tapping and while vibrating the moulds to ensure that powder mass remains constant before and after this stage.

3.3 Sintering

Sintering is a thermal cycle consisting of heating the compacted part at an elevated temperature which is below its melting point. During the process powder particles fuse together, voids between the particles decrease, and eventually a dense solid body is obtained. Typically, bronze powder is sintered at 800 to 1000°C for 20 min in a reducing atmosphere to form a liquid phase between the particles and cooling thereafter ensures that a strong metallurgical bond is formed at the contact points between the powder particles [18]. The moulds filled with powder are then loaded in the tubular hydrogen furnace for sintering. Several trial experiments were carried out in order to optimize the sintering temperature and soaking duration for obtaining ~ 35% density. In the present investigation these experiments were carried out using the -325+400 mesh bronze powder. The most appropriate sintering temperature and soaking duration were found to be 830°C and 30 min to achieve the desired density for the development of the filters. Necessary precautions were taken to extract the components from the mould subsequent to completion of the sintering process. Also, the components were cleaned thoroughly with compressed air to remove any possible remnants of the mould or any other impurities.

Since the filter specifications for necessary compatibility are quite stringent, it was essential to characterize the developed filters prior to testing it with the JT cooler assembly. Thus the characterization and assessment of the filters include viz., adherence to required dimensional accuracy, adherence to projected filter mass, filter porosity and material phase purity. Subsequently, the porous neck filter assembly was tested in conjunction with the JT cooler assembly and a comparative assessment in terms of the cold temperature and the cool down time with filter and without filter condition was carried out.

4. Results and Discussion

The specimen components were sintered and used for density, dimensions measurements and microstructure evaluation. The micro-structural characterization as well as weight, dimension and % porosity measurements were carried out on the neck filter. A porous SEM image of a sintered bronze filter is shown in Fig. 3. During our experimental work, it was calculated that the porosity of our specimen is comparable to one reported in literature [19].

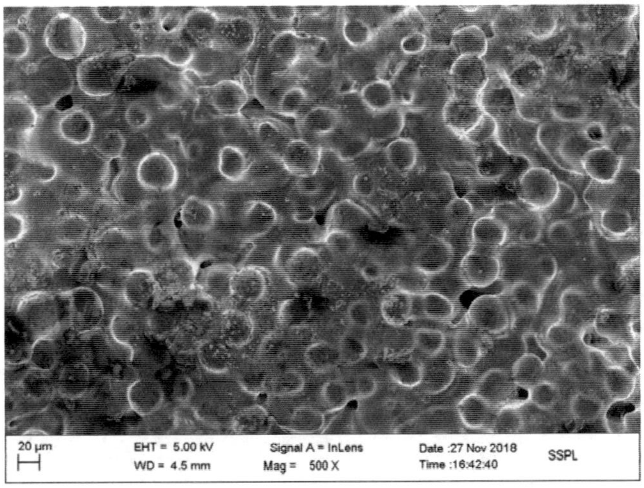

Fig. 3: SEM image of porous sintered bronze filter

The measured dimensions of the bronze filters are given in Table 1. The density of this filter was measured by the Archimedes water displacement method after impregnating the filter with xylene. Table 2 presents the density and the corresponding porosity of the filter.

Table 1: Measured Dimensions of bronze filters samples

Sample number	1	2	3	4
Wt. (g)	0.067	0.069	0.066	0.062
d1(mm)	3.93	3.92	3.94	3.91
d2(mm)	2.46	2.5	2.48	2.41
d3 (mm)	1.84	1.8	1.82	1.85
L (mm)	3.0	3.04	3.02	3.02
T (mm)	0.6	0.6	0.6	0.6

IOP Conf. Series: Materials Science and Engineering **748** (2020) 012013 doi:10.1088/1757-899X/748/1/012013

Table 2: Density and porosity of bronze filters

Sample number	1	2	3	4
W_s, air (g)	0.067	0.069	0.062	0.068
W_{s+x}, air (g)	0.069	0.071	0.064	0.070
W_{s+x}, w(g)	0.057	0.059	0.053	0.058
vol (cm^3)	0.012	0.012	0.011	0.012
•ρ(g/cm^3)	5.58	5.75	5.64	5.67
•ρ(%th)	63.7	65.6	64.3	64.7
Porosity	36.3	34.4	35.7	35.3

An XRD spectrum of the sintered sample, Fig. 4 **[pcpdf -501333],** shows the Copper-Zinc phase of the material with a pure cubic structure. The corresponding SEM images of the neck filter with pore sizes ranging between 10 to a 30 μm are shown in Fig. 5 along with photographs of the developed hardware.

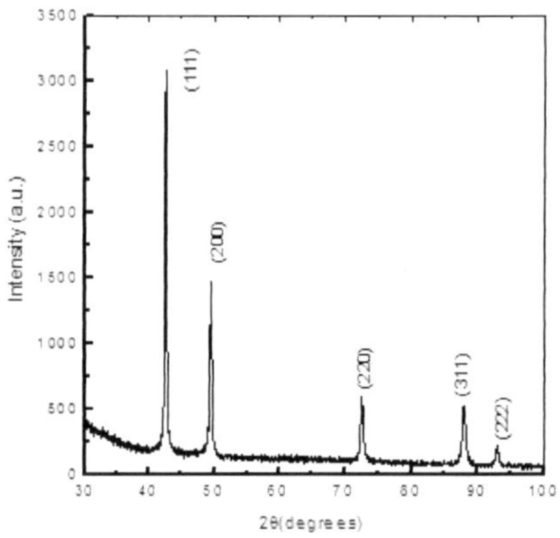

Fig. 4: XRD analysis of sintered filter sample 2

Fig. 5: Photograph of developed neck filter (sample 2) with corresponding SEM image

Further, in order to critically evaluate the porous neck filter it is required to integrate it with the JT cooler assembly. This is quite arduous and should be done with utmost care.

The assembly of the porous neck filter in the JT cooler assembly is shown in Fig. 6(a), which also clearly depicts the relative size of the filter with respect to the JT cooler. JT cooler assembled in glass dewar is shown in Fig. 6(b). The latter assembly was used for checking the flow of nitrogen through the cooler.

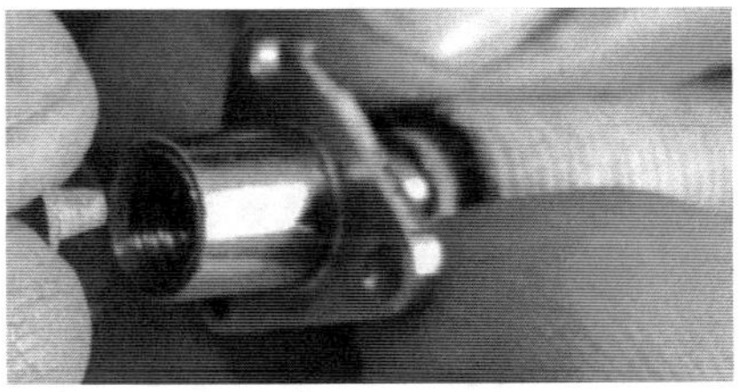

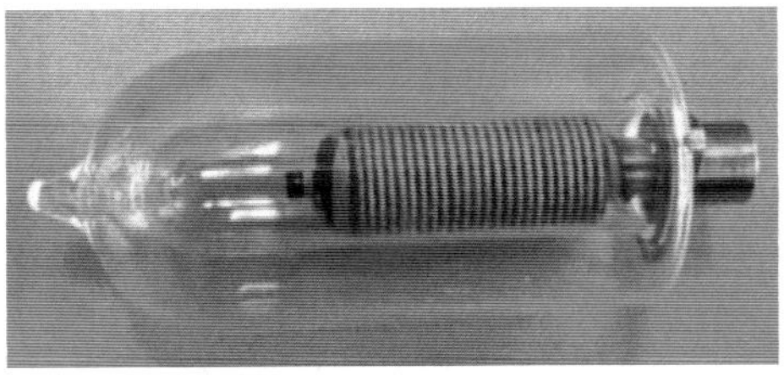

Fig 6: (a) Assembly of Filter in JT cooler (b) JT Cooler assembled in a glass dewar

JT coolers integrated with developed sintered porous filters were finally tested in an infrared detector cryochamber or Dewar for cooling parameters such as *Cool Down Time (CDT)* and achieved cold temperature. All the four sample filters from the lot developed were assembled into different available JT coolers and were tested in a sealed IR detector cryochamber or Dewar for *CDT* which is basically the time required for the Infrared detector packaged in a cryochamber to reach a cryogenic temperature. A single Dewar or Infrared detector cryochamber was used to minimize uncertainties due to variation in quality of vacuum in different Dewars or due to integration of Dewars in general. These tests were carried out in a cooling test set-up the layout of which is shown in Fig. 7.

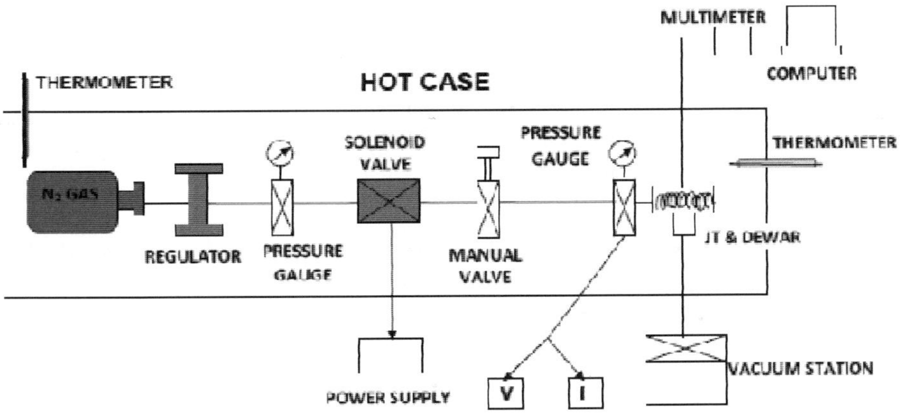

Fig. 7: Cooling Test Set-up of Dewar with JT Cooler

The set up employs nitrogen gas filled in a high pressure bottle typically at a pressure of 31.02 MPa with the help of a booster. The gas pressure is then regulated down to a standard value of 20.68 MPa with a line regulator to carry out all the tests at nearly the same input pressure irrespective of the bottle pressure. The gas flow is initiated by operating a solenoid valve electrically, which then enters the JT cooler after passing through the neck filter producing cooling due to Joule Thomson effect.

The fall in temperature at the header level is then recorded using a RTD (Resistance Temperature Device), type Pt-100 (Make: Omega, Model: PR-17) at the place of device so that the inside temperature can be measured. The results in terms of *CDT,* cold temperature along with the temporal profile of the fall in temperature are correspondingly acquired through a dedicated DAS.

Thereafter, the evaluation of testing of JT cooler with neck filter was carried out. Cold temperature achieved was less than 80K and the cool down time was within 90 secs. Both the results in terms of cool down time (CDT) time and cold temperature achieved were as per required specifications.

5. Conclusions

Infrared (IR) detector cryochambers or Dewars typically operate at cryo-temperatures of ~ 77K and are mostly employed in conjunction with miniature JT cooler assemblies. Usually, JT coolers are beset with issues pertaining to choking and diminished performance in term of the cool down time *(CDT)* and cold temperatures achieved. Hence, there is a clear need for developing customized neck filters with required porosity and filtering capabilities to achieve continuous desired performance in repeated cool down cycles. A sintered porous neck filter with 5μm pore size and filter porosity of nearly 35% has been successfully developed and duly characterised. Since producing such filters is associated with several development criticalities the involved processes have also been briefly discussed Further, porous filter developed was assembled at the warm end of the cooler and the cooler assembly was integrated with the IR cryochamber for in system evaluation. The complete assembly was tested in test setup created in the laboratory and the required cryo temperature was achieved in specified cooldown time. This confirmed the desired qualitative and operational specifications of the developed porous filter.

Acknowledgements

The authors thank Mr. Sushil Kumar and Director of Solid State Physics Laboratory, Delhi-110054, India for the technical discussions and for giving permission to carry out the experimental work presented in this paper.

References

[1] Vincent J D 1990 *Fundamentals of infrared detectors operation and testing* (John Wiley and sons)

[2] Veprik A, Riabzev S, Avishay N, Oster D, Tuitto A 31 May 2012 Linear cryogenic coolers for HOT Infrared detectors. *Proceedings SPIE Infrared Technology and Applications XXXVIII* **8533**

[3] Fraenkel R, Haski J, Mizrahi U, Shkedy L, Schtrichman I 2014 Cooled and uncooled infrared detectors for missile seekers *Proceedings SPIE, Infrared Technology and Applications XL* **9070**

[4] Glaister D S, Donabedian M, Curran D G T, Davis T 1999 An assessment of the state of cryocooler technology for space applications *Proceedings of the IEEE Aerospace Conference*

[5] Wiecek B, 2005 Cooling and Shielding systems for Infrared detectors-requirements and limits *Proceedings of the IEEE Engineering in Medicine and Biology 27th Annual conference*

IOP Conf. Series: Materials Science and Engineering **748** (2020) 012013 doi:10.1088/1757-899X/748/1/012013

[6] Barron R 1985 *Cryogenic Systems* (London : Oxford University Press)
[7] Monteiro W A, Cosandey F, and Bandaru P, July 1997 The effect of Thermomechanical Treatment on the Microstructure of a Cu-Ni-Be alloy *Proceedings of THER-MEC'97* (Wollongong, Australia)
[8] Stadtler W A, 1989 Production of Metallurgy Parts, *Powder Metallurgy,* (Ohio: ASTM) 449-463
[9] Thümmler F & R Oberacker 1993 Introduction to Powder Metallurgy *The Institute of Materials* **ISBN 0-901716-26-X**
[10] Nadkarni *A* Copper Powder Metallurgy Alloys and Composites – *In*: Lee PW, Trudel Y, Iacocca R, German RM, Ferguson BL, Wisen WB, Moyer K, Madan D, Sandreow H 1998 *Powder Metal Technologies and Application* ASM Handbook **7** (OH: ASM International)
[11] Cusano C 1994 Porous metal bearing. In: Monitoring, materials, synthetic lubricants and applications - In: Booser E R *CRC Handbook of Lubricant and Tribology* **III** (New York: CRC Press)
[12] Müller Albano, 1982 Filter elements of highly porous sintered metals *Power Metallurgy International* **14(2)** 73-9
[13] Palfalvi A, Vida-Simiti I, Chicinas I, 1998 Influence of Porosity on the Mechanical Characteristics of Sintered Porous Stainless Sheets *Powder Metallurgy International* **4** 16-9
[14] Getman OI, Chernysher LI 2003 The Porous Structure of a Permeable Nickel Material with a Bimodal Pore ••Size Distribution *Powder Metallurgy and Metal Ceramics* **42** (11-12) 630-7
[15] Bram M, Buchkremer HP, Stöker D Development of Porous Composites Membranes for Microfiltration Devices *Euro PM 2004 Conference Proceedings 183-8*
[16] Podrezov IN Structural Engineering of High-Porous PM Materials *Euro PM 2004 Conference Proceedings* 125-30
[17] *ASM Handbook Powder Metal Technologies and Applications* 1998 (OH: ASM International) **7**
[18] Davis JR 2001 *ASM Specialty Handbook Copper and Copper Alloys* (OH: ASM International) **220**
[19] Neikov OD, Naboychenko SS, Dowson G 2009 *Handbook of Non-Ferrous Metal Powders: Technologies and Applications* (Elsevier)

An IOT based secure transaction system using Micro field communication

Sanjay Kumar [1], Dhananjay Singh [2], Rahul Singh[3], Gagan Gupta[4], Dr. S K Singh[5]

[2]Student, Department of Information Technology, Galgotias College of Engineering & Technology, Gr. Noida, Uttar Pradesh, India
[1,3,4]Assistant Professor, Department of Information Technology, Galgotias College of Engineering & Technology, Gr. Noida, Uttar Pradesh, India
[5]Professor, Department of Information Technology, Galgotias College of Engineering & Technology, Gr. Noida, Uttar Pradesh, India

Email: dhananjay0397@gmail.com, skhakhil@gmail.com

Abstract. The humans have always tried to improve quality of living and technology plays a major role in this cause. Modern technology has been evolving with the emergence of smart technologies. Cash Based transaction involves lot of human effort of withdrawing and then storing the cash. It is also an insecure way of payment involving fears of theft etc... With large amounts of data transfer and e-payment needs, NFC is one of the prime technologies contributing towards fast and secure transmission of electronic signals, data and money. This paper focuses on NFC technology utilisation for a simplified and secure unified transaction system using IOT. Internet of things is a technology of present and future which has its applications in various fields like healthcare, transportation, manufacturing etc… There has already been a lot of research and developments on utilisation of NFC technology for transactions, but we propose a secure wireless transaction system by providing features such as database connectivity and user authentication in the system

Keywords:Near field communication, NFC technology, Rfid, smartcard, Contact less payments.

1. Introduction

Near Field Communication is a set of rules/ protocol which ensures simple wireless connectivity within a small range. This technology is primarily useful in contactless payment system, it can allow payments with mobile devices such as mobile, smartcards etc. It is not just limited to simplified payments transactions, NFC enabled devices can also be used for social networking, e-ticketing and file sharing (such as photo, video, contacts etc...). It is a short range wireless connectivity that uses magnetic field by bringing them within the range of 4cm. It is implemented through electromagnetic induction for simple secure and safe exchange between two devices. It has been successfully implemented for the first time by NOKIA in their smartphones. It operates on the principle similar to that of smart card and RFID tags. The reader starts the communication and retrieves data from the card or the NFC enabled device.

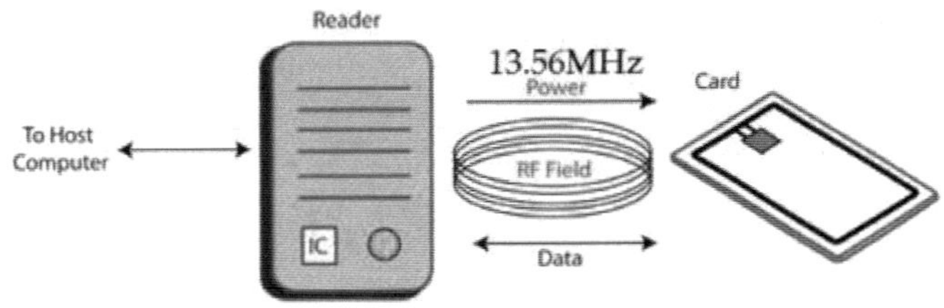

Figure 1 Working of NFC by magnetic field induction

With advantages such as simplicity, security, and ability to connect devices, NFC has vast potential to improve IoT adoption. Both IOT and NFC are technologies slated to expand in the long run, great integration between NFC and IoT can be anticipated.The key applications of NFC card can be depicted through the given diagram:

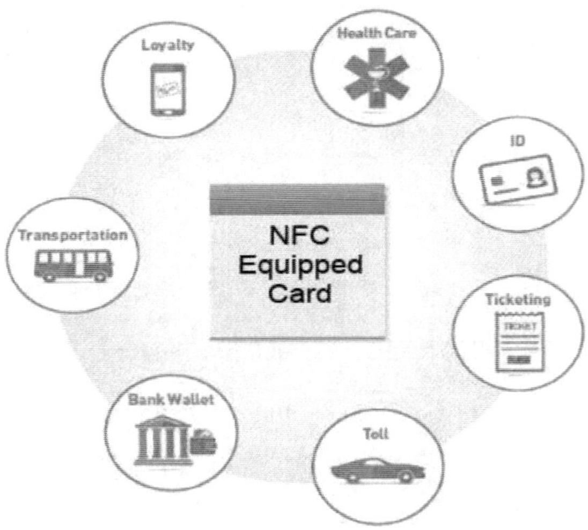

Figure 2 Application of NFC card

2. Operation Modes of NFC

NFC is a short range RF system which transmits data between NFC enabled devices. One device acts as the initiator which is an active device generating magnetic field. It initiates the communication. The other device is the target. It is the passive device responding to the initiator. NFC enabled devices such as NFC Card performs read/write operations from tags and also be utilised a digital storage for NFC readers both at the same timeNFC operates on various modes. The Operation modes of NFC are:

1. Card Emulation Mode
2. Peer to Peer Mode
3. Reader/Writer Mode

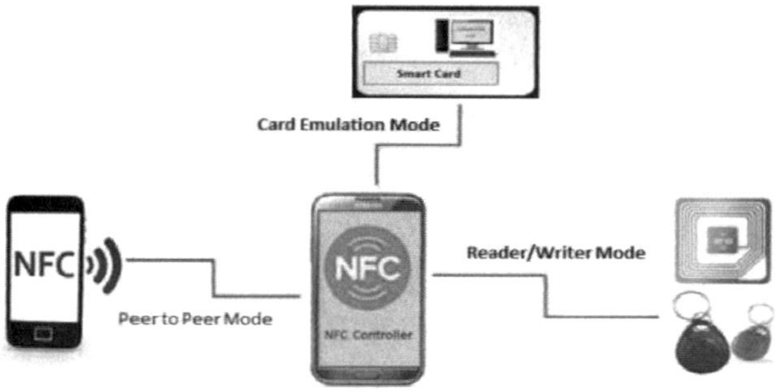

Figure 3 Operating Modes of NFC

2.1Card Emulation Mode:
In this mode the NFC cared stores multiple smart application in the smart card. Card emulator gets rid of the hassle of carrying numerous cards which fatten our wallets and are difficult to maintain. It helps make one unified smart card which caters to variety of our transactional needs.

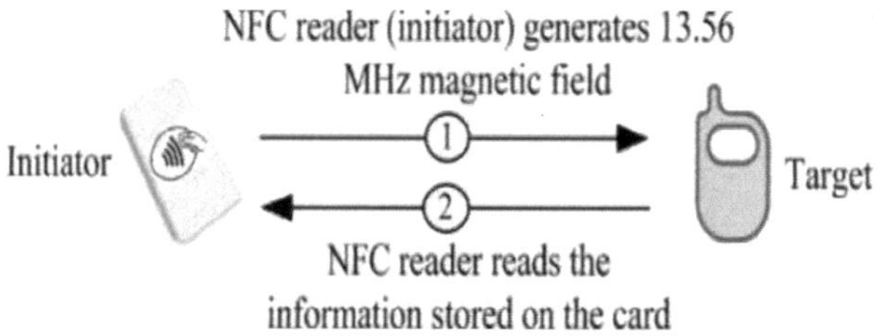

Figure 4 Card Emulation Mode

In this mode the NFC reader generates the field enabling payment and ticketing system. It is compatible to work with the existing smart card systems.

2.2Reader/Writer Mode:
In this mode NFC enabled device performs the role of reader like the smart cards which work without any contact between them. The NFC device can perform the task of reading and writing data into the detected NFC tags i.e. card with read/write operations. It enables communication of an NFC enabled device with the NFC tags. This mode functions in 2 sub modes, Reader mode and Writer mode. In reader mode data is read from NFC tag consisting of required information and program. In writer mode the NFC device such as mobile acts like initiator writing data to the tag. Its application are in Remote shopping, Poster Scanning etc.

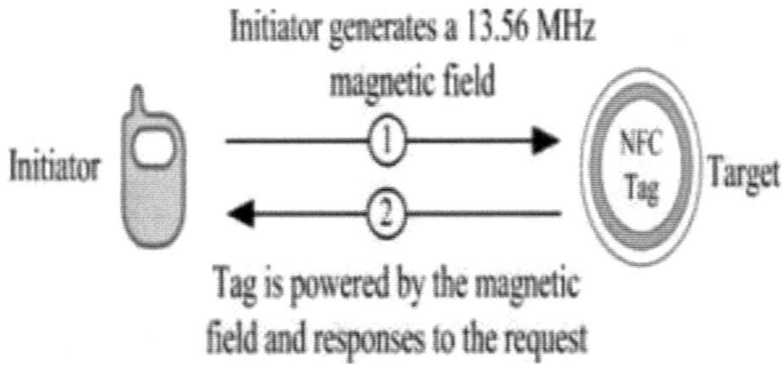

Figure 5 Reader Writer Mode

2.3Peer to Peer Mode:
In this mode the NFC devices are able to transmit and receive data between themselves. It is through a bi-directional channel in which data is transferred one at a time. One device waits until the other transfers data. Its applications are in money transfers, data exchange, Mobile Social i.e. Social Networking etc...

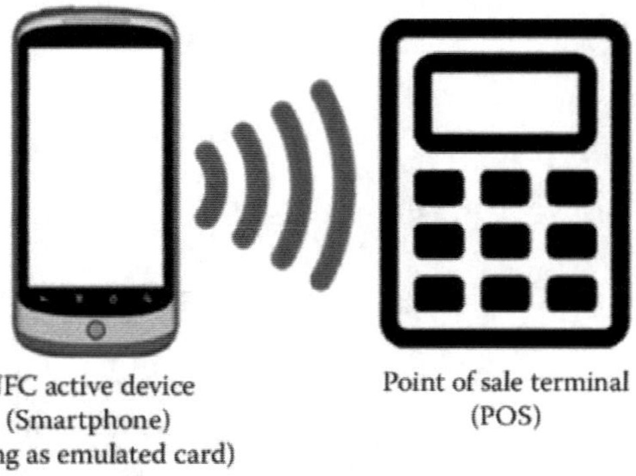

Figure 6 Peer to Peer Mode

3.Current Usage Trends of NFC

NFC is a new yet a very resourceful concept. The concern while developing such system was, would it also have practical application or would it just be limited to a theoretical term. But in recent times there has been an enormous growth in the number of NFC enabled devices.

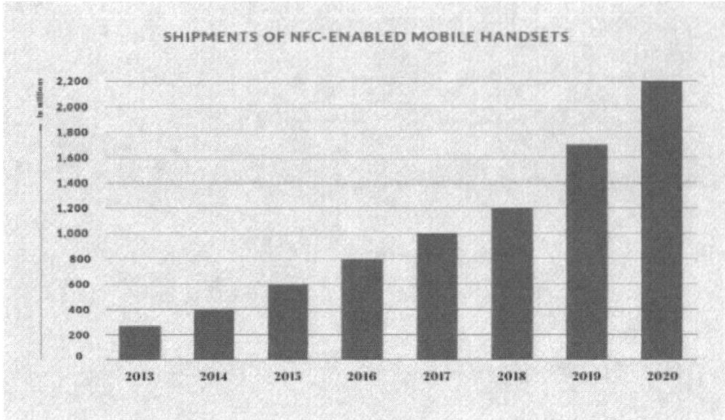

Figure 7 Growth in NFC Enabled Handsets

With several large companies investing in this technology, this technology looks to completely take over the contactless payment market in future. Google, Apple, Microsoft, Master Card are few of the many companies venturing and investing in NFC based payment systems.

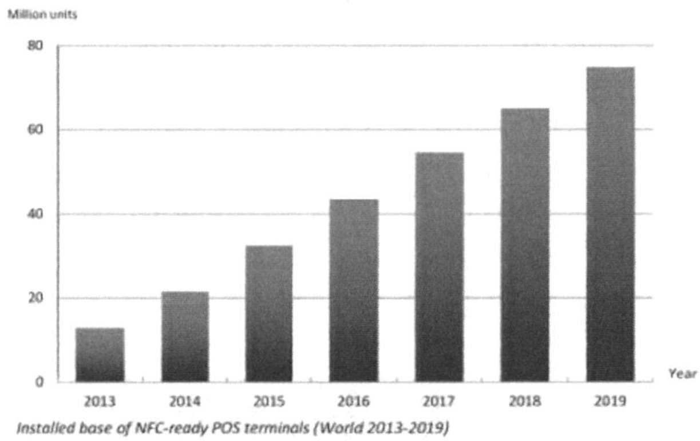

Figure 8 Growth in number of NFC based POS terminals

Some major examples of NFC based contactless cards already in use are:

1. Express Pay
2. Master Card PayPass
3. Visa Paywave
4. RuPay Contactless
5. QuickPassetc...

With so many companies investing in this field, the value of NFC transaction will continue to increase, with projected value by 2020 estimated to be about USD 47 Billion.

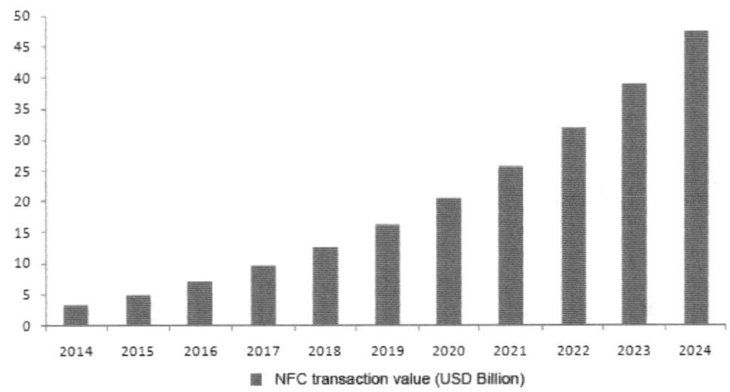

Figure 9 Growth in NFC Transactions Market value

4. Limitations of NFC Card Based Transactions

The previous models had many limitations such as follows:

1. Failure Rate: The NFC cards which are usually made up of plastics are used in harsher environments and as a result of wear and tear they encounter a higher failure rate. One way of overcoming this limitation is to use the card keeping them in an enclosed cover.
2. Privacy: This is another issue faced by the smart card systems. As the card are used in a mass transit system, the transaction operator and the authorities can track individual movements thus posing privacy risks.
3. Use abroad: It is easy to convey data at a quick speed in inland networks but it is not possible when such cards are used abroad.
4. Theft and Fraud: High value transactions are considered safe as they require the user to enter PIN each time the transaction is initiated. Whereas, in low transactions it is not necessarily required. This makes the card being vulnerable to being stolen or frauds.
5. Multiple Card detection: When two or more cards are within the range of the NFC the system may get confused as to which card is intended for the transaction and reader may charge the wrong card. For example in a wallet containing many cards a NFC reader intended for parking entry would detect the apartment entry card and reject the transaction.

5. Implementation& Result

We propose a model which takes into consideration all the drawbacks of previous NFC cards, and have also implemented the proposals in our project. To overcome the multiple card detection issue, we have proposed a unified NFC card which can cater all the transactional requirements irrespective of the bank or the field in which the card is being used. e.g.healthcare, ticketing, identification, payments etc.. To overcome the issue of authentication we propose a system connected to a GSM module. In our system the NFC card sends the card id to Sql database with the help of Arduino which works on scanning card in the MFRC 522 NFC reader setup. This ensures the authentication of the card as each card has unique credentials of the user associated with it. Further, the Arduino UNO board is connected with GSM800 module which provides internet service connectivity to the NFC card. This provides SMS compatibility and as soon as any transaction in the card is processed, an SMS update is sent to the user's registered contact number. This enables both authentication and accountability of the system.

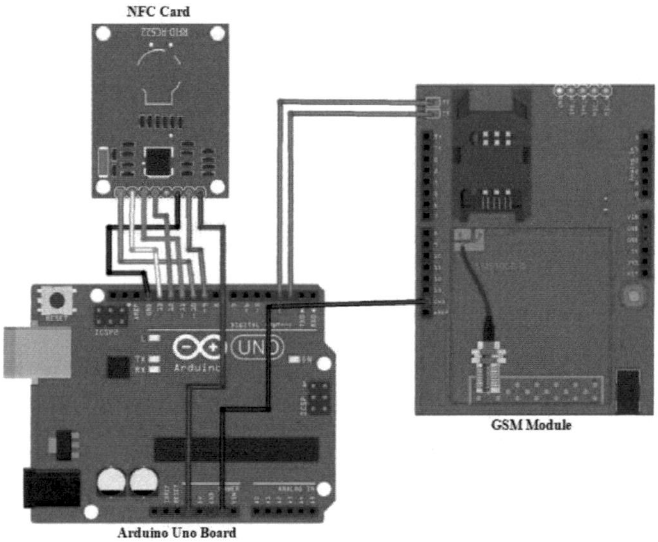

Figure 10 Proposed Model Setup

6. Conclusion

Concluding, we would like to state that NFC is technology which continues to grow day by day with enormous future scope in implementation. Based on the recent trends the day would not be far when the people would completely rely on cashless and contactless modes of payment. Based on our understanding we have proposed a solution but there is still a lot of work which can be done in order to make the implementation of this system more secure and feasible for the common people. One such scope is to extend this NFC Card system to a JAVA based web based project which further can be used for complete transaction and database management over a website secured by username and password credentials. The user can add value, receive reports of previous usage and transactions and also add and delete records. In today's busy world where people have less time, this system is a technology of the future which can help to achieve the target of simplifying the lives of people.

References

[1] Majumder, Alak, ShirshaGhosh, JoyeetaGoswami, and Bidyut K. Bhattacharyya. "NFC in IoT-Based Payment Architecture." (2017).

[2] Anguranak, Taweechai. "NFC Technology Payment System for Android Operation." *Journal of Software*, 2015.

[3] Desai, Ekta, and Mary Grace Shajan. "A review on the operating modes of near field communication." *International Journal of Engineering and Advanced Technology (IJEAT)* 2, no. 2 (2012): 322-325.[4] *Current Trends in Near Field Communication.* 2017. http://nearfieldcommunication.org/current-trends.html (accessed May 18, 2019).

[4] Kumar, Sanjay, AshishNegi, and J. N. Singh. "Semantic Segmentation Using Deep Learning for Brain Tumour MRI via Fully Convolution Neural Networks." Information and Communication Technology for Intelligent Systems. Springer, Singapore, 2019. 11-19.

[5] Brian, A. LarsanAro, L. Arockiam, and P. D. S. K. Malarchelvi. "An IOT based secured smart library system with NFC based book tracking." *International Journal of Emerging Technology in Computer Science & Electronics (IJETCSE)* 11, no. 5 (2014).

[6] Kumar, Sanjay, Gagan Gupta, and KunwarRajat Singh. "5G: Revolution of future communication technology." In *2015 International Conference on Green Computing and Internet of Things (ICGCIoT)*, pp. 143-147. IEEE, 2015.

[7] Monteiro, David M., Joel JPC Rodrigues, and Jaime Lloret. "A secure NFC application for credit transfer among mobile phones." In *2012 International Conference on Computer, Information and Telecommunication Systems (CITS)*, pp. 1-5. IEEE, 2012.

[8] Haselsteiner, Ernst, and KlemensBreitfuß. "Security in near field communication (NFC)." In *Workshop on RFID security*, pp. 12-14. sn, 2006.

[9] Dutot, Vincent. "Factors influencing near field communication (NFC) adoption: An extended TAM approach." *The Journal of High Technology Management Research* 26, no. 1 (2015): 45-57.

[10] Sharif, Rabia, Syed Waqas Ahmad, HirraAnjum, NaveedRamzan, and ShahidRaza Malik. "Effect of infusion time and temperature on decaffeination of tea using liquid–liquid extraction technique." *Journal of Food Process Engineering*37, no. 1 (2014): 46-52.

[11] van der Aalst, Wil MP, and Akhil Kumar. "Xml–based schema definition for support of interorganizational workflow." *Information Systems Research* 14, no. 1 (2003): 23-46.

[12] Gavish, B. and Gerdes, J. (1998) Anonymous Mechanisms in Group *Decision Support Systems* Communication. Decision Support Systems, 23(4), 297-328.

[13] Von Alan, R. Hevner, Salvatore T. March, Jinsoo Park, and Sudha Ram. "Design science in information systems research." *MIS quarterly* 28, no. 1 (2004): 75-105.[14] Mulliner, C. (2009) Vulnerability Analysis and Attacks on NFC-enabled MobilePhones. *In Proceedings of* 2009, International Conference on Availability,Reliability and Security. 695-700.

[14] Strommer, Esko, JouniKaartinen, JuhaParkka, ArtoYlisaukko-oja, and IlkkaKorhonen. "Application of near field communication for health monitoring in daily life." In *2006 International Conference of the IEEE Engineering in Medicine and Biology Society*, pp. 3246-3249. IEEE, 2006.

[15] Tiwari, Naveen Kumar, and Sanjay Khakhil. "Security System for DNS using Cryptography." *International Journal of Computer Applications* 120, no. 17 (2015).

[16] Gupta, Anil Kumar, ShashankShukla, SandeepSaxena, and Sanjay Khakhil. "A Journey towards Workflow Scheduling of Cloud Computing." *International Journal of Computer Applications* 123, no. 4 (2015).

[17] Gupta, Anil Kumar, ShashankShukla, SandeepSaxena, and Sanjay Khakhil. "Bloom filter for elasticity in DBaaS." In *2016 IEEE International Conference on Recent Trends in Electronics, Information & Communication Technology (RTEICT)*, pp. 672-677. IEEE, 2016.

[18] Shukla, Shashank, Anil Kumar Gupta, SandeepSaxena, and Santosh Kumar. "An evolutionary study of multi-objective workflow scheduling in cloud computing." *International Journal of Computer Applications* 133, no. 14 (2016): 14-18.

[19] Shukla, Shashank, Anil Kumar Gupta, Santosh Kumar Upadhyay, and Ashu Gupta. "Wireless Sensor Data Collecting to MobileCloud Device." *International Journal of Computer Applications* 138, no. 11 (2016).

[20] AMAZON, MULTI-OBJECTIVE WORKFLOW SCHEDULING IN. "International Journal OF Engineering Sciences & Management Research."

Application of ELECTRE- I,II methods for EDM performance measures in manufacturing decision making

Rashmi Tiwari*[1], Sharad Agrawal[2], Dinesh Kumar Kasdekar[3]

[1, 2 & 3] Department of Mechanical Engineering,Madhav Institute of Technology and Science, Gwalior-474005(M.P), India.

[1*]tiwarirashmi174@gmail.com

Abstract: Due to the demand of customer point of view requires good performance component with low cost. So, we have to expand the machining characteristics of Al7075 alloy containing 2wt% SiC on EDM. During machining following indicator has to be used for measures EDM performance. These output indicators for EDM are Surface roughness i.e., Ra and Rz, and also Energy consumption (PC). For experimenting, we choose the CCD approach in RSM. In this work, we have to apply a novel approach, multi-attributes decision making to predict the optimum result for the following criteria. In MADM approach using Electre I and II as an application to optimize EDM performance. Electre I and II have decided,based on Entropy weighting percentage. To predict sophisticated optimal attribute that contains the values of criteria (as an EDM performance indicator). Electre I and II offer Rank of alternatives that helps the designer to choose the best decision. At last, the graph shows the comparison of the computational results between Electre I and Electre II.

Keywords: Electric discharge machine (EDM), RSM,MADM, Entropy-Weight method, ELECTRE I & II.

1 Introduction

Day-to-day, raising the demand of new material which can fulfill the demand of these areas like aerospace, defense, and medicalfield requirements. So, due to this requirement, machining of that material is typical task to cut into the different intricate shapes. Due to this machining of that material is not possible with the traditional machining process, because material having property to high hardness as well as strength. For this type of material can be easily machined with nontraditional machining methods. These NT-Machining (Non-Traditional-Machining) techniques have to resolve the problem of machining of these materials causes to affect the material removal rate, and also tool wear rate. So, to solve the issue of above two, causes with following desired output during the machining like good surface finish, dimensional accuracy and low power consumption are achieved with the electrical discharge machining (EDM). EDM machine has also ability to machined die steel and metal matrix composites. So, the performance of EDM machine is continuously trying to improve in the past decades. So, for improving the performance of EDM with the fulfillmeant of requirement of material requirement that have to be mentioned above, study the literature of few researcher to providing a solution is discussed in the below section. HO et. al., [1] has an attempt to study the last 20 years improvement in machining technology, especially in non-traditional machining processes like water jet machining, laser assisted machining, and electrical discharge machining. During this study, researchers was found that for precision requirement, traditional machining not prefer, so that at this time these were to suggest EDM machine is better for this. Tristo et. al., [2] has an attempt to study the industrial machining environment, because of raising the aspect of environment and ecological features. In this study, to suggest the machine to resolve the manufacturing requirement on the basis of environment without affect productivity and quality of goods. So, for this use electrical discharge machining to cut different intricate shape with low power consumption, and low expenses required to produce products with high productivity. Liang et. al., [3] has a study to find that CNC machine was required in manufacturing sectors, because it have a high reliability. So, due to this CNC machine has necessary for the survival of any industry in the competitive market. Shimizu et. al., [4] has an attempt to make the product with machines, reliability has important consideration during machining. So, on the basis of reliability, to produce high performance products according to customer need says that always choose the optimal parameter of the machine. Dubey et. al., [5] has an investigate the performance of power mixed EDM, while chromium particles added in dielectric fluid with copper electrode. During this MRR are the response parameter, for this response uses four intakes. For this performance, use the central composite design approach on the response surface methodology for design the experiment.Kathiresh et. al., [6] has studied to investigate the performance characteristics of EDM such as surface roughness (Ra), TWR, MRR. For this performance characteristic select four intakes such as T_{on}, current, gap-voltage, and T_{off}. While this investigation, experiment was designed by Taguchi L9 Orthogonal array technique and optimized by using GRA. Kar et. al., [7] has attempted to develop the mathematical model for machining aluminium MMC with EDM. For this select response parameter such as radial over cut, MRR, and electrode wear rate and vary the following inputs such as open circuit voltage, discharge current, and T_{on}. This experiment was conducted on the basis of CCD-RSM approach. Nagarjun et. al., [8] has put an attempt to optimize multi-objective responses such as TWR, SR (Ra), and MRR by applying the integrated Fuzzy method with Taguchi. For this use intakes such as discharge current, discharge-voltage, T_{on}, inter-electrode gap. Govindan et. al., [9] has an attempt to provide the suitable service provider by using ELECTRE-1 (determining the impact of reinforced preference) method. For this study, uses Simos procedure and stochastic multicriteria acceptability analysis for various service providers. Wang et. al., [10] has an attempt to remove the uncertainty in accuracy in ranking done by ELECTRE-I method. So, for this ELECTRE-II and ELECTRE-III are used to select the best attributes. Pang et. al., [11] has a studied CNC machine on the basis of reliability design scheme by applying ELECTRE-I method. In this study, weightage of the attributes determined by AHP method. And through this ELECTRE-I method provides an optimal ranking solution that was preferred by designers. Supraja et. al., [12] has an attempt to solve multi criteria decision making, problem of determine best PC framework out of three. So, in this study weight percentage of each criteria was calculated by AHP technique and the solved issue with ELECTRE method after comparing each other and providing a ranking for preference. From the literature survey find a gap, during the study, we have to see that in the manufacturing sector, one response are easily optimized during the process, but when multiple responses are considered process become complex to be optimized. And the material that becomes required in the industry found that aluminium MMC has an ability to solve the problem. During this decision making process have a problem for this solved by using traditional optimization technique. So, here decided after literature study, the objective of the present research is Al MMC i.e., to be used mainly are SiC and alumina for the reinforcement and rarely used B_4C (boron carbide). So. We have to use Al7075+2% wt SiC MMC i.e., fabricated by mechanical stir casting procedure is used in the present study. For this MMC machined by EDM. And this work aim is to develop the correlation between EDM input such as Ton, Toff, current and machining output are surface roughness (Ra and Rz), and also power consumption. And the desired response has to be optimized data set is preferred for machining is used ELECTRE-I and ELECTRE-II method. And the responses after comparing with each alternative and predict optimal data set of input corresponding to the response value for Al7075+2% wt MMC by EDM machine.

2 Equipment used in this experiment

In this work, practical conduct on an EDM machine having their model no. is Electronic Xpert1 die-sinking machine that is based on PLC model no. Electronica PRS-20 controller. The motion of an electrode tool has controlled by a DC servo control to provide too and fro motion toward the workpiece. To achieve good finished surface and high speed of machining, dielectric fluid to be used in EDM. Dielectric fluid used in EDM is Paraffin oil. While performing experiments, positioning of tool electrode and workpiece achieved by dial gauge. During this work, we have to choose '+ve polarity' electrode. For this Copper electrode, circular in size has to be used as a tool, having their dimension 18mm in diameter and 80mm length[13]. The workpiece Al7075 alloy having circular in size and their dimension are 30mm diameter and 20mm thick has to be used. While performing following runs, depth of cut for each sample is 0.2mm. After machining, cleaning and polishing of both workpiece and electrode is necessary. But workpiece has clamped with vice and submerged in paraffin oil. The Al7075 alloy used consist following chemical ingredients shown in Table 1, Prasanna et. al., [14]. For conduct experiment suitable parameters, limits shown as Table 2 and Figure 1 seen the experimental setup.

Measuring the weight of workpiece before and after by using digital weighing balance (citizen), for calculating MRR. And the surface roughness of finished workpiece are measured by using TIME@3100 equipment, that gives both values of Ra, and Rz. Always take three readings in each measure and then average of this take as a final reading of SR And power consumption following the reading of current and voltage note that give directly fromthe control panel, put in formula for evaluation. For machining Al7075 composite in EDM, four process input functions were varied as Ton, Toff, and current, has chosen to study their impact on response parameters

Ra, Rz and PC. The limit of input parameter has decided on the experience of the pilot of the experiment as shown in Table 2.

Table 1. Chemical composition of Al7075

Element	Zn	Cu	Mg	Cr	Si	Ti	Fe	Al
Weight (%)	5.65	1.78	2.51	0.27	0.36	0.19	0.48	Balance

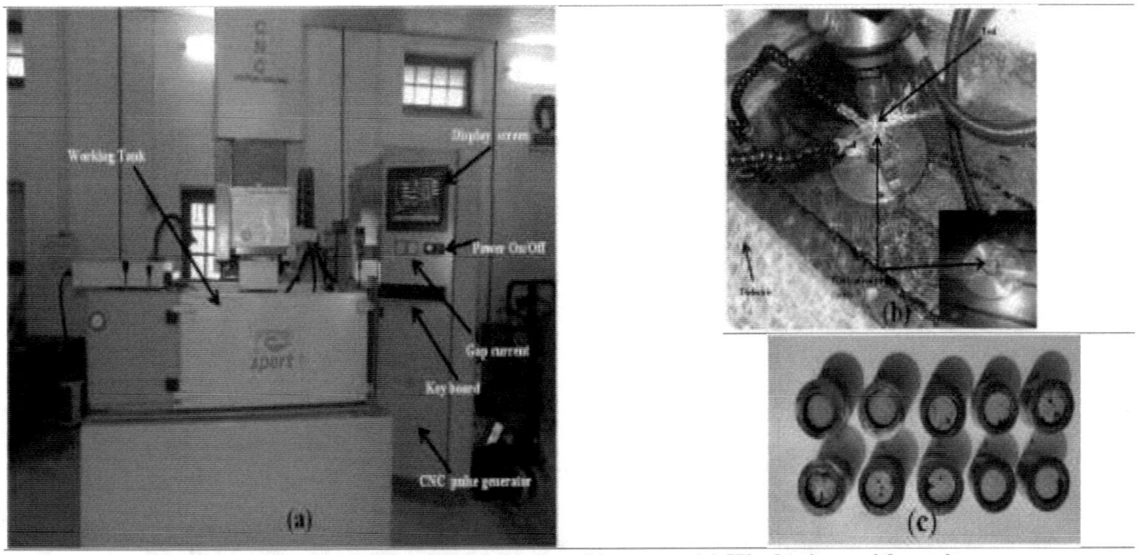

Fig: 1(a) CNC die sinking EDM (b) Sparking zone (c) Work piece with cavity.

Table 2: Factors and their levels

Input Para neters of EDM

S.No	Parameters	Units	Levels		
			L-1	**L-2**	**L-3**
1	Pulse on time (T_{on})	μs	100	150	200
2	Pulse off time (T_{off})	μs	10	15	20
3	Current (C)	Ampere	3	4	5

3. Design of Experiments

Design of Experiments (DOE) has to be used to efficiently and effectively for purposes such as design, planning, and analysis. In DOE, change in inputs called as factors, cause changes in output (as a response). In DOE, the response surface approach used CCD to design the experiment. In a CCD, the face centered design has to be chosen to give a suitable impact of all inputs on the responses. The design of experiments on basis of Table 2 further goes to investigate the significance of the model, perform ANOVA. In DOE software, the regression analysis approach helps to retrieve the response according to independent input parameters. In RSM, dependent regression equation, i.e., the function of independent input parameters as shown in eq. 1:

$$x = f(y_1, y_2, \ldots, y_n) \pm \in, \ldots\ldots\ldots\ldots \text{Eq. (1)}$$

Here, $\in$ denote the error in response y and surface function expressed as is $f(x_1, x_2, x_3, x_n)$ known as response surface.

3.1 Analysis of variance (ANOVA)

ANOVA has to be used for check adequacy of the model. This measures the effect of independent function Ton, Toff, and current on the dependent responses Ra, Rz, PC. The term total sum of squared deviations is formed by adding, sum of squared deviations for each design parameter and the sum of squared error. Statistically, F- test performed to identify that parameter those have significant effect on the response characteristic. Then, the F- value finds for each design parameter is generally F> 4, then it describes the change in design parameter has affected significantly in response. ANOVA has to be used to form a significant model for Ra, Rz and PC are presented in Table 3. On the basis of this result, we have to optimized EDM responses by Electre method.

Table 3 Analysis of variance

ANOVA for Ra					
Source	**DF**	**SS**	**MS**	**F**	**P**
Regression	09	0.75	0.083	12.03	0.0003
Residual error	10	0.069	6.899E-003		
Lack-of-fit	05	0.061	0.012		
Pure error	05	7.750 E-003	1.550E-003	7.90	0.0203
Total	19	0.82			
ANOVA for Rz					
Regression	09	68.26	7.58	13.10	0.0002
Residual error	10	5.79	0.58		
Lack-of-fit	05	5.57	1.11	25.55	0.0014
Pure error	05	0.22	0.044		
Total	19	74.05			
ANOVA for PC					
Regression	9	9.301E+005	1.033E+005	14.97	0.0001
Residual error	10	69038.35	6903.84		
Lack-of-fit	5	69038.35	13807.67		
Pure error	5	0.000	0.000		
Total	19	9.9991E+005			

4. Multi-criteria decision-making (MCDM)

In this section, basic definitions of Entropy- approach, ELECTRE- I,II are presented. Multi-criteria decision-making (MCDM) methodology is popular in now a day for making decisions on realistic applications. Due to MCDM provide qualitative decision making that is efficient in circumstances. While the typical MCDM problem approach is to give a ranking of a finite number of alternatives that contains criteria (as a response) i.e., explicit both to each other simultaneously[10].

4.1 Entropy approach for weight determination

For calculating the weight of criteria, Entropy approach has to be widely used. Decision weights can explain the importance of each criteria, which has to be determined subjective weight on their knowledge, and the objective weight by calculating [15]. The objective weight measurement of criteria calculated by following some steps explained in Table 3.

Entropy Weight Calculation Steps		
Step 1	Data normalization, we choose to follow method: $$G=[g_{ij}]_{mxn} = \frac{X_{ij}}{\sqrt{[\sum_{i=1}^{m} X_{ij}^2]}},\ 1 \le i \le m,\ 1 \le j \le n$$ Here, $$X_{ij} = \begin{cases} X_{ij} & \text{for beneficial attribute} \\ 1/X_{ij} & \text{for non-beneficial attribute} \end{cases}$$	Eq. (2)
Step 2	Calculate decision information entropy value:	

	$E_j = -K\sum_{i=1}^{m} r_{ij} \ln r_{ij},\ 1 \le i \le m,\ 1 \le j \le n,\ \{0 \times \ln(0) = 0\}$ Here, m is the number of evaluation objects, and $K = 1/\ln(m)$	Eq. (3)
Step 3	Calculate difference degree: $G_j = 1 - E_j,\ 1 \le j \le n$	Eq. (4)
Step 4	Calculate entropy weight w: $W_j = G_j / \sum_{j=1}^{n} G_j \quad 1 \le j \le n.$	Eq. (5)

4.2 Electre method

In ELECTRE method evaluating performance criteria that consist both as a qualitative and quantitative. However, this method gives the advantage that exactly focusses the needs of decision maker and provide appropriate results. In ELECTRE methods two major evaluations require i.e., concordance indices and discordance indices for each alternative.During this concordance index can measure hypothetically the strength between alternative A_i and A_k define A_i is better than A_k. And discordance index explains hypothesis that has to be endeavoring to each pair[16].

4.2.1 Electre – I

For solving MCDM problem, ELECTRE I is used. In this ELECTRE method, pairwise comparison of alternatives for both concordance and discordance index.The flow procedure of ELECTRE I and II is explained in Table 4 as:

ELECTRE I and II Calculation Steps		
Step 1	Normalizing the Decision Matrix: $r_{ij} = \bar{x}/\sqrt{\sum_{i=1}^{m} \bar{x}_{ij}^2},\ 1 \le i \le m,\ 1 \le j \le n$ Here $\bar{x}_{ij} = \{ x_{ij},$ if x_{ij} is a benifit index. $1/x_{ij}$, if x_{ij} is a cost index.	Eq. (6)
Step 2	Weighting the Normalized Decision Matrix: $D = [y_{ij}]_{mxn} = g_{ij} \times W_j$ (i=1,2,.....,m; j=1,2,....,n) and $\sum_{i=1}^{n} W_i = 1$	Eq. (7)
Step 3	Decide the Concordance and Discordance Sets: Concordance set: $c_{kl}=\{ j,\ y_{kj} \ge y_{lj} \}$ for J=1,2,3,...,n Discordance Sets: $d_{kl}=\{ j,\ y_{kj} < y_{lj} \}$ for J=1,2,3,...,n	Eq. (8) Eq. (9)
Step 4	Build the Concordance and Discordance Matrices: Concordance Matrix Indices: $c_{kl} = \sum_{j e c_{kl}} wj$ for J=1,2,3,...,n Discordance Matrix Indices: $d_{kl} = \frac{\max_{j \in D_{kl}} \lvert y_{kj} - y_{lj} \rvert}{\max_{j} \lvert y_{kj} - y_{ij} \rvert}$	Eq. (10) Eq. (11)
Step 5	Determine the concordance index matrix: $\underline{c} = \frac{\square 1}{m(m-1)} \sum_{k=1 and k \ne l}^{m} \sum_{l=1 and l \ne k}^{m} c_{kl} f_{lk} = 1,\ if c_k \ge \underline{c}$ otherwise, $f = 0, if c_{lk} < c_k$ _	Eq. (12)
Step 6	Determine the discordance index matrix: $\underline{d} = \frac{\square 1}{m(m-1)} \sum_{k=1 and k \ne l}^{m} \sum_{l=1 and l \ne k}^{m} d_{kl} f_{lk} = 1,\ if c_k \ge \underline{c}$ otherwise, $f = 0, if c_{lk} < c_k$ _	Eq. (13)
Step 7	Calculate the net superior and inferior value:	Eq. (14)

	$C_j = \sum_{k=1}^{n} c(j,k) - \sum_{j=1}^{n} c(k,j), (j \neq k)$; $D_j = \sum_{k=1}^{n} d(j,k) - \sum_{j=1}^{n} d(k,j), (j \neq k)$.	Eq. (15)
Step 8	Determine the Aggregate Dominance Matrix: $e_k = f_k \times g_k$	Eq. (16)

4.2.2 Electre –II

Outranking gives by Electre I, has given two different outranking separately. But in the Electre II can provide ranking by computing pairwise comparison of aggregated relation formed by concordance and discordance. In which alternative $A_i$strongly outranks $A_k$for which examine concordance, pass strong A_i\$$A_i$ and discordance test also for strong pass perform. Note that, here all criteria becomes true for this method[17]. Note that in this method, all criteria are true. From past decades and present MCDM problem point of view, we have to go Electre II method that compensates Electre I limitations and obtained the final best ranking of the alternatives[18].

5 Result & Discussion

5.1 Entropy approach for weight determination

To determine the optimal machining parameters for EDM machine first to calculate the weightage of these criteria's.For those we have to choose an Entropy method for weight calculation. According to Entropy approach, to calculate weighted by criteria's we have to solve following equations from eq 2 to eq. 5 to obtained weight as shown in table 4. After these calculations expelled in Electre method to determine optimal EDM machining parameters.

Table 4: represents Entropy based Weight values.

Criteria	C1	C2	C3
Weights	.332	.333	.335

5.2 Illustrations of Rank Reversals with ELECTRE I and II

In ELECTRE method have exploited for MADM to get found best attribute those having multi-performance criteria. In the present work, the main features of experimental model can be improved by implementing these Electre methods. In this experimental output obtained from EDM should take twenty attributes and having three criteria. In which each criteria are non-benefit index because each criteria becomes minimized. So, in the ELECTRE method, we have to find optimal attributes for those criteria. ELECTRE give a ranking, in which criteria are non-benefit index, so the "smaller-the-better" type. In the ELECTRE require to calculate the weight for each criteria, that have calculated on the basis of the entropy weight method as discussed in earlier sections. After this ELECTRE approach has applied to find the best attribute for those criteria's explained below.

As an illustration the experimental result of EDM has to be applied for analytical approach, in this study. To obtain standard data for designers, Entropy based Electre I and II has more potential to study in EDM application.

5.3 Electre-I (Ra, Rz, PC)

In this region, discuss the obtained results on the basis of their study, for all alternatives that contain sets of criteria. In this MCDM problem is defined as to take 20 alternatives and 3 criteria to find best ranking. Note here all the criteria that has chosen are Non-benefit criteria, so that, the lower score gives better performance. So, in this segment, we have expelled the Electra I method to get found an optimal alternative ranking for these criteria's. The individual alternatives are ranked by the users of ELECTRE-I concerning dominance. In Electra 1 solved by following the procedure that has explained in fig. 2. In this criteria are non-beneficial so, normalization for Electra have used the Eq. 6. After this normalized weight matrix calculated by Eq. 7 then find concordance set index and according to these find concordance indices for concordance according to Eq. 8 and

Eq. 10. Also, find discordance set index and discordance indices for discordance from eq. 9 and eq. 11. Then find $\overline{C}$ and $\overline{D}$ values from eq 12 and Eq. 13 then formed concordance matrix and discordance matrix after this find net superior value from the concordance matrix according Eq. 14. C_j is the difference between the sums of Competitive superiorities of all alternatives, in which the more value defines, the better alternative. From Eq. 15 and net inferior value of the discordance matrix in which smaller becomes better. This is reason behind that the smaller net inferior value gets dominant better alternatives that has achieved large value in net inferior value, then provided ranking both superior and inferior, then to get a final matrix from average to get final ranking.From Table 5, decision matrix contains the performance alternatives A_i which expresses the criteria C_j are as follows: to see that attribute A_7 and A_3 has a better because it has a lower value. So attribute A_7 has optimal ranking according to Electre-I.

When computing net superior value and net inferior value to sort the result as shown in Table 5. In this table compare each pair of alternatives separately for both net superior and net inferior. From Table it becomes clear that the net superior values are lower value for A_{11} alternative, which is giving best ranking (see Table 5). While for net inferior value has A_7 ranked first and A_{10} on last. But due to Electre I for an optimal machining response from EDM machine on the basis of average, two alternatives A_3 and A_7 are best. As the Ranking gives as:

$A_3 = A_7 < A_2 = A_8 = A_{11} < A_4 = A_{16} < A_{14} < A_{15} = A_{18} < A_{12} = A_{17} < A_9 < A_6 < A_1 < A_{10} < A_{20} < A_5 < A_{19} < A_{13}$.

According to these Electre-I result, A_3 and A_7 alternative gives best result for EDM response parameter values at this alternative A_3 are Ra (0.64), Rz (5.37), and PC (250) relative input parameters are Ton (100), Toff (15), and current (4). And alternative A_7 response values Ra (0.45), Rz (4.62), PC (375). These criteria correspond to input parameter Ton (200), Toff (20), current (5). While those parameters are optimized parameter for machining on Al7075 machining in EDM machine were selected, then go to implement problem in Electre-II.

Table 5: Decision Matrix with Their Ranking

Attributes / Criteria	C1	C2	C3	Net superior values	Net Inferior values	Electre –I Final ranking	Electre - II Final ranking
A1	0.63	5.4	375	-1.99	-2.11	9	7
A2	0.42	3.21	300	-14.32	0.61	2	6
A3	0.64	5.37	250	-8.36	3.11	1	5
A4	0.59	4.33	375	-5.98	1.12	3	2
A5	1.15	12	375	13.64	-1.35	12	3
A6	0.86	7.5	400	13.66	1.93	8	4
A7	0.45	4.62	375	-6.98	5.18	1	1
A8	0.72	5.6	300	-3.02	2	2	5
A9	0.47	3.5	750	-4.29	-1.36	7	3
A10	0.56	4.89	750	-0.96	-6.92	10	4
A11	0.27	2.3	225	-19	0	2	7
A12	0.79	5.9	300	2.97	1.3	6	4
A13	1.05	6.06	1125	15.68	-2.8	14	4
A14	0.83	6.2	300	5.63	2.3	4	2
A15	0.59	4.55	375	-4.99	-0.31	5	7
A16	0.76	5.73	300	-0.36	2.17	3	5
A17	0.75	6	300	1.65	1.19	6	5
A18	0.74	5.87	300	-0.36	1.48	5	6
A19	0.69	5.82	750	5.7	-6.41	13	5
A20	0.77	7.4	450	11.67	-1.11	11	2

5.4 Electre-II(Ra, Rz, PC)

In this work, we have to also find a ranking based on Electre II because Electre I has not completely define the ranking of the system define only attribute dominance according to superior and inferior. Due to the following limitations, we have to go to Electre II method. In Electre II some steps same as Electre I as shown in fig. 2. In this calculation from Eq. 6 to Eq. 11 follow after this form E matrix from condition given in Eq. 12 in binary coding 0 and 1. Also, form F matrix according to condition 13 same as binary form. And then form a final matrix G = E*F from Eq. 16 according to Boolean AND gate operation. Then, Ranking assigned on alternatives based on ascending and descending manner, then final ranking provided based on the average of these. From the above Table 5 in which A_7 dominates all other alternatives as seen in it. As the Ranking gives as:

$A_7 < A_4 = A_{14} = A_{20} < A_5 = A_9 < A_6 = A_{10} = A_{12} = A_{13} < A_3 = A_8 = A_{16} = A_{17} = A_{19} < A_2 = A_{18} < A_1 = A_{11} = A_{15}$.

According to these Electre-II result, A_7 alternative gives best result for EDM response parameter values at this alternative are Ra (0.45), Rz (4.62), PC (375). These criteria corresponds to input parameter Ton (200), Toff (20), current (5) are optimized result for machining on Al7075 machining in EDM machine.

The Ranking found from Electre I and Electre II plot on the graph between Ranking and Alternatives as shown in Fig.2. These following results from ELECTRE I and II to give optimized responses according to process parameters. In graph clearly A7 gives the best alternative set would belong in both Electre I and II, and resolve limination of Electre I clearly seen. To fulfill the needs of industries where less power consumption produces a better surface finish for those uses these results.

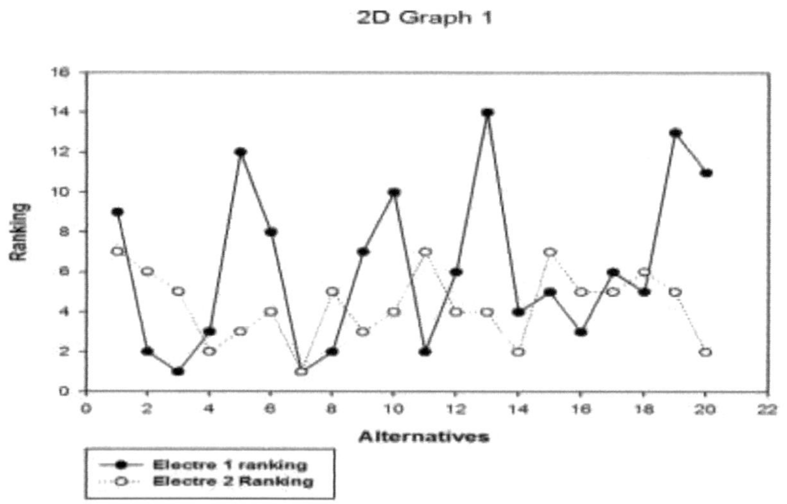

Fig: 2 Plot graph between Alternatives and Ranking according Electre I and II

6 Conclusions

The conclusion of this study is that too obtained,to find the optimal parameters of EDM machinebased on their response criteria by ELECTRE-I and ELECTRE-II methods.For this experiment has the design based on CCD, in RSM.Then firstly, we have to adopt an Entropy approach for the calculation of weights of machining criteria. This Entropy-based ELECTRE-I methodology was used synthetically to rank the following machining alternatives. And then also analyze on the Entropy-based Electre-II method to compensate the Electre-I limitation not provide the final ranking of whole EDM machine to define A_i better than A_k. So, this approach has proposed by this paper has provided choices for the designer to make a product, for this select the best alternative. At this alternative, the corresponding intakes areTon (200), Toff (20), current (5) and optimized response values are Ra (0.45), Rz (4.62), PC (375). And then finally implement this novel approach of the algorithm in the EDM machine to give good prospect. In the above research was motivated to select this problem while machining in EDM generally takes place. In this work, to find an optimal alternative for that performance criterion of EDM corresponding to intake variables, for which model is developed. So, these optimal variables should choose while deciding on machining in EDM. This work is especially for manufactureAircraft rifles. In the future, we will extend our work based on Electre III and IV.

References

[1] Ho K.H and Newman S.T 2003 State of the art electrical discharge machining (EDM). *International Journal of Machine Tools and Manufacture*, ***43***(13), pp.1287-1300.

[2] Tristo G, Bissacco G, Lebar A and Valentinčič J 2015 Real time power consumption monitoring for energy efficiency analysis in micro EDM milling. *The International Journal of Advanced Manufacturing Technology*, ***78***(9-12), pp.1511-1521.

[3] Liang J, Mourelatos Z.P and Tu J 2008 A single-loop method for reliability-based design optimisation. *International Journal of Product Development*, ***5***(1-2), pp.76-92.

[4] Shimizu H, Otsuka Y and Noguchi H 2010 Design review based on failure mode to visualise reliability problems in the development stage of mechanical products. *International journal of vehicle design*, ***53***(3), pp.149-165.

[5] Dubey V and Singh B 2018 Study of Material Removal Rate in Powder Mixed EDM of AA7075/B4C Composite. *Materials Today: Proceedings*, ***5***(2), pp.7466-7475.

[6] Kathiresh P.V.M, SivaKumar K, Dineshkumar S, Sundar S, Vinothbabu N and Surya K 2019 Optimization of machining parameters in EDM of Al7075-SiC-Al2O3 using taguchi method. *Engineering Reports*, ***2***(1), pp.11-16.

[7] Kar C, Surekha B, Jena H and Choudhury S.D 2018 Study of Influence of Process Parameters in Electric Discharge Machining of Aluminum–Red Mud Metal Matrix Composite. *Procedia Manufacturing*, ***20***, pp.392-399.

[8] Nagaraju N, Venkatesu S and Ujwala N.G 2018 Optimization of process parameters of EDM process using fuzzy logic and Taguchi methods for improving material removal rate and surface finish. *Materials Today: Proceedings*, ***5***(2), pp.7420-7428.

[9] Govindan K, Kadziński M, Ehling, R and Miebs G 2019 Selection of a sustainable third-party reverse logistics provider based on the robustness analysis of an outranking graph kernel conducted with ELECTRE I and SMAA. *Omega*, ***85***, pp.1-15.

[10] Wang X and Triantaphyllou E 2008 Ranking irregularities when evaluating alternatives by using some ELECTRE methods. *Omega*, ***36***(1), pp.45-63.

[11] Pang J, Zhang G and Chen G 2011 ELECTRE I Decision Model of Reliability Design Scheme for Computer Numerical Control Machine. *JSW*, ***6***(5), pp.894-900.

[12] Supraja S and Kousalya P 2016 ELECTRE Method for the selection of best computer system. *Indian Journal of Science and Technology*, ***9***(39), pp.1-5.

[13] Kasdekar D.K, Parashar V and Soni P 2016 Optimization of Machining Parameters in Electro Discharge Machining of Al6061-Cu-SiC-Graphite Metal Matrix Composite. In *Materials Science Forum* (Vol. **860**, pp. 61-64). Trans Tech Publications.

[14] Prasanna P, Sashank T.V, Manikanta B and Aluri P 2017 Optimizing The Process Parameters Of Electrical Discharge Machining On AA7075-SiC Alloys. *Materials Today: Proceedings*, ***4***(8), pp.8517-8527.

[15] Kasdekar D.K and Parashar V 2015 MADM Approach for Optimization of Multiple Responses in EDM of En-353 Steel. *International Journal of Advanced Science and Technology*, ***83***(1), pp.59-70.

[16] Afshari A.R, Mojahed M, Yusuff R.M, Hong T.S and Ismail M.Y 2010 Personnel selection using ELECTRE. *Journal of Applied Sciences(Faisalabad)*, ***10***(23), pp.3068-3075.

[17] Figueira J, Greco S and Ehrgott M. eds 2005 *Multiple criteria decision analysis: state of the art surveys* (Vol. **78**). Springer Science & Business Media.

[18] Milani A.S, Shanian A and El-Lahham C 2006 Using different ELECTRE methods in strategic planning in the presence of human behavioral resistance. *Advances in Decision Sciences*, *2006*.

Study of hybrid photovoltaic thermal systems

Ashish Saurabh[1,3], Deepali Atheaya[1], Anil Kumar[2]

[1]Department of Mechanical and Aerospace Engineering,
Bennett University, Tech Zone – II, Greater Noida - 201310, UP, India
[2]Department of Mechanical Engineering,
Delhi Technological University, Delhi
[3]Corresponding Author, email: as6723@bennett.edu.in

Abstract. Sun is the primary source of renewable energy. It is abundant, inexhaustible and clean. It plays a very important role in the present energy crisis. Solar energy can be harnessed by hybrid photovoltaic thermal system to generate power and heat. These devices generate thermal and electrical energy simultaneously. Hybrid photovoltaic thermal systems have high efficiency. There is ample scope in this area as much work remains to be done. The hybrid system has huge potential in India where the availability of solar energy is spread throughout the country. The paper focusses on the study of hybrid photovoltaic thermal systems.
Keywords: Photovoltaic thermal; Hybrid solar; PVT performance.

1. Introduction

Climate change, global warming and environmental pollution have forced the world to look for clean energy technology. These problems can be solved by using renewable energy from sun and wind. All renewable power sources have their primary source in Sun. The motivating fact is that solar insolation rate received by earth is greater than the demand of fossil fuels. The earth intercepts about 1.8×10^{11} MW power from sun. This is greater than the present rate of all energy consumption. [1][2] Solar power is abundant, and its conversion is also less costly. Different methods for collecting solar energy in the usable form are being developed. Some methods focus on the use of solar PV, whereas others on thermal conversion. Solar PV works on energy of electrons excited by solar photons. Solar thermal depends on energy of water heated by sunlight. This hot water can be used for heating or electricity generation. [3][4][5] Solar cell converts only a fraction of solar energy (14-20%) to electricity through photovoltaic effect. Remaining energy is dissipated as heat energy. The efficiency of silicon solar cell decreases 0.45 % per °C . Panel temperature can reach 70 °C in hot summer days. This means panels will produce up to 20 % less power compared to their rated power capacity at 25 °C. To overcome this, hybrid solar technology was developed. A hybrid solar concentrator system captures this waste energy and thus removes the waste heat from solar module. This also increases the efficiency of the combination remarkably. [6][7] The individual silicon cell has an efficiency of 25% whereas efficiency of solar PVT system can reach up to 70%. [8]

2. History of Solar PVT system

The history of solar PV began in 1839 after the breakthrough discovery of photovoltaic effect by Alexandre-Edmond Becquerel, a French physicist. Solar cells are made up of silicon which captures the photons emitted by sun and generate voltage.[9] The photoconductivity of selenium was then discovered in in 1873 by English electrical engineer Willoughby Smith. The first solar cell was designed in New York by Charles Fritts. It was done by coating selenium with a thin cover of gold. The efficiency of solar cell at the time of invention was mere 1% compared to the efficiency of modern solar cells at 20%. Silicon was found to be more efficient than selenium in 1953 by Bell Laboratories. And in this way production of solar cells took off commercially. The demand for solar power increased as the oil prices rose in 1970.This led to more research in the area and further reduction of cost. [10] The history of solar thermal dates to 1839. There are records of solar heating in United States. Captain John Ericsson invented solar power engine in 1864. Solar Stirling Engine was driven by a parabolic concentrator. [11] In 1913,the first parabolic trough system was made for generating steam in Maadi, Egypt. The studies in PVT technology started in 1970. PVT technology has become an attractive energy solution for buildings as they require less space and provide more efficiency. [12] Hybrid PVT systems combine both the above idea.

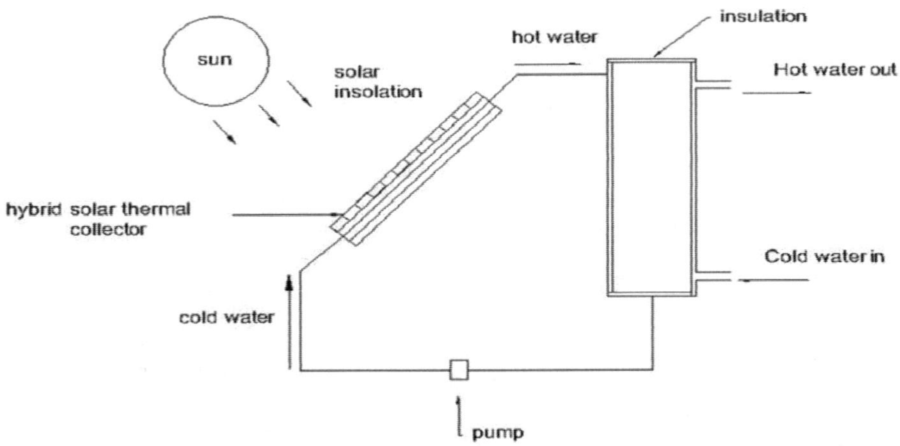

Figure 1. Solar thermal collector system

3. Hybrid solar PVT system

Hybrid photovoltaic solar concentrator is a combination of solar cell and solar thermal collector. The model of hybrid solar thermal collector system is shown in Fig. 1 above. The cold water is supplied from household to the hybrid collector where it gets heated from the waste heat from the hybrid system. Combining thermal and photovoltaic system helps reducing space, on site electricity generation and thermal generation. The detailed view of PV collector is shown in Fig. 2. The cover glass is used for protection and minimizing heat loss. Eva encapsulant is used to protect PV layer from shock and vibration. Thermal insulation is also provided beneath the PV layer to reduce the heat loss from the surface. Fuentes et al. [13] founded that the combined efficiency of hybrid PV/T system can reach 80% during the day. Bhattarai et al. [14] analysed a hybrid solar PVT collector and found that energy savings of hybrid PVT is more compared to conventional PV. Dupeyrat et al. [15] showed 79 % thermal efficiency and 8.7% electrical efficiency for hybrid solar PVT. The electrical efficiency of solar cells was found to be 15% for 90% of solar insolation absorbed. Remaining was rejected as heat. [16] The study on liquid type PVT systems were first done between 1978-1981. [17] Lovvik and Bergene [18] studied the effect of cooling on solar cell. Temperature in the range of 60 – 80 °C were found common for solar cell and cooling resulted in increased electrical efficiency of about 10-30 %.

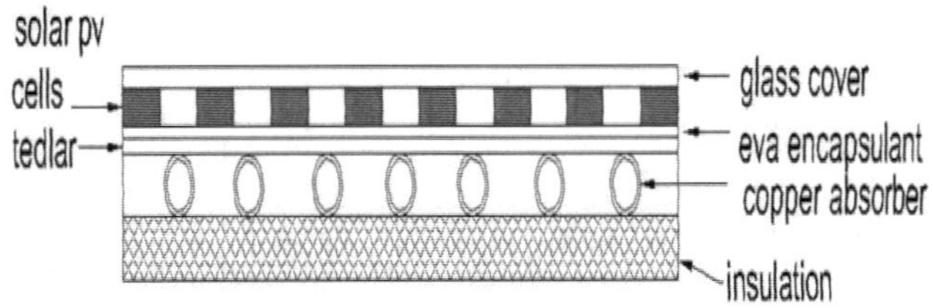

Figure 2. Detailed view of PV collector

4. Application of Hybrid PVT

Solar hybrid PVT liquid collectors have been used for domestic hot water application. Hybrid PVTs have been also integrated with buildings. Chow et al. [19] concluded that building integration for PVT is more popular than that of water application. A 20 kWp PVT with air as thermal fluid is located at Mantaro Library in Spain. Concordia University, Montreal also developed a PVT system with 24.5 kW and 75 kW electrical and thermal power respectively. Beijing Olympic village also installed a PVT system with 10 kW and 20 kW of electrical and thermal power generation respectively. Fang et al. [20] concluded that the integration of hybrid PVTs with building is going to be most focussed area in the world. Global solar thermal capacity grew from 89 × 10^6 m^2 (2000) to 675 × 10^6 m^2 (2017) in China. 3.78 × 10^{10} kg of oil and 1.2 × 10^{11} kg of CO_2 was saved due to this. Denmark, Sweden, Austria, Greece and Spain have large scale solar thermal plants since 1980. China has also recently installed a large no. of such systems.

4.1 Applications of Hybrid Photovoltaic Thermal Systems:

The major application of Hybrid PVT system is found in building integration. These systems can be used for pool water heating (25 – 35 °C), space heating (up to 60 °C) and domestic hot water. [21] Space heating and DHW constitute for nearly 50 % of heat demand. The hybrid systems also allow for the most efficient utilization of roof space. The total installation cost is also lesser when compared to individual installation of solar thermal unit and solar PV unit separately. The hybrid system can be also used in the following areas:

i. Building integrated photovoltaic thermal – BIPVT capacity has increased due to increase in grid connected photovoltaics. BIPVT is generally mounted on rooftop of building. The orientation of BIPVT should be such that solar insolation can be accessed. Irradiance and PV module temperature are important parameters for such system.

ii. Power generation – Hybrid PVT system can be used for power generation. Also, it was found that PVT plant is capable to produce more power than conventional plant. But the cost would be more. [22]

iii. Agriculture – PV pumps are most suitable application when plenty of sunshine is available. Hybrid PVT collector system with air can be also used for agriculture purpose or crop drying. Fterich et al. [23] experimentally studied a solar drier equipped with solar PV/T air collector. The hybrid PVT drier shortened the drying time and increased the efficiency. Barnawal and Tiwari [24] performed grape drying using hybrid photovoltaic greenhouse drier at solar energy park, New Delhi.

iv. Street lights – Hybrid system can be used for lighting streets, yards, compounds and industrial roads.

v. Concentrating PVT – These systems have been used in hospitals and schools. Swedish manufacturer Absolicon installed concentrating PVTs in a private hospital in Harnosand for power. The Cogenra building installation at the university of Arizona Tech Park supplies 191 kW and 36 kW thermal and electric power respectively. [25]

vi. Solar distillation – The overall thermal efficiency of hybrid solar PVT still designed Kumar and Tiwari [26] was found to be 20 % more than the conventional solar still. The electrical efficiency of the former was also found to be higher. Pounraj et al. [27] found 38% efficiency rise for Peltier based hybrid PVT active solar still as compared to conventional PV still.

5. Performance of Hybrid Solar PVT

Based on working fluid, there are two main types of hybrid collector. These are water-based and air-based hybrid collectors. Air has some advantage over water since it does not freeze and there is no damage if leakage occurs. But the heat capacity and conductivity of air is lower. PVT air collectors are generally used in buildings. The overall performance of the hybrid system is the sum of thermal efficiency and electrical efficiency. [28]

$$\eta_{PVT} = \eta_{th} + \eta_{PV} \tag{1}$$

5.1 Thermal Performance

5.1.1 Solar module temperature

$$t_{mo} = t_{amb} + (\text{NOCT} - 20) \times \frac{G}{800} \tag{2}$$

where NOCT the temperature of module in an environment with solar irradiation of 800 W/m^2, 20°C ambient temperature and wind speed of 1 m/s. The heat can be removed from the cell by different modes of heat transfer. The heat removed can be utilized for different purposes. The thermal efficiency of hybrid system and other parameters can be obtained as:

$$\eta_{th} = \frac{\dot{m} \times c_p \times (t_{out} - t_{in})}{a_{pvt} \times G} \tag{3}$$

The loss of heat from the reservoir to the surrounding is:

$$Q_{loss} = \int U_t (t_{in} - t_{amb})\, dt \tag{4}$$

Figure 3. Solar thermal systems on houses in Salzburb, Austria [29]

5.2 Electrical efficiency

The electrical efficiency is given as:

$$\eta_{el} = (f') \frac{i_{sc}.U_{oc}}{G.S} \quad (5)$$

$$f' = \frac{P_{max}}{I_{sc}.U_{oc}} \quad (6)$$

The energy conversion efficiency of the PVT system is:

$$\eta_{PVT} = \frac{w+q}{h} = \eta_{el} + \eta_{th} \quad (7)$$

5.3 Energy saving efficiency

The energy-saving efficiency of the pvt system is given by

$$E_{sa} = \frac{\eta_{el}}{\eta_{power}} + \eta_{th} \quad (8)$$

6. Losses in solar thermal system

Losses in solar thermal system are mainly optical and thermal. The solar thermal system efficiency keeps changing due to changing solar and weather condition. [30]

i. Cosine loss – These losses occur due solar radiation not being perpendicular to reflector.
ii. Shading and blocking – Losses within the system due to blockage by individual elements
iii. Reflectance – The reflector does not allow all the incident light to pass through it
iv. Cleanliness – Dirt and dust from atmosphere accumulate on the surface of reflector and decrease efficiency.
v. Incident angle modifier (IAM) – Optical parameters generally degrade with increase in incident angle.
vi. End losses – Some part of receiver does not receive radiation from solar collector.
vii. Intercept – These losses occur due to geometrical inaccuracies.
viii. Shielding by bellows – These losses occur due to bellows shielding.
ix. Transmittance of glass – These losses are depending on the nature of glass material.
x. Absorbance of receiver – The receiver surface coating absorbs some radiation.
xi. Thermal losses – This is caused by convection and radiation.
xii. Thermal losses due to piping – These losses occur due to conduction and convection.

7. Conclusion

There has been huge research in the field of hybrid system since 1970. Hybrid PVT system transforms solar insolation into usable heat and power. The hybrid system is a combination of solar PV and thermal collector. Solar PV converts solar energy into electric power whereas thermal collector removes waster heat from the solar PV. This results in the increase in the efficiency of the hybrid system. The electrical efficiency of solar PV is about 10% without cooling. The conversion efficiency of hybrid PVT systems can reach up to 80 %. The hybrid system can be very useful for developing country like India which has huge potential of solar energy spread throughout the country.

IOP Conf. Series: Materials Science and Engineering **748** (2020) 012016 doi:10.1088/1757-899X/748/1/012016

Nomenclature

η_{el}	electrical efficiency
η_{power}	efficiency of electrical power generation
t_{out}	outlet temperature of collector (°C)
t_{in}	inlet temperature of collector (°C)
c_p	specific heat (J/kg/K)
η_{th}	thermal efficiency
IAM	incident angle modifier
NOCT	nominal operating temperature (°C)
SRE	standard reference environment
U_t	overall heat transfer coefficient of reservoir (W/m^2K)
q	heat (J)
E_{sa}	energy saving efficiency
s	surface area of the unit (m^2)
G	solar irradiance (W/m^2)
U_{oc}	open circuit voltage (V)
η_{PV}	photovoltaic efficiency
eva	Ethylene-vinyl acetate
$\dot{m}$	mass flow rate (m^3/hr)
G	Irradiance on collector surface (W/m^2)
PVT	photovoltaic thermal
PV	photovoltaic
S	heat absorber plate area (m^2)
t_{amb}	ambient temperature (°C)
t_m	temperature of module (°C)
h	energy input (J)
w	mechanical work (J)
i_{sc}	short circuit current (A)
f'	fill factor
P_{max}	Maximum power (W)
a	area (m^2)
η_{power}	electric power generation efficiency
Q_{loss}	heat loss (J)

Subscripts

el	electrical
out	outlet
in	inlet
th	thermal
mo	module
amb	ambient
max	maximum
avg	average

References

[1] Hannah L 2015 Mitigation: Reducing Greenhouse Gas Emissions, Sinks, and Solutions Climate Change *Renewable Energy* **2(1)** 379-402

[2] Paridaa B, Iniyanb S and Ranko G 2011 A review of solar photovoltaic technologies *Renewable and Sustainable Energy Reviews* **15(3)** 1625-1636

[3] Thomas N 2003 Field Guide to Appropriate Technology *Energy* 157-275

[4] Bhatia S 2014 Photovoltaics Energy resources and their utilisation. *Advanced Renewable Energy Systems* **31(6)** 1-31

[5] Belz K, Kuznik F, Werner K, Schmidt T and Ruck W 2015 Thermal energy storage systems for heating and hot water in residential buildings Woodhead Publishing Series in Energy *Advances in Thermal energy storage system* **12(5)** 441-465

[6] Abdulyan M and Chiasson A 2017 Thermal Study of Hybrid Photovoltaic-Thermal (PVT) *Energy Procedia* **141(3)** 102–108

[7] Du D, Darkwa J and Kokogiannakis G 2002 Thermal management systems for Photovoltaics (PV) installations: A critical review *Solar Energy* **97(6)** 238–254.

[8] Adria E. Brooks 2014 Solar Energy: Photovoltaics. Future Energy Improved, Sustainable and Clean Options for our Planet **4(3)** 383-404

[9] http://www.pvresources.com/en/introduction/history.php

[10] https://www1.eere.energy.gov/solar/pdfs/solar_timeline.pdf

[11] Kongtragool B and Wongwises S 2003 A review of solar-powered Stirling engines and low temperature differential Stirling engines *Renewable Sustainable Energy* **7(4)** 131–54

[12] Aidroos D, Hasimah B, Rahman A, Zaidi W, Omar W and Obaid S 2015 Historical development of concentrating solar power technologies to generate clean electricity efficiently *Renewable and Sustainable Energy* **41(3)** 996-1027

[13] Fuentes J, Vivar M, Casa J and Aguilera J 2018 An experimental comparison between commercial hybrid PV/T and simple PV systems intended for BIPV, *Renewable and sustainable Energy Reviews* **93(1)** 110-120

[14] Bhattarai S, Ohb J, Euh S, Kaflea G and Kima G 2012 Simulation and model validation of sheet and tube type photovoltaic thermal solar system and conventional solar collecting system in transient states *Sol Energy Matter Sol Cells* **103(5)** 184–93

[15] Dupeyrat P, Ménézo C, Wirth H and Rommel M 2011 Improvement of PV module optical properties for PV-thermal hybrid collector application *Sol Energy Mater Sol Cells.* **6(1)** 2028–36

[16] Dupeyrat P, Menezo C and Fortuin S 2015 Study of the thermal and electrical performances of PVT solar hot water system *Energy and buildings* **68(5)** 751-755

[17] Hendrie S 1982 Photovoltaic/thermal collector development program. Final report

[18] Bergene T and Lovvik O 1995 Model calculations on a flat-plate solar heat collector with integrated solar cells *Sol. energy* **55 (6)** 453-462

[19] Chow T, Tiwari G and Menezo C 2012 Hybrid solar: A review on photovoltaic and thermal power integration *International Journal of Photoenergy* 220-227

[20] Fang X and Li D 2013 Solar photovoltaic and thermal technology and applications *Renewable and Sustainable Energy Reviews* **23** 330-340

[21] Mellor A et al. 2018 Roadmap for the next-generation of hybrid photovoltaic-thermal solar energy collectors *Solar Energy* **174** 386–398

[22] Gagliano A et al. 2019 Comparative assessments of the performances of PV/T and conventional solar plants *Journal of Cleaner Production* **219** 304-315

[23] Fterich et al 2018 Experimental parametric study of a mixed-mode forced convection solar dryer equipped with a PV/T air collector *Solar Energy* **171** 751–760

[24] Barnwal P and Tiwari G 2008 Grape drying by using hybrid photovoltaic-thermal (PV/T) greenhouse dryer: An experimental study *Solar Energy* **82** 1131–1144

[25] Good C et al 2015 Hybrid Photovoltaic-thermal Systems in Buildings – A Review. *Energy Procedia* **70** 683-690

[26] Kumar S and Tiwari A 2010 Design, fabrication and performance of a hybrid photovoltaic/thermal (PV/T) active solar still *Energy Conversion and Management* **51** 1219–1229

[27] Pounraj et al. 2018 Experimental investigation on Peltier based hybrid PV/T active solar still for enhancing the overall performance *Energy Conversion and Management* **168** 371-381

[28] Radziemska E 2009 Performance Analysis of a Photovoltaic-Thermal Integrated System *International Journal of Photoenergy* 80-233

[29] https://www.iea-shc.org/Data/Sites/1/publications/Solar-Heat-Worldwide-2018.pdf
[30] Hermann et al 2017 The Performance of Concentrated Solar Power (CSP) Systems Modelling, Measurement and Assessment *Solar Energy* 115-165

Ranking of Barriers to Green Manufacturing Implementation in SMEs Using Best-Worst Method

Dilip Kushwaha[1] and Faisal Talib[2]

Department of Mechanical Engineering

Zakir Husain College of Engineering & Technology

Aligarh Muslim University, Aligarh-202002, India

E-mail: [1]dilipkush1992@gmail.com; [2]ftalib77@gmail.com

Abstract. Nature has provided us with limited resources on the earth, but the requirement is unlimited, one is fulfilled another will crop up. Modern manufacturing scenario and rough consumption of natural resources create huge challenges in front of human beings in the form of environmental degradation and wastage of material. The purpose of this study is to identify the barriers to green manufacturing (GM) from selected Small and Medium Enterprises (SMEs) in Delhi and NCR region and rank them as per their priority. An extensive literature review together with expert's opinion was done to identify the GM barriers. A newly developed approach known as Best-Worst method (BWM) was adopted for the ranking of identified GM barriers. This study identifies 21 barriers to GM implementation which are arranged into 5 main barriers for well understanding. Further, the findings of this study show that, “Lack of new technology facilities and processes”, “Lack of experts related to new technology”, “Lack of expertise training programs”, “High initial capital cost to implement GM” and “Weak Legislation” are top-five barriers that lead to difficulties in GM implementation in SMEs while least important observed barrier is “Not enough and understandable information”. Therefore this study provides knowledge about the green manufacturing barriers and their ranking most severe to less severe that is most important to less important and this study also provides a systematic way to sort out the barriers in GM implementation according to their priority. In the end, the study provides conclusions and some implication of the study as well as the scope for future research.

Keywords: Green Manufacturing, barriers, BWM, SMEs.

1. Introduction

Numerous researches have been done during the past few decades to control environmental degradation and several suggestions were proposed. The manufacturing sector is one of the major sector that is responsible for this degradation. Environmental and natural resource degradation is a huge challenge in front of human beings. In manufacturing sector, Small and Medium Enterprises (SMEs) are mostly responsible for the generation of industrial pollution as they are in huge numbers and still growing at a rapid pace. Therefore, government is intended to focus and assist them by developing guidelines for sustainable green environment. Still SMEs failed to act responsively due to growing new manufacturing scenario. Hence, the requirement for new methodologies emerges to endure this relentless challenge and support aggressiveness. To overcome these challenges, Green Manufacturing (GM) becomes a necessary and reliable manufacturing strategy as it can successfully save the quality of the environment and reduce the manufacturing wastes and enhance recycling (Prasad and Sharma, 2014). Therefore, GM is an idea of eco-friendly manufacturing effort which acts like eliminating the waste and pollution, recycling and reuse of materials, by re-defining the existing production process or system through research and process design. GM goals are also for conserving natural resources for future generations. Therefore, GM can be implemented in two different ways (Dani, 2018)

1. "The manufacturing of green products, particularly those used in renewable energy systems and clean technology equipment of all kinds."
2. "The greening of manufacturing – reducing pollution and waste by minimizing natural resource use, recycling and reusing waste, and reducing emissions."

Green products and green production techniques are competitive weapons. GM has also become a vehicle for long-term job creation in India. According to Goswami and Chaudhury, 2017, investment in green structure and sustainable power sector can develop around six million skilled and 2.5 million un-skilled jobs till 2020. The Union Ministry of New and Renewable Energy (MNRE) have announced a target of installing 175 giga watts (GW) of renewable energy capacity by March 2022. The Government of India has set a target of introducing 60 GW of wind control limit by 2022, against which 35 GW limit has just been introduced (MNRE, 2019). One major sector of green manufacturing is in the metals industries which can exceed expectations as in recycling and reusing waste materials. Several organizations and producer's are focused to forms set up for the group, reuse, and recycle scrap material i.e. gathering scrap in assigned canisters or boxes and reusing these materials, it additionally improve producer's main concern, support representative inspirations, and show social duties to the outside world (Dani, 2018). SMEs and micro and medium enterprises (MSMEs) represent 95 percent of the complete modern industrial movement in India and can assume an essential job creation.

From the above discussion it is clear that the GM implantation is mandatory for developing nations like India and SMEs are the backbone of these countries. Further, SMEs are the driving force behind the dynamic growth of any economy, therefore, they are working tirelessly to implementing GM strategies since effective implementation of GM will lead to gaining a competitive advantage over other and sustain in long run (Mathiyazhagan et al., 2014). But, SMEs face a lot of hindrances in implementing GM in their organizations. Thus, there is the strong need for SMEs to identify and categories GM barriers and rank them utilizing a robust multi-criteria decision-making (MCDM) technique so that these barriers can eliminate one by one according to their ranking order. Keeping in view on the above mentioned points, this study has the following objectives.

1. To identify the GM barriers in SMEs through literature review and expert's opinion.
2. To categorize the identified GM barriers into different criterion and sub-criterion.
3. To rank the identified GM barriers of SMEs using best-worst method (BWM).

With these objectives keeping in mind, this research aims in identifying and ranking the barriers to GM implementation in SMEs with regards to an emerging economy like India. An extensive literature review together with expert's opinion has been done to identify barriers to GM implementation and from a robust MCDM technique called BWM created by Rezaei (2015) has been used to rank these barriers.

The remainder of this paper is organized in the following manner. In Section 2, an extensive literature review is done for the identification of possible barriers to GM implementation in SMEs. Section 3, discusses the research methodology. In section 4, result and analysis were done according to steps discussed in BWM. Section 5, presents the detail discussion on the present study. Section 6, presents the conclusions highlights the limitations of this study and discusses possible future scope.

2. Literature Review

Green manufacturing (GM) is a modern manufacturing strategy, recently it's become profitable due to eco-friendly manufacturing activities, and have kept on still developing. GM involves the change of mechanical tasks in three different ways, (1) utilizing Green vitality, (2) creating and selling Green items and (3) utilizing Green procedures in business activities (Bhattacharya et al. 2011). Rapid consumption of natural resources, growing energy expenditure, increasing customer awareness about environmentally conscious products and the requirement for consistency with ecological enactment through the advancement of green procedures prompted the development of a GM strategy worldview (Gandhi et al., 2017). However, Indian SMEs are still struggling to implement GM strategies. SMEs have poor records when it comes to implementation of GM strategies. Therefore to finding the causes behind this, broad survey has been accomplished to explore the barriers to GM implementation in SMEs. Mittal et al. (2013) identified drivers and barriers to GM implementation in two dissimilar economies - a developed economy and a rising economy and have been approved by statistical analysis. Mittal and Sangwan (2014), in this paper the authors have identified 12 GM barriers and prioritize them according to Fuzzy theory based on ecological, social and financial points of view. Ghazilla et al. (2015), identified 39 drivers and 64 barriers which encourage and restrain the execution of GM implementation in Malaysian SMEs. Kumar et al. (2016), identified that GM strategies are difficult to actualize in Indian situation because of the presence of barriers, such as lack of top management commitment, expertise training program and fund. Majumdar and Sinha (2018), identified 12 important barriers to green supply chain management in Southeast Asian nations. Piyathanavong et al. (2019), identify 10 barriers related to the implementation of GM in Thai manufacturing strategies. These various studies and some other important studies conducted for GM implementation are summarised in Table 1.

Table 1. Previous GM studies found in literature review

Researchers and years	Major findings	Used Methodology
Mittal et al. (2013)	In this study, drivers and hindrances to GM have been approved utilizing a factual examination of the information gathered through the poll study. This research has given a wide viewpoint on the drivers and hindrances to GM in two unique economies - a developed economy and a rising economy.	statistical analysis (Standard Deviation, Independent T-Test)

Mittal and Sangwan (2014)	In this paper, the authors have identified 12 GM barriers and prioritize them according to Fuzzy theory based on natural, social and financial viewpoints.	Fuzzy TOPSIS
Ghazilla et al. (2015)	Total 39 drivers and 64 barriers were identified and analysed for GM in Malaysian SMEs. The drivers were classified into 7 categories and the obstructions are arranged into 8 categories.	Literature review and Delphi method
Kumar et al. (2016)	They conducted a study to identify the barriers of GLSPD and 21 barriers are identified from the literature review and then an ISM technique is applied. Further, MICMAC analysis was also done and found that 7 barriers are distinguished as driver hindrances, 9 as a reliant, five obstructions as linkage and no boundary as independent.	Literature review ISM approach
Gandhi et al. (2017)	They conducted a study to determine the drivers of incorporated lean and green manufacturing are chosen through writing survey and specialists' sentiments. In this investigation, 15 normal drivers are distinguished and among them 'top administration duty' is the most basic driver in the fruitful usage of LM and GM practices.	Fuzzy TOPSIS, Fuzzy SAW and Borda Methods
Xia et al. (2018)	The authors categorize barriers of Green technology adoption for enterprises into competence and psychology and then the skill class is investigated with three basic gatherings including individual, procedure, and acquirement and psychology is also sub-categorized into three groups as Individual barriers, Group barriers, and Social barriers.	AHP and Fuzzy set theory
Majumdar and Sinha (2018)	In this paper, the authors identify 12 important barriers to green supply chain management in southeast Asian nations. ISM is utilized to change over the impression of space specialists into an unmistakable organized guide which portrays the logical relationship among these barriers.	Literature Review,S urvey Questionn aire ISM
Piyathanavong et al. (2019)	Authors identify 10 barriers for the execution of Green Manufacturing in the Thai manufacturing area and this paper grows our insight in the field of tasks and ecological manageability.	Survey Questionn aire

After examining the recent papers on barriers to green manufacturing implementation, it is found that there is a lack of proper research in the context of the barriers to GM implementation in Indian SMEs, which should tell the barriers to GM implementation in a systematic way and there ranking based on their weights with the help of a robust and reliable MCDM technique.

Based on gaps identified, an extensive literature review (2010 to 2019) is done and many (around 43) barriers are identified and from several rounds of discussion with experts 21 are identified as most important barriers related to GM implementation in SMEs in Delhi and NCR region.

2.1 *Most Important Identified Barriers to GM Implementation in SMEs and their Categorization*

For well understanding it is necessary to categorize these 21 most important identified barriers into their proper related areas which will be known as main barriers. It is observe that these barriers can be categorised into five main barriers i.e. Management/organization related barriers (MO), Law / Legislation related barriers (LG), New technology related barriers (NT), Financial and other Fears related barriers (FF) and Information related barriers (IN). Element of main barriers will be known as sub-barriers. In this way schematic categorization is done according to their nature which is shown in Table 2

Table 2 . Main and Sub-barriers to GM Implementations and their References

Main Barriers	Sub-barriers	References
Management /organization related barriers (MO)	Lack of expertise training program towards GM in SMEs (MO1)	Panizzolo et al. (2012), Mathiyazhaganet al. (2013)
	Lack of top management Commitment (MO2)	Murillo-Luna et al. (2011), Mittal et al. (2013), Kumar et al. (2015)
	Undeveloped organizational GM culture and absence of inspiration and support (MO3)	Govindan et al. (2014)
	Lack of green manufacturing thinking (MO4)	Reyes et al. (2014)
	High hesitation to convert traditional practices to GM or Difficulties in changing positive GM mentalities into activities (MO5)	Zhu et al. (2012a, b), Mittal et al.(2013), Mathiyazhagan et al. (2013), Mangla et al.(2017)
Law / Legislation related barriers (LG)	Weak Legislation (LG1)	Brammer et al. (2012), Zhu et al. (2012), Mittal et al. (2013)
	Uncertain Future Legislation (LG2)	Singh et al. (2012), Bey et al. (2013), Ghazilla et al. (2015), Mittal and Sangwan (2015),
	The absence government Support and policies to GM implementation (LG3)	Luthra, et al. (2015), Ghazilla et al. (2015), Hojnik and Ruzzier (2016),
	Lack of Monitoring, Controlling, supporting and direction from administrative specialists on GM (LG4)	Murillo-Luna et al. (2011), Ghazilla et al. (2015)
New technology related barriers (NT)	Lack of new technology facilities and processes to support GM in SMEs (NT1)	Murillo-Luna et al. (2011), Ghazilla et al. (2015)
	Lack of experts related to new technology for GM implementation in SMEs (NT2)	Ghazilla et al. (2015)
	Risk of implementing new technology (NT3)	Ghazilla et al. (2015)

Financial and other Fears related barriers (FF)	Competition and uncertainty (FF1)	Mudgal et al. (2010), Luthra, et al. (2013), Johansson and Sundin (2014)
	Uncertain Benefits from GM implementation (FF2)	Mittal et al. (2013)
	Poor supplier commitment towards GM or Lack of green supplier (FF3)	Mathiyazhagan et al. (2013), Luthra,et al. (2015b), Kumar, et al. (2015)
	High initial capital cost to implement GM (FF4)	Mudgal et al. (2010)
	Absence of adaptability to change over to GM-based frameworks (FF5)	Mudgal et al. (2010),Ghazilla et al.(2015)
	Fear of failure (FF6)	Jinzhou (2011)
Information related barriers (IN)	Inappropriate identification of GM activities, Data Collection, and Retrieval System (IN1)	Johansson and Sundin (2014), Luthra, et al. (2015)
	Difficulties in obtaining GM/GT information for potential improvements (IN2)	Murillo-Luna et al.(2011), Ghazilla et al.(2015)
	Not enough and understandable information (IN3)	Mangla et al., (2017)

3. Research Methodology

For the ranking of GM barriers, the following methodology is followed. Initially, the GM barriers are identified through extensive literature review together with expert's opinion. Through a detailed literature review, many(around 43) barriers were identified and for sorting them again discussion will be done among 6 experts (three from academic and three from SMEs) with questionnaires in which they rated all the 43 barriers on 5 points Likert scale as per their knowledge and experience. In which those barriers which were scored 2 or less than 2 on the Likert scale by at least 2 experts was removed for the next analysis. From this process, 21 GM barriers were remains as most important and finally, categorized these GM barriers into five main criterions which are called main GM barriers and elements of main barriers are called sub-barriers. After that, for the ranking of these GM barriers, a new MCDM technique called BWM (Best-Worst method) is used, which is given by J. Rezaei (2015). There are many MCDM techniques available (Talib and Rahman, 2015; Talib et al., 2011) like Analytical Hierarchal Process (AHP), Analytical Network Process (ANP), Multi-Attribute Utility Theory (MAUT), Simple Multiple Attribute Rating Technique (SMART), Technique for order preference by Similarity to Ideal Solution (TOPSIS), Weighted Product Model (WPM), Weighted Sum Method (WSM), VIseKriterijumska Optimizacija I Kompromisno Resenje (VIKOR), and Preference ranking organization method for enrichment evaluation (PROMETHEE) method, etc. to rank the GM barriers by calculating weights of the barriers but BWM is a very strong MCDM technique for pairwise comparison. BWM performs fundamentally superior to AHP as for the consistency proportion, and the other assessment criteria like-least infringement, complete deviation, and congruity (J. Rezaei, 2015). The Best-Worst method is a recently developed robust optimization technique. The

BWM is a multi-criteria decision-making method that utilizations two vectors of pairwise comparisons to determine the weights of criteria. First, the best (for example most attractive, most significant), and the worst (for example least attractive, least significant) criteria are identified by the decision-maker, after which the best criterion is compared to the other criteria [BO] and the other criteria to the worst criterion [OW] using a no. on (1-9) scale (Rezaei-2015).
After that, a non-linear minmax model or linear minmax model is used to identify the weights such that the maximum absolute difference between the weight ratios and their corresponding comparisons is minimized. For the solution of non-linear minmax model, Interval Analysis is required which gives multiple optimal solutions and linear minmax optimization model gives unique solution from linear programming where it can be solved using any commercial software package (Best-Worst Method-Solver on Microsoft Excel solver)
Recently BWM has been used in a lot of contexts and areas all over the world among the researchers and industrialist such as Energy efficiency in buildings (Gupta et. al., 2017), Airports evaluation and ranking model(Shojaei et.al., 2017), Assessing the social sustainability of supply chains(Ahmadi et.al., 2017), Evaluating firms' R&D performance (Negin and Rezaei, 2017), Efficiency of university-industry PhD projects (Negin and Rezaei, 2016), A group multi-criteria decision-making(Safarzadeh et.al., 2018), Measuring the relative importance of the logistics performance index indicators (Rezaei, et.al., 2018), etc.

3.1 Steps used to obtain the weights of GM barriers using BWM.

In BWM Rezaei proposed the following five steps.
Step 1. Identification of barriers to GM implementation in SMEs for analysis.
In this step, we identify and finalized the GM barriers through extensive literature review and experts opinion.
Step 2. From the finalized GM barriers identify the best (B) (e.g. the most desirable, the most important) and the worst (W) (e.g. the least desirable, the least important) GM barriers or decision criteria based on the decision-maker(s)/expert(s) opinion. Table 3 represents the best and worst criteria according to different expert's choices.
Step 3. Experts are determining the preference of the best decision criterion (B) over all the other decision criteria, using a 9-point scale (numbers between 1 and 9; in which for no.1: *B* is equally important to *j*, and for no.9: *B* is extremely more important than *j*). The result is a best-to-others (BO) vector as follows.

$A_B = (a_{B1}, a_{B2}, a_{B3}, \ldots\ldots, a_{Bn}),$

Whereas $_{Bj}$ indicates the preference of the best criterion (B) over any criterion j and it is clear that $a_{BB} = 1$.

Step 4. Now experts are also determining the preference of all the decision criteria over the worst criterion (W), using a 9-point scale (numbers between 1 and 9; in which for no.1: *j* is equally important to *W*, and for no.9: *j* is extremely more important than *W*), which results in the others-to-worst (OW) vector as follows.

$A_W = (a_{1w}, a_{2w}, a_{3w}, \ldots\ldots\ldots, a_{nw})^T,$

Where a_{jW} indicates the preference of any criterion j over the worst criterion W and it is clear that $a_{WW}= 1$.

Step5. Find the optimal weights ($w_1^*, w_2^*, w_3^*, \ldots\ldots w_n^*$) for all the GM barriers.

To obtain the weights of GM barriers we consider a linear model of BWM because we want to increase the efficiency of the decision-making process along with simple calculation and to obtain a unique optimal solution.
The optimal weights should be determined such that the maximum absolute differences for all *j* can be minimized for

$$\{ | w_B - a_{Bj} w_j |, | w_j - a_{jw} w_w | \}$$

or equivalently:

$$\text{Min max}_j \{ | w_B - a_{Bj} w_j |, | w_j - a_{jw} w_w | \}$$

s.t.

$$\sum_j w_j = 1$$

$$w_j \geq 0, \text{ for all } j \qquad (1)$$

The problem (1) is rewritten as a linear optimization problem as follows:
Min ξ^L

s.t.

$$| w_B - a_{Bj} w_j | \leq \xi^L, \text{ for all } j$$

$$| w_j - a_{jw} w_w | \leq \xi^L, \text{ for all } j$$

$$\sum_j w_j = 1$$

$$w_j \geq 0, \text{ for all } j \qquad (2)$$

The problem (2) is a linear optimization problem with a unique solution, values for optimal weights ($w_1^*, w_2^*, w_3^*, \ldots w_n^*$)and the optimal objective function value ξ^L are obtained by solving it.
ξ^L or Ksi, is the consistency, it values close to zero show a high level of consistency of the pairwise comparisons provided by the decision-maker(s)/expert(s) (Rezaei, 2015)
For more accurate results we use excel solver for calculating the weights of GM barriers and consistency of GM barriers. Best and Worst barriers identified by three SMEs managers and three academic experts are represented in Table 3.

Table 3. Expert's Chosen for the selection of GM barriers

Green manufacturing main barriers	Sub-barriers	No. of experts Considered as best	No. of experts Considered worst
Management/organization related barriers (MO)	MO1	4	
	MO2	1	
	MO3	1	
	MO4		5
	MO5		1
Law / Legislation related barriers (LB)	LG1	5	
	LG2		4
	LG3		2
	LG4	1	
New technology related barriers (NT)	NT1	6	
	NT2		
	NT3		6
Financial and other Fears related barriers (FF)	FF1		5
	FF2		1
	FF3	1	
	FF4	4	
	FF5	1	
	FF6		
Information related barriers (IB)	IN1	6	
	IN2		1
	IN3		5

4. Results and Analysis

The analysis was done according to the steps discussed in BWM. For calculating the weights of GM barriers and consistency of GM barriers Best-Worst Method-Solver is to be used. Table 4 represents the pairwise comparison of the main GM barriers and from Table 5 to 9 represents the pairwise comparison of sub-barriers to GM implementation. Figures show their weights respectively. Table 10 shows the Cumulative weights of all the main barriers and sub-barriers to GM implementation and the Global weight of sub-barriers are calculated by multiplying the sub-barriers weights with their respective main barriers weights. Finally, according to the global weights of all the sub-barriers, their ranking has been done. Highest value of the weight gets the first rank and the lowest weight is in the last position e.g. in Table 10, NT1 gets the maximum global weight i.e. 0.239 so that it obtains the first rank and IN3 get minimum global weight i.e. 0.004 so it obtains the last rank. The **Ksi** shows to what extent the results are reliable, the **Ksi** value closer to zero is better e.g. management/organization related barriers have minimum Ksi i.e. 0.066 this represent that in this category the barriers are closer to equivalent importance with respect to worst barriers and that's why we can say that, this category barriers have been high consistency to each other.

4.1 Determination of the optimal weights of main GM barriers using BWM The experts individually made pairwise comparisons and suggested the preferences of the "Best" over other barriers and the preferences of other barriers over the "Worst" using number from 1 to 9

which led to the formulation of BO and OW vectors as shown in Tables 4 and figure-1 shows their weight calculated from equation (2) or from Best-Worst Method-Solver on Microsoft Excel and their consistency respectively.

Table 4 . Comparison of main GM barriers

BO	Management/organization related barriers (MO)	Law / Legislation related barriers (LG)	New technology related barriers (NT)	Financial and other Fears related barriers (FF)	Information related barriers (IN)
Best barrier: NT	2	4	1	3	7
OW				Worst barrier: IN	
Management/organization related barriers (MO)				6	
Law / Legislation related barriers (LG)				4	
New technology related barriers (NT)				8	
Financial and other Fears related barriers (FF)				5	
Information related barriers(IN)				1	

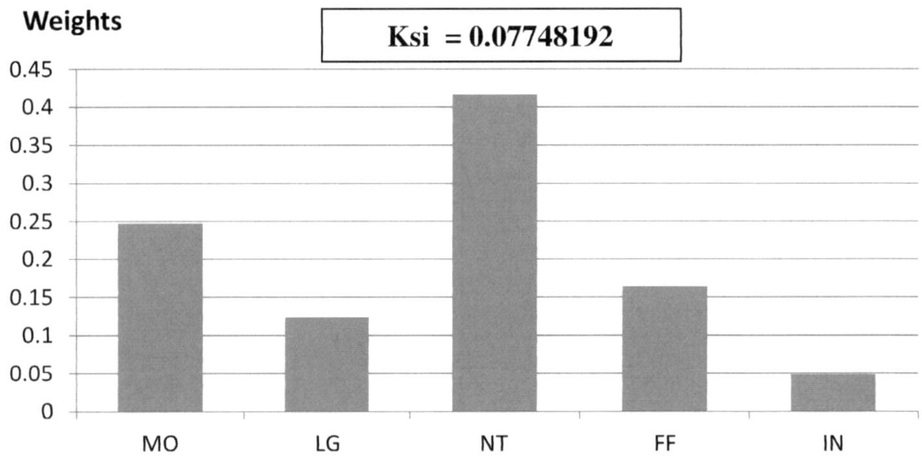

Figure 1. Weights of Main Barriers

It is quite evident from figure-1 that the value of Ksi which represents the consistency of comparisons is very close to zero indicating highly reliable weights of the barriers. Figure-1 further reveals that "New technology related barriers (NT)" and "Information related barriers (IN)" are the most important and the least important barriers, respectively.

4.2 Determination of the relative weights of the sub-category enablers using BWM

For the determination of the relative weights of the sub-barriers, experts individually identified the "Best" and the "Worst" barriers and indicated the preference best to other and other to worst using a number between 1 and 9 leading to the generation of BO and OW vectors. These vectors were used in Equation (2) which was solved to produce optimal weights of the sub-barriers and their consistency (Ksi) for each sub-barrier of each main barrier. Pairwise comparison of different sub-barriers are represented in their respective Tables 5 to 9 and their calculated weight with their consistency (Ksi) are represented in figure from (2-6). All tables and figures can be understood as discussed above.

Table 5. Pairwise comparison for Management/organization related barriers

BO	MO_1	MO_2	MO_3	MO_4	MO_5
Best barrier: MO1	**1**	**4**	**3**	**9**	**2**
OW	**Worst barrier: MO4**				
MO_1	**8**				
MO_2	**4**				
MO_3	**5**				
MO_4	**1**				
MO_5	**3**				

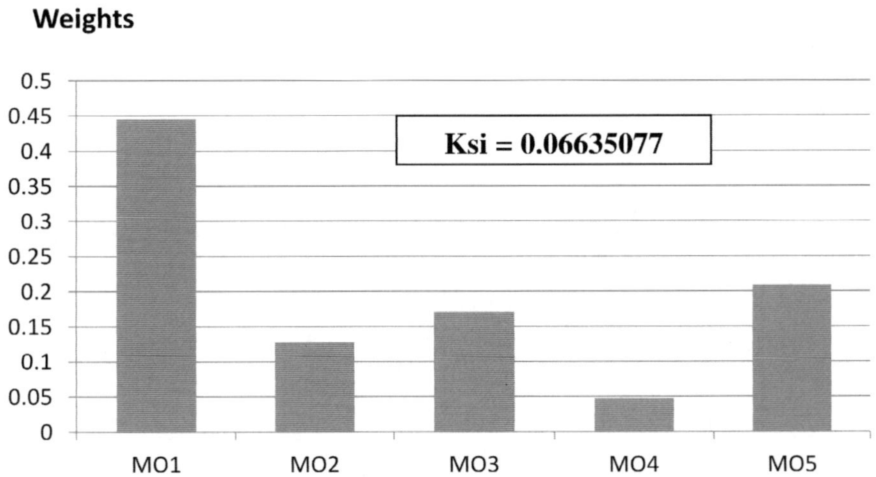

Figure 2. Weights of Management/organization related barriers

Table 6. Pairwise comparison for Law / Legislation related barriers

BO	LB_1	LB_2	LB_3	LB_4
Best barrier: LG1	**1**	**8**	**5**	**2**
OW	**Worst barrier: LG2**			
LG_1	**7**			
LG_2	**1**			
LG_3	**4**			
LG_4	**5**			

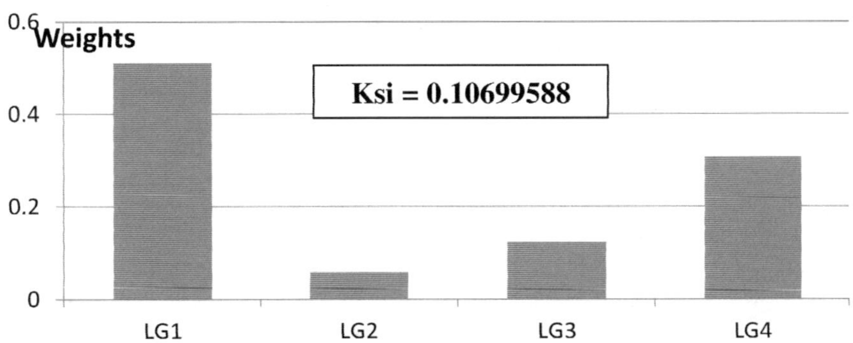

Figure 3. Weights of Law / Legislation related barriers

Table 7 . Pairwise comparison for new technology-related barriers

BO	NT_1	NT_2	NT_3
Best barrier: NT1	**1**	**2**	**5**
OW			**Worst barrier: NT3**
NT_1			**5**
NT_2			**4**
NT_3			**1**

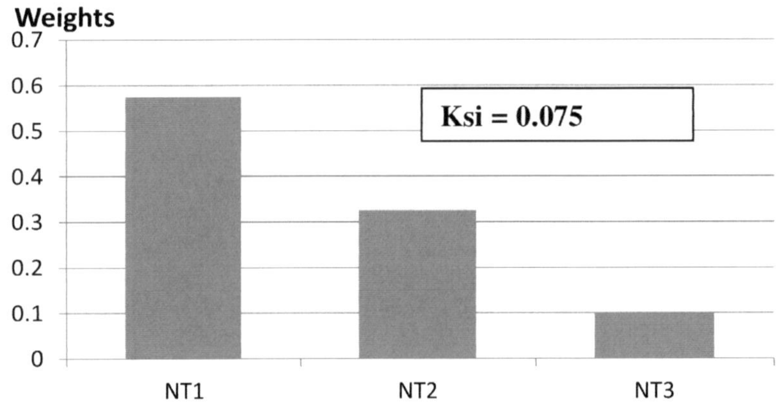

Figure 4 . Weights of new technology-related barriers

Table 8. Pairwise comparison for Financial and other Fears related barriers

BO	FF_1	FF_2	FF_3	FF_4	FF_5	FF_6
Best barrier: FF4	**9**	**4**	**3**	**1**	**2**	**5**
OW					**Worst barrier: FF1**	
FF_1					**1**	
FF_2					**3**	
FF_{13}					**5**	
FF_4					**8**	
FF_5					**7**	
FF_6					**4**	

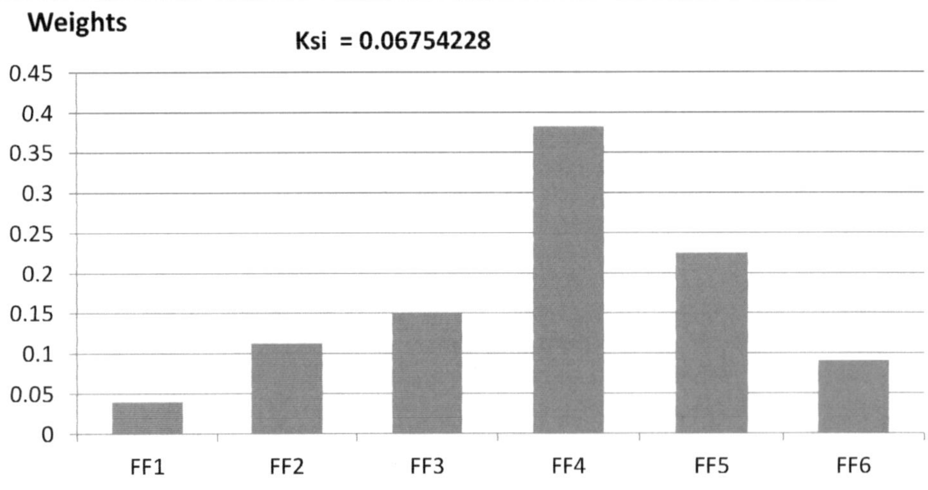

Figure 5. Weights of Financial and other Fears related barriers

Table 9. Pairwise comparison for Information related barriers

BO	IB_1	IB_2	IB_3
Best barrier: IN1	**1**	**3**	**5**

OW	Worst barrier: IN3
IN_1	**6**
IN_2	**5**
IN_3	**1**

Ksi = 0.18181818

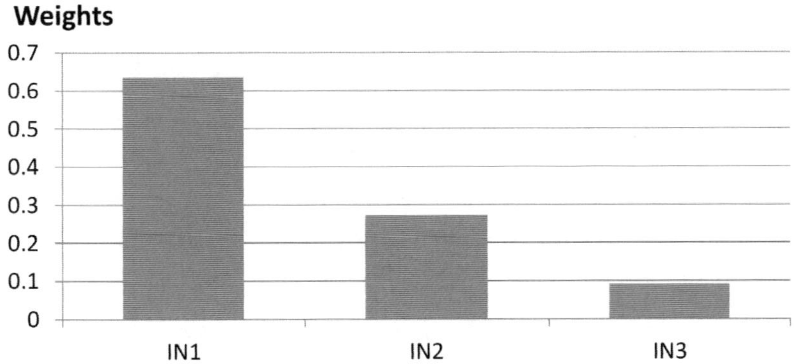

Figure 6. Weights of Information related barriers

4.3 Determination of the global weights of the sub-barriers

Table 10 shows the global weights of all sub-barriers to GM implementation. The Global weights of sub-barriers are calculated by multiplying the sub-barriers weights with their respective main barriers weights. Finally, according to the global weights of all the sub-barriers, their ranking has been done. Highest value of the global weight gets the first rank and the lowest global weight is in the last position i.e. NT1 gets first rank and IN3 gets last rank.

Table 10. Cumulative table for weights of the main barriers and the sub-barriers of GM implementation

Main barriers	Weighs of main barriers	Sub - barriers	Local Weighs of sub-barriers	Global weight of sub-barriers	Ranking
Management/organization related barriers (MO)	0.247	MO1	0.445	**0.110**	**3**
		MO2	0.128	0.032	11
		MO3	0.171	0.042	8
		MO4	0.047	0.012	18
		MO5	0.209	0.052	6

Law / Legislation related barriers (LG)	0.123	LG1	0.510	**0.063**	**5**
		LG2	0.058	0.007	19
		LG3	0.123	0.015	16
		LG4	0.309	0.038	9
New technology related barriers (NT	**0.416**	NT1	0.575	**0.239**	**1**
		NT2	0.325	**0.135**	**2**
		NT3	0.100	0.042	7
Financial and other Fears related barriers (FF)	0.165	FF1	0.039	0.006	20
		FF2	0.113	0.019	14
		FF3	0.150	0.025	13
		FF4	0.383	**0.063**	**4**
		FF5	0.225	0.037	10
		FF6	0.090	0.015	15
Information related barriers (IN)	0.048	IN1	0.636	0.031	12
		IN2	0.272	0.013	17
		IN3	0.091	0.004	21

5. Discussion

Best–Worst method is applying for ranking of barriers to green manufacturing. Table 10 shows the weights of whole main barriers as well as sub-barriers, the rankings are obtained on their respective global weights. Total five main barriers were concluded and among them, New-technology related barrier (NT) is ranked first through analysis. This result shows the past research conformance (Xia et al., 2018) In which they found that the new green technology application was generally thought for green manufacturing. therefore, they study the technical compatibility with its operational system consisted of competence and psychology section and established an identification framework and indicators system of barriers in the way of green technology adoption. Another research shows that insufficient knowledge about the available new technology choices and limited access to green literature is the top-ranked barrier which hinders the implementation of GM in the industry (Mittal and Sangwan, 2014). Some other researches show that top basic obstructions which impede the execution of GM in SMEs have insufficient R&D support because of the absence of assets and specialized ability concerning green practices (Ghazilla et al., 2015). Majumdar and Sinha (2018), is also indicated that The intricacy of the green procedure and framework plan which is a technology-related hindrance is the most elementary obstruction which prevents the GM implementation in material and attire store network of Southeast Asia. In this way, these results demonstrate the importance of our study and make them valid. Management/organization related barriers (MO) obtain the second rank among the main barriers. This result shows the past research conformance with Gandhi et al., (2017), in which they observed that top administration responsibility is prescribed as the most significant driver for the usage of GM in Indian SMEs. Government of India has just begun management development program (MDP) as one of the segments of the national manufacturing competitiveness program (NMCP) to improve the decision-making ability of managers of SMEs and to enhance profitability/productivity. Financial and other Fears related barriers (FF) obtain the third rank among main GM barriers. Therefore, financial support is necessary for small organizations to GM implementation because it has High initial capital cost (Gandhi et al., 2017). Financial and other Fears related barriers are also due to Weak market positions for GM-based products or Low Customer Demand is generally found among customers, they do not aware the benefit of GM based product, they think that the GM based product is costly (Mudgal et al., 2010). But cooperation between advertisement and political inspiration can change purchase and utilization, the behaviour of consumers'. These combined efforts are more effective towards green attitude; possibly these combine efforts could create a better environment for all consumers. These combine efforts extensively affected consumers' purchase, usage, and recycling intentions and attitude. (Kidwell et. al. 2013)

Among 21 sub-barriers, Lack of new technology facilities and processes to support GM (NT1) is ranked first and second rank is also related to new technology its indicate that for GM implementation our concentration is to be on developing new technology and to fulfil the required experts related to new technology(Ghazilla et al.(2015). Lack of expertise training program is the third most important sub-barriers which is also the most important barrier in the Thai manufacturing organization (Piyathanavong et al., 2019). After implementing GM it may become a major barrier (Panizzolo et al., 2012; Mathiyazhagan et al., 2013). High initial capital cost to implement GM (FF4) is the fourth major sub-barrier, this barrier plays an extensive role when we want to initiate implementing GM in our organization (Mudgal et al., 2010). Weak Legislation (LG1) is the fifth major sub-barrier to GM implementation, 'weak legislation' in terms of complete absence of environmental laws or complex and ineffective environmental legislations play effective role towards dirty manufacturing specially in Indian SMEs so the Government ought to likewise guarantee the uniform natural enactment in all states/areas of the nation to prevent organizations from moving the messy manufacturing to places with remiss ecological performance (Mittal, et al., 2013).

6. Conclusions, Limitations and Future Research

The manufacturing sector is one of the vital components for developing the national economy. Therefore every country is engaged in developing the manufacturing sector but the limited natural resources and environmental degradation pose very tough challenges in front of human beings. To overcome these challenges a new manufacturing strategy was developed which is known as GM. However, the implementation of GM in SMEs is delayed due to barriers. Neither there are any systematic arrangements of these GM barriers nor any powerful method is applied for their ranking, is presented in any literature. To eliminate these barriers, a systematic understanding and their prioritization are required.

To address these gaps, this study has developed a broad framework to identify GM barriers in SMEs and through using a strong MCDM technique named as Best-Worst method is used to rank these barriers. The structure was created with the assistance of extensive literature review and experts opinion. Total twenty-one (21) barriers are identified which are categorized into five (5) main barriers according to their nature and for well understanding. Then Best-Worst analysis is done on these barriers to rank them. The results of this analysis show that "Lack of new technologies facilities and processes" is found to be most important barrier followed by "Lack of experts related to new technology", "Lack of expertise training program", "High initial capital cost" and "Weak Legislation" related barriers. Therefore, the studies show that technological development is mandatory for implementing GM. The execution of GM is conceivable just with teamed up endeavours of government and industry in a vital manner by alleviating the GM barriers.

Although this study is done on a broad level it has several limitations. The barriers originated from various manufacturing divisions. This may have biased the results, therefore, it should be possible to see the effects of the current examination with comparative manufacturing segmentation. In this study, BWM is used for ranking GM barriers but there are other MCDM techniques like VIKOR, AHP, Fuzzy TOPSIS, MAUT, SMART, PROMETHEE, etc. can also be used to think about the outcomes for any progressions. Further, this study can be fertilized by taking a large number of SMEs and to use statistical tools for analyse the findings. The sensitivity analysis can be directed to break down the effect of variation in the weights of the barriers.

References

[1] Mudgal R K, Shankar R, Talib P, and Raj T 2010 Modelling the Barriers of Green Supply Chain Practices: An Indian Perspective International Journal of Logistics Systems and Management **7(1)** 81–107.

[2] Jinzhou W 2011 Discussion on the relationship between green technological innovation and system innovation Energy Procedia **5** 2352–2357.

[3] Murillo-Luna J L, Ayerbe C G and Torres P R 2011 Barriers to the adoption of proactive environmental strategies Journal of Cleaner Production **19** 1417-1425.

[4] Bhattacharya A, Jain A, and Choudhary A 2011 Green Manufacturing Energy Products and Processes bcg com

[5] Mathiyazhagan K, Govindan K, Haqand A N and Geng A 2013 An ISM Approach for the Barrier Analysis in Implementing Green Supply Chain Management Journal of Cleaner Production **47** 283–297.

[6] Mittal V K, Geed P, Herrmann C and Sangwan K S 2013 Comparison of Drivers and Barriers to Green Manufacturing: A Case of India and Germany 20th CIRP International Conference on Life Cycle Engineering Singapore 97-102.

[7] Johansson G and Sundin E 2014 Lean and Green Product Development: Two Sides of the Same Coin? Journal of Cleaner Production **85** 104–121.

[8] Kidwell B, Farmer A and Hardesty D M 2013 Getting liberals and conservatives to go green: political ideology and congruent appeals journal of consumer research **40(2)** 350-367.

[9] Prasad S and Sharma S K 2014 Lean and Green Manufacturing Concept and its Implementation in Operations Management International Journal of Advanced Mechanical Engineering **4** 509-514

[10] Reyes J A G, Jacques G W, Lim M K, Kumar V and Rocha-Lona L 2014 Lean and Green-synergies, Differences, Limitations, and the Need for Six Sigma In Advances in Production Management Systems Innovative and Knowledge-based Production Management in a Global-local World Berlin Springer Berlin Heidelberg 71-81

[11] Luthra S, Kumar S, Garg D and Haleem A 2015 Critical Success Factors of Green Supply Chain Management for Achieving Sustainability in Indian Automobile Industry Production Planning & Control **26 (5)** 339–362

[12] Ghazilla R A R, Sakundarini N, Rashid S H A, Ayub N S, Olugu E U and Musa S N 2015 Drivers and barriers analysis for green manufacturing practices in Malaysian SMEs: A Preliminary Findings 12th Global Conference on Sustainable Manufacturing Procedia CIRP **26** 658 – 663

[13] Mittal V K and Sangwan K S 2017 Prioritizing Barriers to Green Manufacturing: Environmental, Social and Economic Perspectives, Variety Management in

Manufacturing Proceedings of the 47th CIRP Conference on Manufacturing Systems Procedia CIRP **17** 559- 564.

[14] Talib F and Rahman Z 2015 Identification and prioritization of barriers to total quality management implementation in service industry: an analytic hierarchy process approach The TQM Journal **27(5)** 591-615.

[15] Kumar S, Luthra S, Govindan, Kumar N and Haleem A 2016 Barriers in green lean six sigma product development processes: an ISM approach, Production Planning & Control doi: 10.1080/09537287 1165307

[16] Hojnik J and Ruzzier M 2016 Drivers of and barriers to eco-innovation: a case study International Journal of Sustainable Economy **8 (4)** 273–294.

[17] Mittal V K, Sindhwani R, Kasariya V, Salroo F, Sangwan K S and Singh P L 2017 Adoption of integrated Lean-Green-Agile strategies for modern manufacturing systems 24 CIRP conference on life cycle engineering procedia CIRP 61 463-468

[18] Gandhi N S, Thanki S J and Thakkar J J 2017 Ranking of Drivers for Integrated Lean-Green Manufacturing for Indian Manufacturing SMEs Journal of Cleaner Production doi: 10.1016/j.jclepro.2017.10.041

[19] Mangla S K, Govindan K, and Luthra S 2017 Prioritizing the barriers to achieve sustainable consumption and production trends in supply chains using fuzzy analytical Hierarchy Process Journal of Cleaner Production **151** 509–525.

[20] Gupta H and Barua M K 2018 A framework to overcome barriers to green innovation in SMEs using BWM and Fuzzy TOPSIS Science of the Total Environment **633** 122–139

[21] Piyathanavong V, Kumar V and Maldonado-Guzman G 2019 The Adoption of Operational Environmental Sustainability Approaches in the Thai Manufacturing sector Journal of Cleaner Production doi: https:// doi.org/10.1016 /j.jclepro. 2019.02.093

[22] Xia D, Zhang M,Yu Q, and Tu Y 2019 Developing a framework to identify barriers of Green technology adoption for enterprises, Resources, Conservation & Recycling, **143** 99–110.

[23] Rezaei J 2015 Best-worst multi-criteria decision-making method Omega, **53** 49–57

[24] Salimi N and Rezaei J 2016 Measuring the efficiency of university-industry Ph.D projects using the best-worst method Scientometrics 109:1911–1938 doi 10.1007/s11192-016-2121-0.

[25] Rezaei J 2016 Best-worst multi-criteria decision-making method: Some properties and a linear model Omega **64** 126–130

[26] Shojaei P, Haeri S A S and Mohammadi S 2018 Airports evaluation and ranking model using Taguchi loss function best-worst method and VIKOR technique Journal of Air Transport Management **68** 4-13

[27] Ahmadia H B, Kusi-Sarponga S and Rezaei J 2017 Assessing the social sustainability of supply chains using Best Worst Method, Resources, Conservation and Recycling **126** 99–106

[28] Salimi N and Rezaei J 2017 Evaluating firms'RandD performance using best-worst method Evaluation and Program Planning https://doi.org/10.1016/j.evalprogplan201710.002

[29] Safarzadeh S, Khansefid S and Barzoki M R 2018 A group multi-criteria decision-making based on best-worst method Computers and Industrial Engineering **126** 111–121

[30] Rezaei J, Van Roekel W S and Tavasszy L 2018 Measuring the relative importance of the logistics performance index indicators using Best Worst Method, Transport Policy **68** 158–169

[31] Talib F, Rahman Z and Qureshi M N 2011 Prioritising the practices of total quality management: an analytic hierarchy process (AHP) analysis for the service industries Total Quality Management and Business Excellence **22(12)** 1331-1351

[32] GM definition Dani-2018,https://www.meadmetals.com [Assessed on 25/03/2019]

[33] MNRE, 2019. Ministry of New and Renewable Energy. https://mnre.gov.in. [Assessed on 25/03/2019]

Modelling Drivers of Coordination Risk in Production System

R Shakya[1] and P Chauhan[2]
JSS Academy of Technical Education, Noida, Uttar Pradesh, India

[1]shakya.rishabh8@gmail.com
[2]prashantchauhan@jssaten.ac.in

Abstract:
The production system process involves the coordination of a wide range of dissimilar and interconnected activities that's why it involves more risk and uncertainty. When an organization coordinate with a differently located organization, it is very critical to established relations with new suppliers and sometimes it may bring about coordination and language problems. Coordination risk is the major risk in the entire production system because many suppliers are supply raw material to the organization. Coordination risks are very volatile and also often more difficult to observe, so they may go unnoticed. The main objective of this paper is to identify and understand the mutual interaction among various drivers of coordination risk which affects the performance of production system. To this effect, authors have identified various drivers through extant review of literature. From this information, an integrated model using interpretive structural modelling (ISM) for drivers of coordination risk in production system is developed and the structural relationships between these drivers are modelled. Further, MICMAC analysis is done to analyse the independent power and dependency of drivers which shall be helpful to managers to identify and classify important conditions and to reveal the direct and indirect effects of each condition of coordination risk on production system.

Keywords: Coordination risk, Production system, Interpretive Structural Modelling (ISM), MICMAC analysis

1. Introduction

Production system is defined as "The procedure, approaches or planning which contains all functions required to gather the inputs, process and produce finished product" or it can also be defined as the method of converting input to output. Production system utilizes infrastructure, funds, materials and labour to produce finished product. Production system has three main components i.e., Input, Conversion Process and Output [1].

The production system process consists of risk and uncertainty because it involves the coordination of a wide range of dissimilar and interconnected activities. Coordination between any organization in a new location with the main organization in other country is very complicated. When an organization moves to a different location, it is very critical to established relations with new suppliers and sometimes it may bring about coordination and language problems. Risk of wrong decision for coordination of new organization creates a risk of misuse by partners, increased level of dependency and loss of technical experience, sudden shift in supplier and customer demand. Risks related with supply chain and coordination are another category recognized in international production literature [2].

Coordination risk exist when there is a miscommunication among employees of organization. A coordination risk occurs when there is necessary to coordinate about the process system but organization may not be able to attain [3]. Efficient network coordination among suppliers depends upon information sharing that it may increase the efficiency of production system [4]. Coordination risk is "the risk of failure of one player's investment due to the probable absence of complementary investments by other players in different phases in the supply chain" [5]. When decision makers ignore the optimal system target lack of coordination arises because they have incomplete information.

Coordination risk happens when decision taken by persons may contribute to a collective result and the decision rules followed by person may be not known [2]. Contributing factor of coordination risk are demand and supply uncertainty, such uncertainty may arise from external sources. Inter-organizational information systems (IOS) is a system which is used within supply chain management (SCM) to improve business processes and to enable better working relations between two organizations. Coordination risk contribute to deviation from optimal behaviour. Coordination risk arise from two factors:

(1) A lack of common information.
(2) A lack of trust between two organizations [2].

Mutual knowledge problems and trust are the two main elements of coordination risk. Experimental studies indicate that how failures of mutual knowledge problems can affect coordination risk. The other element is trusting that other person will truly carry out the optimal ordering policy [2].

Due to lack of coordination in planning problems are created in urban freight transport [6]. Even when the technology advancements, the face-to-face communication is still the most effective way of communicating with others. Therefore, distribution of the teams near located workspaces, will increase the coordination and communication risk [7]. Coordination means two or more groups work together skilfully. In other words, the groups are responsible for specific tasks but work interdependently for mutual target [8]. Uncertainties in supply and demand is due to lack of coordination between two members. Coordination between persons help in dealing inter-dependencies and decreasing uncertainties [9].

According to above literature, many researchers have been created various number of models, methods to study the impact of coordination risks on production system. In this research relationship among several types of coordination risks is found. Due to the complex nature of coordination system, to analyse the inter-relationship among various risks is very difficult [10].

In this research, various methods such as FBN, TNT, FMEA are used to study the interrelationships (see, e.g., [11]; [12]; [13]).

Interpretive structural modelling (ISM) is used by the author. It is a well-established methodology for identifying relationships among specific items.

The main objectives of this research are:
1. Classify drivers of coordination risk related to production system.
2. Create the relationship between drivers using ISM.
3. Suggest a structural model for drivers of coordination risk of production risk.
4. Classify the drivers into several categories using MICMAC analysis.

2. Interpretive Structural Modelling (ISM)

ISM is an interactive learning process. From this method we can define how items or risks are related. According to relationship among items, an overall structure is taken out from the set of items. After modelling and finally overall structure is represented in a diagraph model [14]. It is used to identify interrelationships among various parameters which describe an issue or a problem [15].

In ISM a set of dissimilar directly and indirectly related variables are arranged into a systemic model. The existence of directly or indirectly related variable makes the structure of system more complex. It then becomes difficult to deal with this type of system. Hence, a methodology is required which helps to identify a structure within a system [16].

Four symbols are used to represent the relationship between variables (p and q) are shown below:

V = Factor p influence factor q
A = Factor q influence factor p
X = Factor p and q influence each other
O = p and q are unrelated

This data is indicated in the arrangement of binary matrix called initial reachability matrix (IRM). If variable p reaches variable q, then the entry is 1 and if variable p does not reach variable q, then entry is 0 [16].

3. Identification of drivers of coordination risk in production system

There are many factors of coordination risk. Few factors with literature support are discussed as follows:

Table 1: List of drivers of coordination risk

S.No.	Factors	Definition	References
1	Incomplete information flow	Incomplete information flow due to communication losses and unnatural resources.	(Canzian et al., 2013), (Keyhani & Chatterjee, 2012), (Ozdemir-Gungor et al., 2018), (Keyhani & Annaswamy, 2012), (Lindblad et al., 2018)
2	Lack of information sharing	Sharing of information leads to appropriate decision making within organization.	(Alawamleh & popplewell, 2011), (Somuyiwa et al., 2011), (Xu & Chen, 2016), (Tavella & Hjortso, 2012)
3	Loss of communication	Communication is essential to any organization. Uncertainty is due to less communication.	(Alawamleh & popplewell, 2011), (Dietrich et al., 2013), (Safronova, 2011)
4	Lack of top management commitment	Lack of top management commitment will lead to a failure to meet objectives.	(Alawamleh & popplewell, 2011), (Rahman et al., 2013), (Lodgaard et al., 2016), (Kumar, 2014)
5	Mutual knowledge problems	A lack of mutual knowledge problems seems to be the main obstacle to communication.	(Croson et al., 2014), (Edwards, 2011), (Thomas et al., 2014), (Angeletos & Lian, 2018), (Ingram & Maye, 2016), (Dugnas, 2017)
6	Lack of trust between partners	Trust that occurs between partners tells how much partners believe in others.	(Croson et al., 2014), (Alawamleh & Popplewell, 2011), (Jagd, 2010), (Samantra et al., 2016)

4. ISM methodology

4.1 Contextual relationship establishment among risks

In this research 8 risks are identified. To analyse the risks, a relationship of 'reaches to' form is used. It means one risk influence another risk. From this method, a contextual relationship is obtained.

Few specialists, from several organizations associated to production system were consulted to support in making the contextual relationships between the risks. SSIM is developed by using following four symbols which are used to indicate the relationship between variables (p and q) [31].

1.V is denoted from risk p to risk q (means risk p influence risk q).
2.A is denoted from risk q to risk p (means risk q influence risk p).
3.X is denoted for both direction relations (means risks p and q influence each other).
4.O is denoted for no relation between the risks (means risks p and q are unrelated) [31].

4.2 Final reachability matrix with driving power and dependence power (FRM)

After finding the relationship among risk the final reachability matrix is developed.

Table 2: FRM

S.No.	Factors	1	2	3	4	5	6	Driving Power
1	Incomplete information flow	1	1	1	0	0	1	4
2	Lack of information sharing	1	1	1	0	0	1	4
3	Loss of communication	1	1	1	0	0	1	4
4	Lack of top management commitment	1	1	1	1	0	1	5
5	Mutual knowledge problems	1	1	1	0	1	1	5
6	Lack of trust between partners	1	1	1	0	0	1	4
	Dependence	6	6	6	1	1	6	

4.3 ISM based level of risk

By using partitioning method those elements whose reachability and the intersection sets are same indicate at highest level.

Table 3: ISM based level of risk

S.No.	Factors	Level
1	Incomplete information flow	I
2	Lack of information sharing	I
3	Loss of communication	I
4	Lack of top management commitment	II
5	Mutual knowledge problems	II
6	Lack of trust between partners	I

5. ISM model development

ISM model is developed after applying all the steps of ISM.

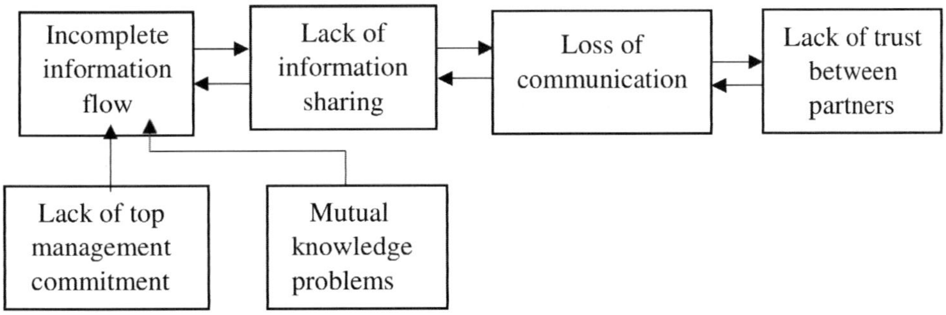

Figure 1: The level of coordination risk is shown by ISM model

6. MICMAC analysis

Matrice d'Impacts croises-multiplication appliqu´ean classment (cross-impact matrix multiplication applied to classification) is referred to as MICMAC. The aim of MICMAC analysis is to analyse the dependence power and driving power of factors. The principle of MICMAC is based on multiplication properties of matrices [20]. According to their dependence power and driving power, the factors are classified into four categories as follows [32].

(1) Autonomous risks: This type of risks has weak driving power as well as weak dependence.
(2) Linkage risks: This type of risks has strong driving power as well as strong dependence.
(3) Dependent risks: This type of risks has strong dependence power but weak driving power.
(4) Independent risk: This type of risks has strong driving power but weak dependence power [33].

Figure 2: indicates the results of MICMAC analysis.

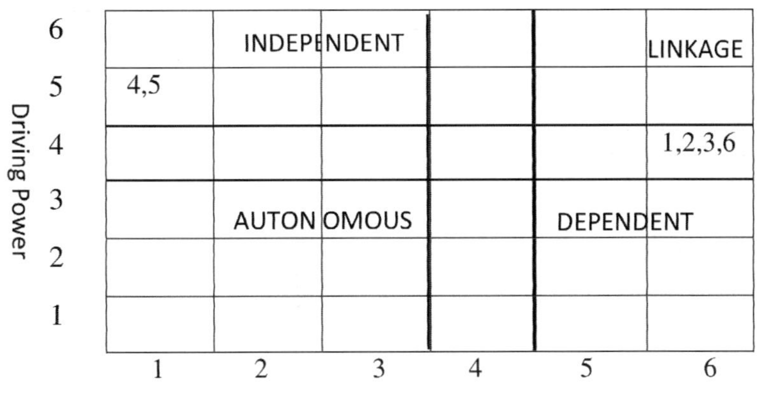

Figure 2: Diagram of Dependence and Driving power

7. Conclusion

In this research, authors identified 6 factors i.e. Incomplete information flow, Lack of information sharing, Loss of communication, Lack of top management commitment, Mutual knowledge problems, Lack of trust between partners that may affect the performance of production system. ISM model and MICMAC approach is used to determine the relationship between them. They classify the risks under four categories namely independent, linkage, dependent and autonomous. The final results attained with the help of ISM. Some of the results are as follows:

(a) From figure.2, it is observed that two drivers Lack of top management commitment (4), Mutual knowledge problems (5) have strong driving power and are less dependent on other drivers. Therefore, these two drivers are considered as the root causes for all drivers.

(b) It is observed that four driver's Incomplete information flow (1), Lack of information sharing (2), Loss of communication (3), Lack of trust between partners (6) are considered as the most important drivers. Management of organisation should recognize the dependence of these drivers on other drivers.

8. References

[1] Edwards A 2011. Building common knowledge at the boundaries between professional practices: Relational agency and relational expertise in systems of distributed expertise. *International Journal of Educational Research, 50(1), 33-39.*

[2] Croson R, Donohue K, Katok E and Sterman J 2014. Order stability in supply chains: coordination risk and the role of coordination stock. *Production and Operations Management, 23(2), 176-196.*

[3] Bass M Herbsleb J D and Lescher C 2009. A coordination risk analysis method for multi-site projects: Experience report. *In 2009 Fourth IEEE International Conference on Global Software Engineering (pp. 31-40). IEEE.*

[4] Alawamleh M and Popplewell K 2011. Interpretive structural modelling of risk sources in a virtual organization. *International Journal of Production Research, 49(20), 6041-6063.*

[5] Tavella E and Hjortso C N 2012. Enhancing the design and management of a local organic food supply chain with soft systems methodology. *International food and agribusiness management review, 15(1030-2016-82765), 47.*

[6] Samantra C, Datta S, Mahapatra S S and Debata B R 2016. Interpretive structural modelling of critical risk factors in software engineering project. Benchmarking: *An International Journal, 23(1), 2-24.*

[7] Wang M, Sun L and Li M 2018. Risk Analysis for FPSO Vapor Cloud Explosion Considering Domino Effect. In ASME 2018 37th International Conference on Ocean, Offshore and Arctic Engineering (pp. V003T02A058-V003T02A058). *American Society of Mechanical Engineers.*

[8] Cao N, Zhang Z, To K M and Ng K P 2008. "How are supply chains coordinated? An empirical observation in textile-apparel business", *Journal of Fashion Marketing & Management, Vol. 12 No. 3, pp. 384-97.*

[9] Arshinder K A and Deshmukh S G 2009. "A framework for evaluation of coordination by contracts: a case of two-level supply chains", *Computers & Industrial Engineering, Vol. 56, pp. 1177-91.*

[10] Chauhan P, Kumar S and Sharma R K 2015. Modelling drivers of political risk in offshore outsourcing. *European Journal of Business and Management, 7(9), 34-44.*

[11] Tenhunen T 2010. *Challenges in Scaling Agile Software Development.*

[12] Xu Y and Chen M 2016. Improving Just-in-Time manufacturing operations by using Internet of Things based solutions. *Procedia CIRP, 56, 326-331.*

[13] Singh R K, Garg S K and Deshmukh S G 2007. Interpretive structural modelling of factors for improving competitiveness of SMEs. *International Journal of Productivity and Quality Management, 2(4), 423-440.*

[14] Sumberg J and Sabates-Wheeler R 2010. Linking agricultural development to school feeding.

[15] Keyhani A and Annaswamy A M 2012. A new automatic generation control with heterogeneous assets for integration of renewables. *In 2012 IEEE PES Innovative Smart Grid Technologies (ISGT) (pp. 1-6). IEEE.*

[16] Lindblad F, Magnusson B, Luu, A and Ragnarsson J 2018. Information Flow Optimization: Enabling Standardisation towards Modular Building Methods of Wood- Building Solutions. *International Journal of Innovation, Management and Technology, 9(6), 232-239.*

[17] Canzian L, Xiao Y, Zame W, Zorzi M and van der Schaar M 2013. Intervention with complete and incomplete information: Application to flow control. *IEEE Transactions on Communications, 61(8), 3206-3218.*

[18] Keyhani A and Chatterjee A 2012. Automatic generation control structure for smart power grids. *IEEE Transactions on Smart Grid, 3(3), 1310-1316.*

[19] Ozdemir-Gungor D, Kıdak L B and Ercan Y 2018. Formal communication channels in a state hospital: a qualitative study. *iioab journal, 9(6), 48-55.*

[20] Somuyiwa A O, Adebayo I T and Akanbi T A 2011. Supply chain performance: an agile supply chain driven by information system (IS) capabilities. *British Journal of Arts and Social Sciences, 1(2), 125-135.*

[21] Dietrich F, Löchte C, Jeschke S and Raatz A 2013. An agent-based concept for planning and control of the production of carbon fibre reinforced plastics aircraft structures with mobile production units. *In Automation, Communication and Cybernetics in Science and Engineering 2011/2012 (pp. 607-621). Springer, Berlin, Heidelberg.*

[22] Safronova K 2011. The hypothesis of a new approach to lean production into the context of lean thinking. *Fulfilling the Worldwide Sustainability Challenge: Strategies, Innovations, and Perspectives for Forward Momentum in Turbulent Times, 868-873.*

[23] Rahman N A A, Sharif S M and Esa M M 2013. Lean manufacturing case study with Kanban system implementation. *Procedia Economics and Finance, 7, 174-180.*

[24] Lodgaard E, Ingvaldsen J A, Aschehoug S and Gamme I 2016. Barriers to continuous improvement: perceptions of top managers, middle managers and workers. *Procedia CIRP, 41, 1119-1124.*

[25] Kumar A 2014. The challenges to the implementation of lean manufacturing. *International Journal of Engineering Science & Advanced Technology, 4(4), 307-312.*

[26] Thomas K A, DeScioli P, Haque O S and Pinker S 2014. The psychology of coordination and common knowledge. *Journal of personality and social psychology, 107(4), 657.*

[27] Angeletos G M and Lian C 2018. Forward guidance without common knowledge. *American Economic Review, 108(9), 2477-2512.*

[28] Ingram J and Maye D 2016. Niche knowledge systems-challenging or invigorating the AKS? *An analysis of the Permaculture community in England.*

[29] Dugnas K 2017. Operational planning and control in eto-companies: thoughts on improving of existing performance measurement practices in the industry 4.0 context.

[30] Jagd S 2010. Balancing trust and control in organizations: towards a process perspective. *Society and Business Review, 5(3), 259-269.*

[31] Attri R, Dev N and Sharma V 2013. Interpretive structural modelling (ISM) approach: an overview. *Research Journal of Management Sciences, 2319, 1171.*

[32] Sharma H D, Gupta A D 1995. The objectives of waste management in India: a future inquiry. *Technological Forecasting and Social Change, 48, 285–309.*

[33] Chidambaranathan S, Muralidharan C and Deshmukh S G 2009. Analyzing the interaction of critical factors of supplier development using Interpretive Structural Modelling—an empirical study. *The International Journal of Advanced Manufacturing Technology, 43(11-12), 1081-1093.*

Synthesis and Characterization of Yttrium based Luminescent Materials

Vartika S.Singh and S.V.Moharil

Department of Physics, RTM Nagpur University, Nagpur 440 033, India

E-mail: vartikasingh1710@gmail.com

Abstract. $NaYF_4$ is well known and most promising upconversion yttrium based fluoride phosphor till date. Fluoride synthesis is rather tricky due to their peculiar physico-chemical properties. Fluorides are highly susceptible to hydrolysis. This is detrimental to the luminescence. Conventional method for preparing complex fluoride compounds is solid state reaction which is a tedious task. In recent years, many new methods such as wet chemical synthesis have been suggested for these novel materials. These new syntheses provide more milder and efficient ways of materials preparation. Here, preparation of two luminescent hosts $NaYF_4$ and KYF_4 was attempted via Wet-Chemical synthesis. Among these, $NaYF_4$ could be successfully prepared, whereas synthesis of KYF_4 was partially successful. Formation of these compounds was confirmed by various measurements. Photoluminescence was obtained for Ce^{3+} doped $NaYF_4$.

Keywords: Fluorides, Phosphors, Photoluminescence, Synthesis

1. Introduction

In the recent years number of yttrium based fluorides are reported as solid state lasers, upconversion and VUV materials. The examples include fluorides of the type MYF_4 where M= Li,Na,K etc.

Pol and Rao[1] have described the synthesis of the compound $NaYF_4$ through a chemical reaction using NaF and YCl_3 as starting materials. More recently, $NaYF_4$ has been also synthesised via few new novel methods such as hydrothermal [2], solvothermal [3] etc. Interesting phosphors result when small amounts of rare earth activators are incorporated into the $NaYF_4$.
$NaYF_4$:Nd^{3+} is known to be potential host since tunable laser action in infrared region is reported in 1969 [4]. It is one of the efficient material for infrared to visible upconversion when activated with Yb^{3+},Er^{3+} since 1972 [5,6]. Many studies have been carried out in recent years. $NaYF_4$:Yb,Er [7-16] has emerged as a most desirable upconversion fluoride material.
Other efficient UC materials are $NaYF_4$:Tm^{3+}, Yb^{3+} [17-28], $NaYF_4$:Yb^{3+},Ho^{3+} [29-31], $NaYF_4$:Yb^{3+},Tm^{3+}/Er^{3+}/Eu^{3+}[32], $NaYF_4$:Yb^{3+}[33], $NaYF_4$:Nd/Yb [34,35] and $NaYF_4$:Ce^{3+},Tb^{3+} [36,37]. Also, interesting properties have been observed in nanocrystals of doped $NaYF_4$ [38-41]. $NaYF_4$ is also useful in bioimaging [42].

The structure of the fluorides $NaLnF_4$ (Ln=Y–Lu) was first described by Burns in 1965[43] and subsequently confirmed[44]. At ambient temperature and pressure, $NaYF_4$ exists in two polymorphs [45], the hexagonal type structure and the cubic (fluorite type) one, depending on the synthesis method. The space group for $NaYF_4$ is P-6 and the unit cell dimensions are a = 5.967 A° and c = 3.523 A°. Each unit cell contains 1.5 formula units, so that the compound may be written as $NaLn(Na1/2Ln1/2)F_6$ to indicate that three cation sites exist, one occupied fully by Ln^{3+} (1a), one occupied randomly by ½ Na^+, ½ Ln^{3+} (1f) and the third occupied fully by Na^+ (2h). Sites 1a and 1f both have C3h symmetry and are nine fold coordinated, whereas site 2h has C3 symmetry. The rare earth ion should be accommodated in both sites 1a and 1f , the number of Ln^{3+} ions in site 1a and 1f being twice that in site 1f. Crystal structure of $NaYF_4$ is given in the **Figure 1**.

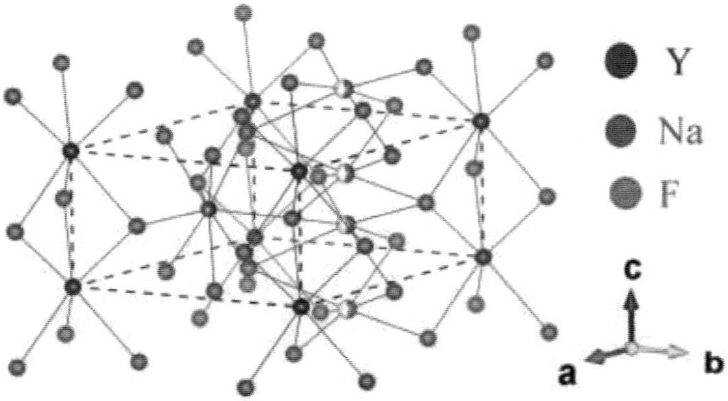

Figure 1: Crystal structure of Hexagonal-type $NaYF_4$.

The alkali metal-rare earth fluoride systems generally have complex phase diagrams[46] which make it difficult to obtain well-crystallized single-phase materials. The transition in $NaYF_4$ between hexagonal and cubic phases occurs at 691°C.

2. Experimental

$NaYF_4$ was synthesized by using wet chemical synthesis. All the reagents used were of the Analytical Reagent grade. Freshly prepared aqueous solution of NaF was used. Chloride salts of Y_2O_3 and $Ce_2(CO_3)_3$ were prepared by dissolving in dil. HCl. NaF was taken in 10 time's excess than the stoichiometric ratio. Both the solutions were heated to 85°C. At 85°C, solution of YCl_3 and $CeCl_3$ was added slowly into the solution of NaF, yielding $NaYF_4$ precipitate. The resultant precipitate was filtered, washed and dried under the fan. The dried powdered was collected and used for subsequent experiments. KYF_4 was prepared in a similar way.

X-ray diffraction patterns were recorded on a Philips PANalytical X'pert Pro diffractometer. Photoluminescence spectra were recorded on a Hitachi F-4000 spectro-fluorimeter with spectral slit width of 1.0 nm in the spectral range 220-700 nm. Scanning electron microscope images were taken on the SEM,VEGA MV2300T/40 operated at 20.0kV voltage.

3. Results and Discussion

3.1. X-ray diffraction analysis

Figure 2 shows the XRD pattern for $NaYF_4$ prepared by wet chemical synthesis. It was found that the XRD pattern of ICDD 77-2042($NaYF_4$) is identical to that of ICDD 89-5167(YOF). Therefore, it raised a confusion regarding the precipitated sample that whether $NaYF_4$ is formed or always YOF precipitates during the wet chemical synthesis

Similar, type of confusion was raised when X-ray diffraction pattern of KYF_4 (our sample) prepared by same wet chemical synthesis was compared with ICDD 79-1688(KYF_4) and ICDD 89-5167(YOF) (**Figure 3**).

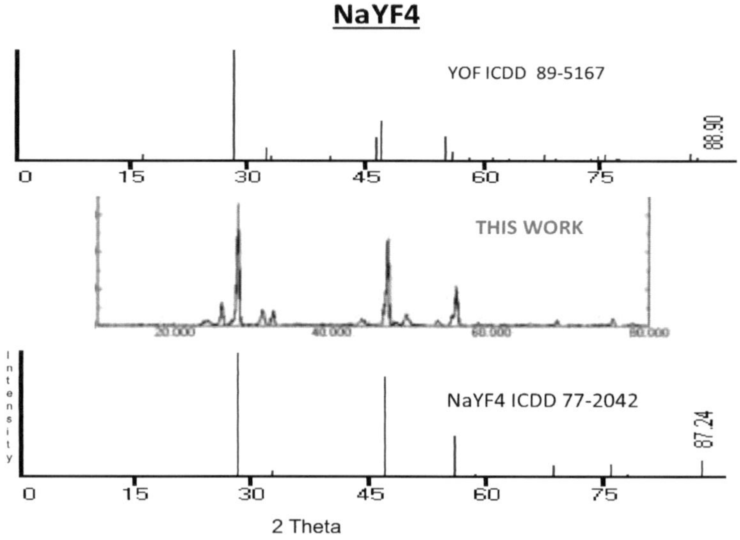

Figure 2: XRD pattern of $NaYF_4$ prepared by wet chemical synthesis, compared with ICDD 77-2042($NaYF_4$) and ICDD 89-5167(YOF).

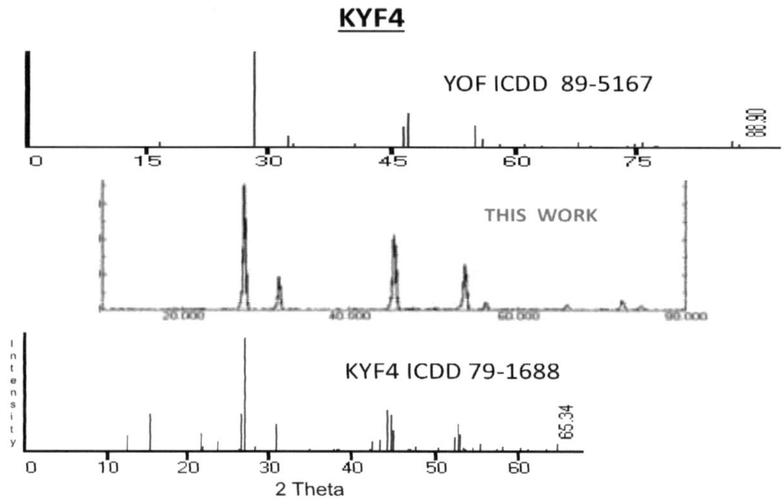

Figure 3: XRD pattern of KYF_4 prepared by wet chemical synthesis, compared with ICDD 79-1688(KYF_4) and ICDD 89-5167(YOF).

3.2. EDAX and SEM

Therefore, Energy dispersive x-ray analysis (EDAX) and scanning electron microscope (SEM) studies were carried out on both the compounds.

Figure 4 represents the SEM image together with the EDAX spectrum of $NaYF_4$. **Table 1**, shows the weight% and atomic% of the constituent elements, respectively. When composition of elements in $NaYF_4$ compound were calculated separately, the values obtained for different constituting elements are Na = 16.66 At%, Y = 16.66 At% and F = 66.67At%, which are found to be approximately nearer to the values observed during EDAX analysis (values given in **table 1**). Therefore, formation of YOF phase can be ruled out since, oxygen is present but in very less amount. This could be due to formation of some oxides at the surface of the material during the measurements. Ce is present in 0.06 At% in the sample.

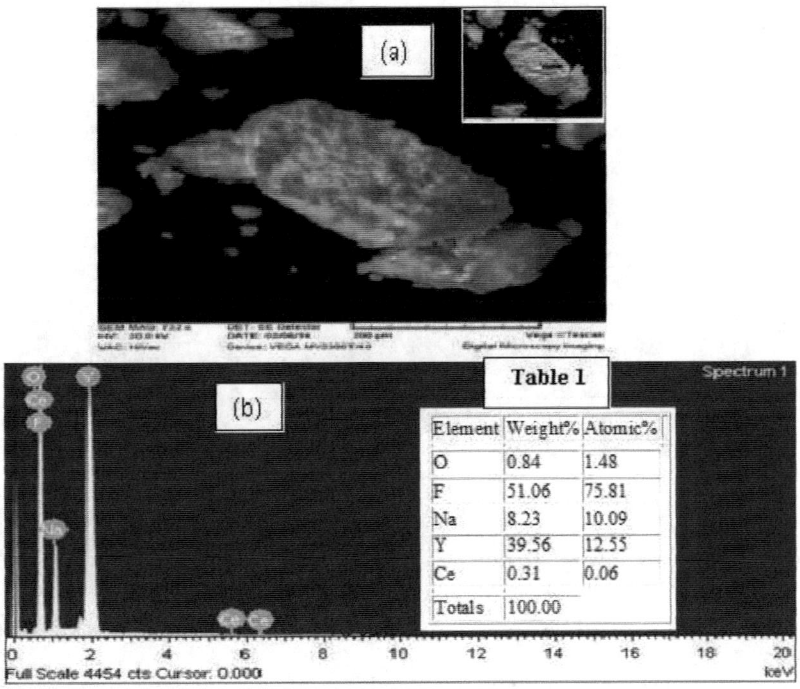

Table 1

Element	Weight%	Atomic%
O	0.84	1.48
F	51.06	75.81
Na	8.23	10.09
Y	39.56	12.55
Ce	0.31	0.06
Totals	100.00	

Figure 4: represents the (a) SEM image and (b) EDAX spectrum of $NaYF_4$.
Table 1 gives the experimental values of weight% and atomic% of the Na, Y and F present in our sample.

Hence, it can be said that $NaYF_4$ has been formed in our experiment.

Figure 5 shows the SEM micrograph and EDAX spectrum of KYF_4 prepared by wet chemical synthesis.

Table 2, shows the composition of various elements as their atomic percent. When composition of elements in KYF_4 compound were calculated separately, the values obtained for different constituting elements are K = 16.66 At%, Y = 16.66 At% and F = 66.67At%. But according to the values given in the **table 2**, oxygen (O) and potassium (K) are present approximately in same concentration, indicating that may be both KYF_4 and YOF are coexisting in our sample. Na is present as an impurity but in a very small amount.

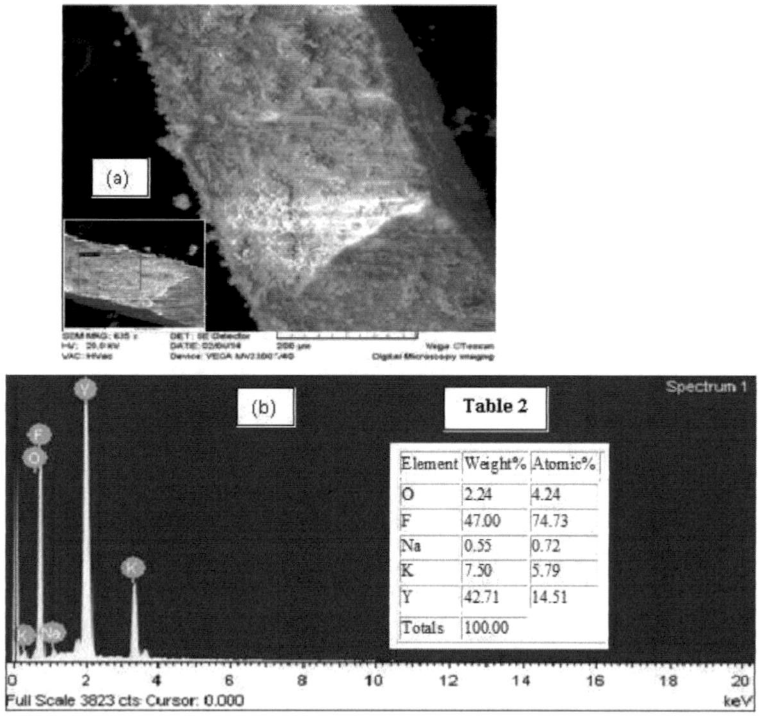

Table 2

Element	Weight%	Atomic%
O	2.24	4.24
F	47.00	74.73
Na	0.55	0.72
K	7.50	5.79
Y	42.71	14.51
Totals	100.00	

Figure 5: represents the (a) SEM image and (b) EDAX spectrum.
Table 2 shows the observed weight% and atomic% values of the K,Y and F present in our sample.

Hence, a very good match between our sample $NaYF_4$ and ICDD file 77-2042 is seen. Thus confirming the formation of hexagonal $NaYF_4$.
Therefore from the above results, it can be concluded that $NaYF_4$ can be prepared by using simple wet chemical synthesis easily but it's difficult to synthesize pure KYF_4 in the same way.
SEM images show the formation of irregular particles of different shapes and sizes.

3.3. Photoluminescence in $NaYF_4$:Ce^{3+}

Figure 6 represents the Ce^{3+} luminescence in $NaYF_4$. A broad emission band peaking at 315 nm is observed for 254 nm excitation. The excitation spectrum consists of a peak at 254 nm and a shoulder at 288 nm. Peaks of the curves appears little disturbing due to very high PL intensity. As prepared Ce-doped sample showed luminescence. Luminescence study on KYF_4 was not attempted due to the doubt regarding the formation of phase pure compound.

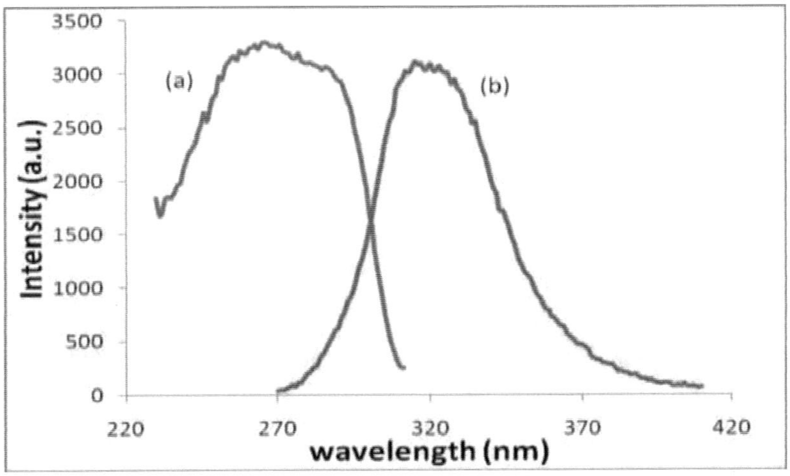

Figure 6: PL spectra of $NaYF_4$:Ce(1%). (a) excitation at λ_{em}=315nm and (b) emission at λ_{ex}=254nm.

4. Conclusions

Two well known host phosphors $NaYF_4$ and KYF_4 were attempted via simple Wet-Chemical process. XRD studies showed that XRD pattern of both the compounds matches well with an another compound YOF which raised doubt regarding the successful formation of desired phases. Hence in order to eliminate this confusion, further studies such as EDAX and SEM were also done. The EDAX confirmed the formation of phase pure $NaYF_4$ whereas KYF_4 was contaminated with YOF.

As prepared Ce^{3+} doped $NaYF_4$ showed luminescence, hence avoiding the need of high temperature and time consuming reduction processes. Emission has been recorded at 315 nm for 254 nm excitation with very high intensity. PL measurements are in good agreement with the previous studies.

References

[1] Pol P. G. and Rao V. J. 1973 *Ind. J. Pure Appl. Phys.* **11** 886.
[2] Chen Z and Fu M-L 2018 *Mater. Res. Bull.* **107** 194.
[3] Tanga Z, Liua Q, Lia J, Wua X, Zhan S, Nie G, Hu J, Hu S, Xi Z, Wu S, Zhang Y, Shi L and Liu Y 2019 *J. Lumin.* **206** 21.
[4] Bagdasarov Kh. S., Kaminsky A. A. and Sobolev B. P. 1969 *Kristallographiya* **13** 779.
[5] Kano T. K., Yamamoto H. and Otomo Y. 1972 *J. Electrochem. Soc.* **119** 1561.
[6] Menyuk N., Dwigth K. and Pierce J. W. 1972 *Appl. Phys. Lett.* **21** 159.
[7] Zhu H, Lin M , Jin G, Lu T J and Xu F 2017 *J. Lumin* **185** 292.
[8] Jia H, Chen Z, Chen P, Xu C, Liu C, Zhao Z, Liu X and Qiu J 2017 *J. Am. Ceram. Soc.* **100** 697.
[9] Xiao Q, 2017 *J. Am. Ceram. Soc.* **100** 1011.
[10] GaoY, Hu Y, Zhoua D and Qiu J 2017 *J. Euro. Ceram. Soc.* **37** 763.
[11] Fan S, Wang S, Xu W, Li M, Sun H and Hu L 2017 *J. Mater. Sci.* **52** 869
[12] Cong T , Ding Y, Liu J, Zhao H and Hong X 2016 *Mater. Lett.***165** 59
[13] Ding M, Chen D, Ma D, Dai J, Li Y and Ji Z 2016 *J. Mater. Chem. C* **4** 2432
[14] Li X, Xue Z and Liu H 2016 *J. Alloys Compd.* **681** 379

[15] Palo E, Pihlgren L , Tuomisto M, Laihinen T, Hyppänen I , Kankare J , Lastusaari M, Soukka T , Swart H C. and Hölsä J 2016 *Opt. Mater.* **59** 49
[16] Tuomisto M , Palo E , Laihinen T , Hyppänen I , Lastusaari M, Swart H C. and Hölsä J 2016 *Opt.Mater.* **59** 115
[17] Kramer K.W., Biner D.A., Frei G., Gudel H.U., Hehlen M.P. and Luthi S.R. 2004 *Chem. Mater.* **16** 1244
[18] Chen Z and Fu M-L 2018 *Mater. Res. Bull.***107** 194
[19] Han Q , Gao W , Zhang C, Mi X , Zhao X, Zhang Z, Dong J and Zheng H 2018 *J. Alloys Compd.* **748** 252
[20] Yan H, Chen X, Song H, Wang X, Shi Z and Lin Q 2017 *Mater. Lett.* **187** 101
[21] Lei L, Wu R, Zhou J, Zhang S, Xiao Z, Zhang J and Xu S 2017 *Mater.Lett.***189** 35
[22] Misiak M, Bednarkiewicz A and Stręk W 2016 *J. Lumin.* **169** 717
[23] Bi X, He G, Di W and Qin W 2016 *Mater. Lett.* **173** 187
[24] Yang Y, Zhou P, Xu W, Xu S, Jiang Y, Chen X and Song H 2016 *J. Mater. Chem. C* **4** 659
[25] Feng Z , Xia H , Wang C , Zhang Z, Jiang D, Zhang J, He S , Tang Q, Sheng Q, Gu X, Zhang Y, Chen B and Jiang H 2016 *J. Alloys Compd* **680** 26
[26] Misiak M, Stręk W, Arabasz S and Bednarkiewicz A 2016 *J. Lumin.* **178** 34
[27] Liu L, Qin F, Zhao H, Lv T , Zhang Z and Cao W 2016 *J. Alloys. Compd.* **684** 211
[28] Pokhrel M , Valdes C and Mao Y 2016 *Opt Mater* **58** 67
[29] Gao W, Sun Z, Han Q, Zhang J, Yan X, Ge H and Dong J 2018 *Mater. Res. Bull.* **108** 10
[30] Pilch A. , Wawrzynczyk D. , Kurnatowska M. , Czaban B., Samoc M., Strek W. and Bednarkiewicz A. 2017 *J. Lumin.* **182** 114
[31] Shen J-W, Lu J, Tu J, Ouyang X and Li H 2016 *J. Mater. Chem. C* **4** 4905
[32] Yao W., Tian Q., Liu J., Wu Z., Cui S., Ding J., Dai Z. and Wu W. 2016 *J. Mater. Chem. C* **4** 6327
[33] Wang X , Xu T, Cai P, Vu T and Jin Seo H 2017 *J. Alloys. Compd.* **691** 530
[34] Zhou N, Xu B, Gan L, Zhang J, Han J and Zhai T 2017 *J. Mater. Chem. C* **5** 1591
[35] Xiaobin D , Fangtian Y , Hongshang P and Shihua H 2016 *J.Rare Earths* **34** 863
[36] Pavel E. , Marinescu V., Lungulescu M. and Sbarcea B. 2018 *Mater. Lett.* **210** 12
[37] Ding M, Lu C , Chen L and Ji Z 2018 *J. Alloys. Compd.* **763** 85
[38] Zheng B, Lin L, Huang L, Feng Z, Zhuanga L, Wang Z and Zheng Z 2018 *Mater. Res. Bull.* **101** 199
[39] Ye S, Xiao P, Liao H, Li S and Wang D 2018 *Mater. Res. Bull.* **103** 279
[40] Du P , Luo L and Su Yu J 2018 *J. Alloys Compd.* **739** 926
[41] Zhuang J, Lei B, Zhang H and Liu Y 2016 *Mater. Lett.* **178** 260
[42] Yang D, Cao C, Feng W, Huang C and Li F 2018 *J.Rare Earths* **36** 113
[43] Burns J. H. 1965 *Inorg. Chem.* **4** 881
[44] Zakaria D., Mahiou R., Avignant D. and Zahir M. 1997 *J. Alloys.Compd.* **257** 65
[45] Martin N., Boutinaud P., Mahiou R., Cousseins J. C. and Bouderbala M. 1999 *J. Mater. Chem.* **9** 125
[46] Thoma R. E. 1966 "The Rare Earth Halides," Progress in the Science and Technology of the Rare Earths, Vol. 2, p. 90, L. Eyring, Editor, Pergamon Press.

Design and analysis of double wishbone suspension system

Pranav Upadhyay[1,2], Mrinal Deep[1], Aryan Dwivedi[1], Ashutosh Agarwal[1], Pikesh Bansal[1] and Pradeep Sharma[1]

[1]Department of Mechanical Engineering, ABES Engineering College, Ghaziabad, India.
[2]Corresponding Author
E-mail: upad.pranav1202@gmail.com

Abstract. Suspension system serves purpose of providing stability to vehicle which results improving comfortability of riding. It provides traction control and steering stability. This paper proposes a method for designing double wishbone suspension system for electric solar vehicle with damper to lower wishbone. Recent trends in suspension system have focused on improving innovations in ride quality and handling of vehicles. It maintains a viable spectrum of economic growth by reducing costs and increasing efficiency simultaneously. Theoretical analysis is conducted using Lotus and Solid works to design double wishbone suspension system. The analysis is done for tadpole geometry car with double wishbone in front and swing arm in the rear wheels. Results in the paper are going to improve vehicle handling characteristics by controlling camber, toe angles and A-arm's length. This is accomplished by changing the length of A-arm and the ground clearance and attach angle with the chassis by hit and trial method and improving possible traction and maneuverability. Results of the paper are presented in form of graph as a result of simulation conducted.

1. Introduction

The suspension system of a vehicle is generally designed to maintain traction with the road. Suspension are designed to meet the following requirements: (1) Ride comfort (2) traction and (3) Handling. Suspensions generally fall into either of two groups: Dependent and Independent Suspensions. Double wishbone suspension helps in controlling the motion of the wheel throughout suspension travel, controlling such parameters as camber angle, caster angle & toe angle. In this suspension research we are focusing on the A arm design and have proposed a design with geometry optimization. The geometry of the suspension is Damper to lower wishbone. For designing A-arms we have used Solid-Works (2018) in which we are able to do different material analysis and testing to make the A- arm light weight and economically feasible while satisfying the driving condition. Calculation of stiffness of shocker was done to get angle between A-arm and shocker best feasible for our ESVC 2019.

2. Literature Review

X. L. Bian et al. [1] has researched on Roberson and Wittenburg (RW) methodology of multi-body dynamic analysis. To simulate the actual working conditions they consider wheel bouncing to the steering performance & employed two weight functions for their model. A passenger vehicle and group of desirable design variables was an example for their procedure examination.

Vinod Cherian et al. [2] have worked on non-linear model of a double wishbone suspension and found a procedure in which we can find the effect on transmission, tire distribution forces acting on wheel spindle & chassis by variation in suspension parameters. The equations of motion are derived with the help of Lagrange multiplier method, and solved numerically using MATLAB.

J. S. Hwang et al. [3] has worked on how to find hard points position with the help of kinematic design of double wishbone which would increase stability of a vehicle.

Ramon Sancibrian et al. [4] have done an attempt of using multi objective dimensional synthesis technique on the kinematic design of double-wishbone suspension systems in vehicles. The Jacobian matrix was obtained by using exact differentiation constructed on gradient determination by synthesis method.

N.Vivekanandan et al. [5] have a main objective to design double wishbone. Material section was done on the basis of carbon content, tensile strength, yield strength, hardness & cost. According to their analysis AISI 1040 was best for their wishbones. Geometric roll center was located by suspension geometry and principles of the instant centre of rotation.

R. C. Silva et al. [6] aims to propose analytical equations implemented computationally in MATLAB which were set to evaluate the double wishbone suspension. Simple concepts of trigonometry were used to develop their formula. Predefined points Cartesian coordinates (x, y, z) which are attained from the CAD design are given as input to formula which will give us the caster, camber angles.

Damodaran, P et al. [7] have presented an optimal dimensional synthesis of 5 link suspension mechanism in terms of length of the 5 links and positions of the ten spherical joints, which are installed on the vehicle chassis and wheel carrier. The influence of the complaint joints on the dynamic behaviour of the suspension system is approached. Matlab optimization tool box is used for this optimization. The aim was to take the motion of the body to the wheel carrier with respect to the vehicle chassis as close as possible to a vertical translation.

Shijil P et al. [8] have done a dynamic study of the vehicle when driven on an offroad racetrack conditions were analyzed on the parameter dynamic and static suspension system of an ATV. The performance of the ATV is affected by many parameters. This paper work is focused on design and analysis, determination, optimization of suspension systems and their assembly best performance. The aim was to identify and optimize the parameters which affects the dynamic performance of suspension systems.

Arun Y Patil et al. [9] have researched on an upper A arm which was designed so as to help in accommodating a coil spring shock absorber in ATVs without altering the motion ratio and providing more flexibility in choosing an angle correction factor. Solidworks is used for CAD model. The virtual simulation and structural analysis were performed on ANSYS Workbench. Several iterations were carried out to optimize the proposed design using the software in terms of strength factor, stress concentration and weight.

2.1. Literature Review Conclusion

From literature survey above it is evident that various softwares are available to conduct analysis on suspension geometry and to conduct structural analysis of A-arms. Out of these softwares we found out that a combination of Solid Works and Lotus will suit our purpose. The analysis conducted by all the researchers has been done for various configuration of their vehicles. We are conducting analysis for tadpole geometry solar vehicle being manufactured for ESVC-2019. Analysis results for various configurations of A-arms are presented in paper and final optimized results for Lotus geometry are show cased in this paper. Our objective is to reduce weight and optimize geometry in order to maintain stability of vehicle and maintain traction with road.

3. Stiffness of shocker

The weight transfer calculation was done according to driving conditions to calculate stiffness of shocker mounted on lower A-arm for absorbing loads and maintaining traction with road. Shocker stiffness is defined as amount of deflection or travel, shocker will show when force is applied at different driving conditions.

4. Selection of material for A-Arm

For manufacturing of A-arms simulation of the A-arms is done on Solid works with different materials to make the A-arm economical and light weight. The list of different materials is enlisted. The simulation of the A-arms is performed by using Solid works (2018) to find the stress distributed on the A-arms.

Properties of different Materials:

Table 1. Properties of materials.

Properties	AISI1020	AISI4130	AA6063 T6	AA6061 T6
Yield Strength	351.57 MPa	460MPa	214MPa	276 MPa
Ultimate Strength	420 MPa	670MPa	241 MPa	310 MPa
Density	7900kg/m3	7850kg/m3	2700kg/m3	2700kg/m3
Young's Modulus	205GPa	205GPa	68.9GPa	70GPa
Moment of Inertia	$2.027*10^{-8}m^4$	$2.027*10^{-8}m^4$	$1.263*10^{-7}m^4$	$1.263*10^{-7}m^4$

The total weight of the vehicle is 250Kg(assumed) including the driver's weight, which is taken by including the Factor of Safety i.e. that the design is safe.

From calculations (refer pg. 13) we calculated that the maximum force on the front wheel will be at the time of braking condition.

Total mass transfer on each front wheel = 110 kg (refer pg. 10)

Total Weight on wheel = $110 + 20\%\ of\ total\ weight$

$= 110 + 50 = 160\ kg$

All the simulation is being performed at 1600N of force on different material with different dimensions. Result is show in the following table 2.

5. Result and Discussion

Figure 1-3 represents the analysis of A-arm with requisite boundary conditions for AISI 4130 and 1.25mm, 2.00 mm, 2.50 mm thickness of pipe. The results of FOS for Von mises criteria and maximum stress are presented in table 2. Figure 4-6 shows the analysis of A-arm with requisite boundary conditions for AISI 1020 and 1.25mm, 2.00 mm, 2.50 mm thickness of pipe. The results of FOS for Von mises criteria and maximum stress. Similarly, figure 7-9 represents the analysis of A-arm for material AA 6061-T6 and 1.25mm, 2.00 mm, 2.50 mm thickness of pipe. Also, the tests are done on AA 6063-T6 in figure 10-12 for the analysis of A-arm with requisite boundary conditions and 1.25mm, 2.00 mm, 2.50 mm thickness of pipe. The results of FOS for Von mises criteria and maximum stress are presented in table 2. All the above-mentioned pipes are taken as experimental setup and are analyzed to obtain optimized results for concluding the final material to be selected for fabrication of the control arm (commonly called as A-Arm). The simulations are done on Solidworks-2018 software and the simulation results have been shown below in fig. 1-12. The control arm has been fixed at the points where it is attached directly to the chassis and the load of 1600 N is applied at the points which are fixed directly to the suspension of the vehicle and the wheel assembly point.

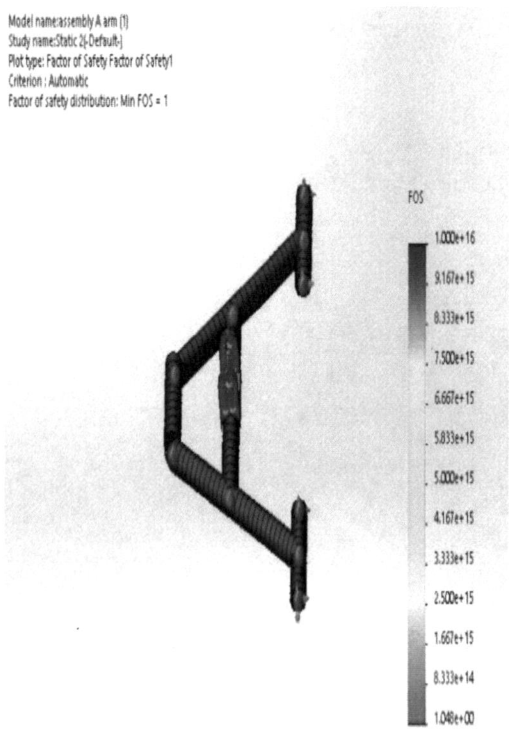

Fig 1. AISI 4130 1inch 1.25mm.

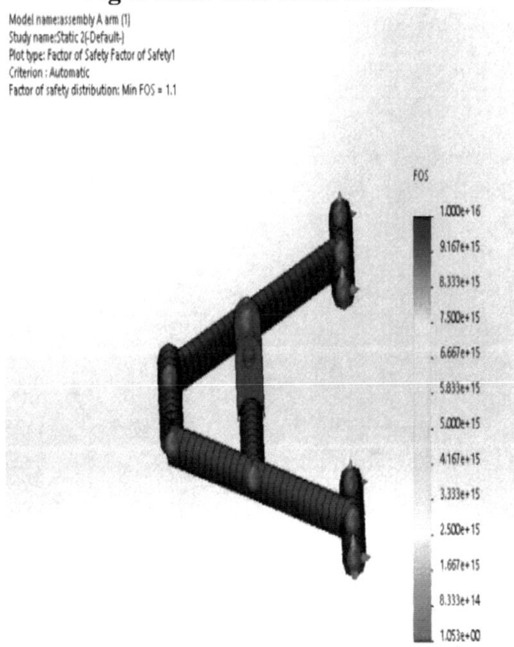

Fig 2. AISI 4130 1inch 2mm.

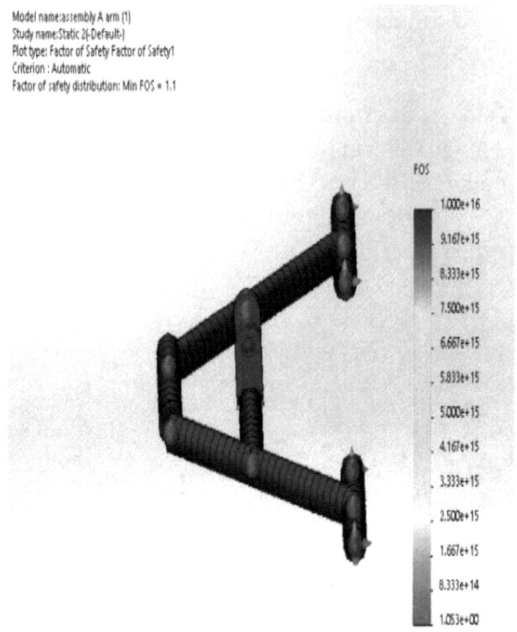

Fig 3. AISI 4130 1-inch 2.5mm.

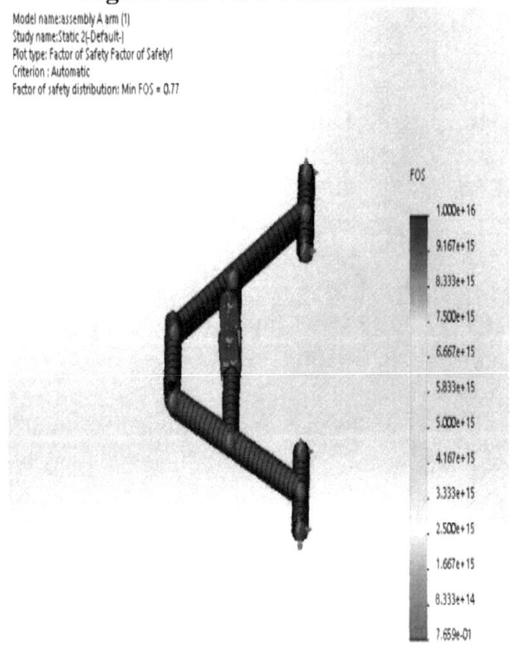

Fig 4. AISI 1020 1-inch 1.25mm.

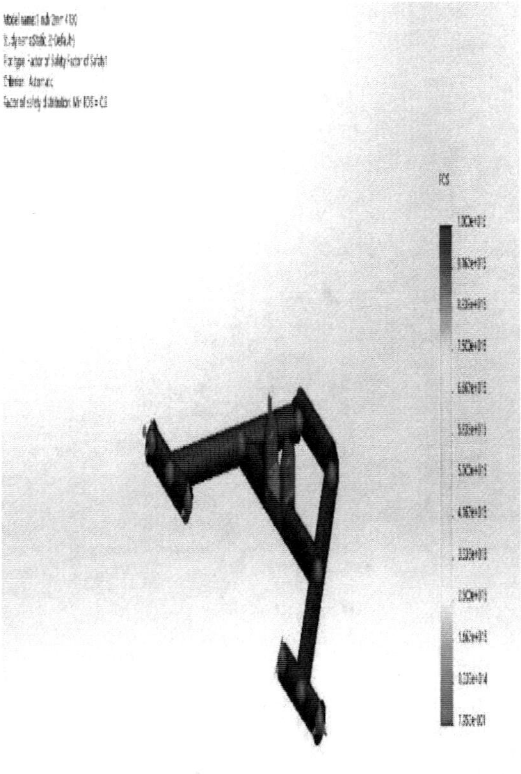

Fig 5. AISI 1020 1-inch 2.00mm.

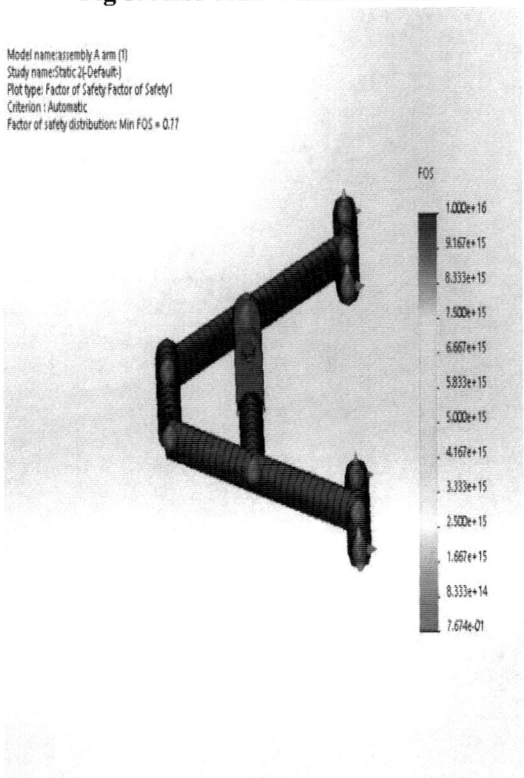

Fig 6. AISI 1020 1-inch 2.00mm.

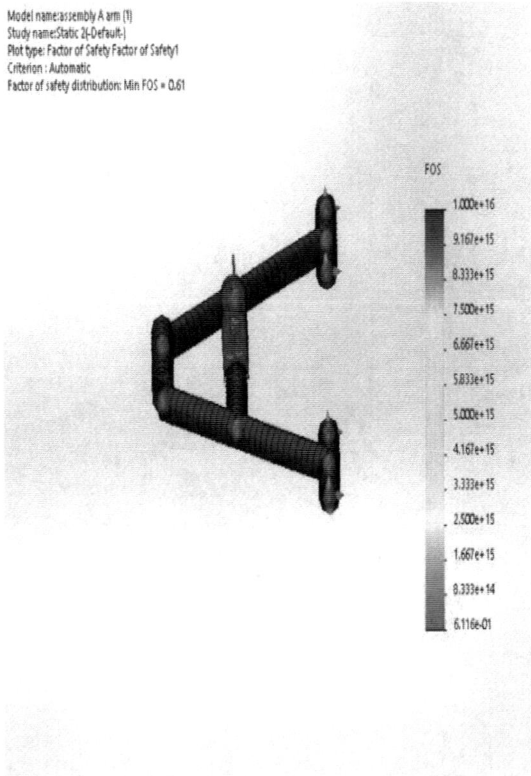

Fig 7. A-6061T6 1-inch 1.25mm.

Fig 8. A-6061T6 1-inch 2mm.

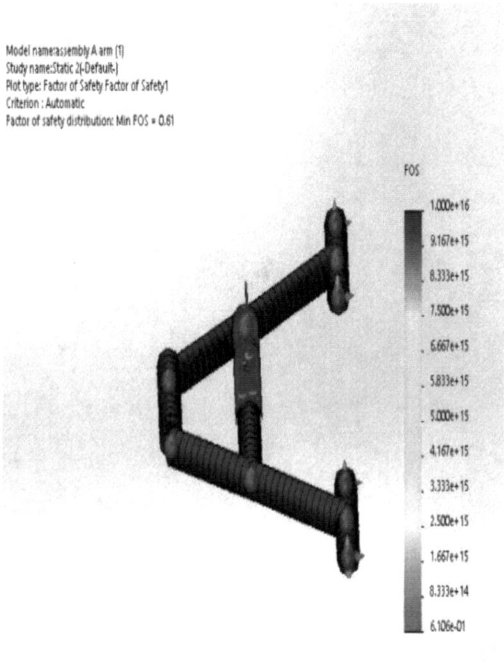

Fig 9. A-6061T6 1-inch 2.5mm.

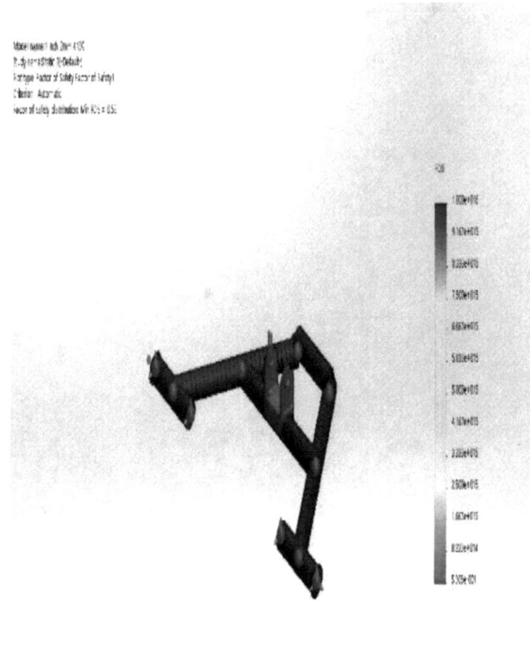

Fig 11. A-6063T6 1-inch 1.25mm.

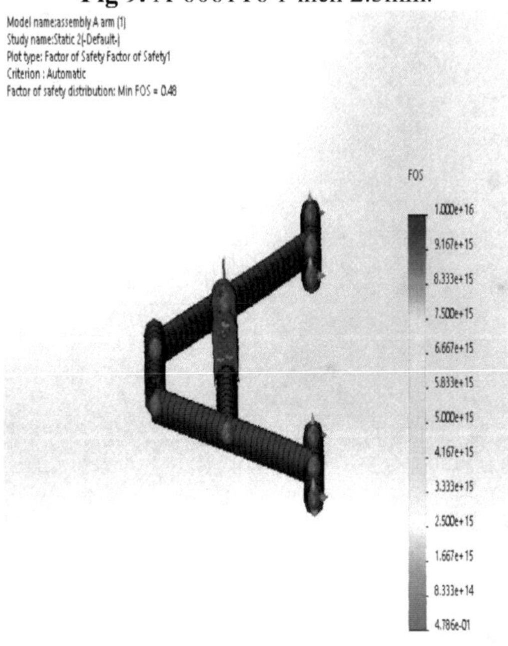

Fig 10. A-6063T6 1-inch 1.25mm.

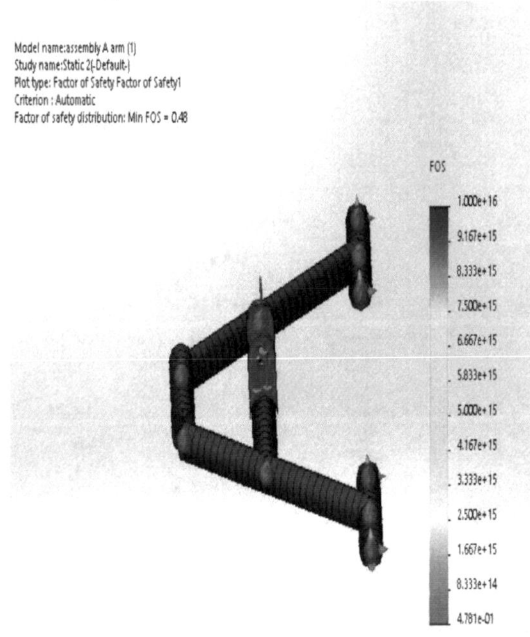

Fig 12. A-6063T6 1-inch 1.25mm.

Table2. Results of Analysis.

Figure No.	Pipe Thickness	FOS	Stress (N-m)
		AISI-4130	
1	1.25mm	1.0	4.388×10^8
2	2.00mm	1.1	4.344×10^8
3	2.50mm	1.1	4.600×10^8
		AISI-1020	
4	1.25mm	0.77	4.590×10^8
5	2.00mm	0.80	3.445×10^8
6	2.50mm	0.77	4.581×10^8
		AA-6063-T6	
7	1.25mm	0.48	2.097×10^8
8	2.00mm	0.53	2.150×10^8
9	2.5mm	0.48	3.308×10^8
		AA-6061-T6	
10	1.25mm	0.61	2.496×10^8
11	2.00mm	0.68	2.750×10^8
12	2.50mm	0.61	3.504×10^8

6. Conclusion

After analyzing the data it has been found that AISI 4130 is best suited for A-Arm material as for same FOS weight of component is optimized when we use AISI 4130. But for this application further analysis led to conclusion that pipe of outer 1 inch and 2 mm thickness is best suited as FOS for this will be 1.8 as per analysis. From the data its predicted to use AISI 4130 inch outer diameter and 2 mm thickness for this application.

7. References

[1] Bian X L, Song B A and Walter R 2004 Optimization of steering linkage and double-wishbone suspension via R-W multi-body dynamic analysis *Research in Engg., Vol. 69(1)* p 38-43

[2] Cherian V, Jalili N and Haque I 2004 *Development of a Non-Linear Model of a Double Wishbone Suspension for the Characterization of Force Transmission to the Steering Column and Chassis* ASME 2004 International Mechanical Engineering Congress and Exposition Paper No. IMECE2004-59922 p 775-780

[3] Hwang J S , Kiml S R and Han S Y 2007 *Kinematic design of a double wishbone type front suspension mechanism using multi-objective optimization* 5th Australasian Congress on Applied Mechanics,10-12 December 2007 p 788-793

[4] Sancibrian R, Garcia P, Viadero F , Fernandez A and De-Juan A 2010 Kinematic design of double-wishbone suspension systems using a multi objective optimization approach *Int. J. of Vehicle Mechanics and Mobility Vol. 48(7)* p 793-813.

[5] Vivekanandan N, Gunaki A , Acharya C, Gilbert S and Bodake R 2014 Design Analysis and Simulation of Double Wishbone Suspension System *Int. J. of Mech. Engg., Vol 2 (6)* p 1-7

[6] Silva R C, Nunes M A A, Oliveira A B S and Andrade G O 2015 *Estimation of the Geometric Parameters of a Front Double Wishbone Suspension Based on Geometry Formulation* ASME 2015 Int. Mech. Engg. Congress and Exposition Vol 12 Paper No. IMECE2015-50012

[7] Dhamodharan P, Mohamed Raafiq N and Madhu S 2015 A Review Paper on Multilink Suspension Mechanism with Compliant Joints *Int. J. of Applied Engg. Research Vol. 10(33)* p 25491-25493

[8] Shijil P, Vargheese A, Devasia A, Joseph C and Jacob J 2016 Design and Analysis of Suspension System for an all Terrain Vehicle *Int. J. of Sci. & Engg. Research Vol. 7(3)* p 164-190

[9] Sunal S, et al 2018 *Design modification and analysis of upper A-arm of an ATV suspension system* NAFEMS India Regional Conference on Engineering Modelling, Analysis, Simulation and 3D-Printing (NIRC 2018)

Optimization of chassis for a solar powered vehicle

Pranav Upadhyay[1,2], Sachin Kumar Sharma[1], Gautam Kumar[1], Pikesh Bansal[1] and Pradeep Sharma[1]

[1]Department of Mechanical Engineering, ABES Engineering College, Ghaiziabad, India
[2]Corresponding Author
E-mail: upad.pranav1202@gmail.com

Abstract. This study aims at designing and analysis of roll cage for an electric solar vehicle fabricated for Imperial Society for Innovative Engineers ESVC (Electric Solar Vehicle Competition) 2019. The vehicle will be powered by electric motor and charging will be done by solar panels which makes the design eco-friendly. This study focuses on the development of an optimized roll cage structure which will include the compactness of frame design, light weight and driver's safety considering all the design ergonomics and the driver's safety measures. Optimization is based on reducing weight, strength considerations and ergonomics consideration. Variation of materials, pipe diameters and 3-D structure is implemented during analysis. Analysis of roll cage is conducted using SolidWorks and loading conditions are taken as per standards. Three-wheel tadpole configuration is used for analysis purpose. Results are presented in form of graphs and SolidWorks analysis outputs.

1. Introduction

1.1. Literature Review

The objective of the study is to design, develop and fabricate the roll cage of an electric solar vehicle for ESVC 2019 in accordance with the rules specified by the ISIE. It is designed to compensate the swelling demand of the conventionally powered vehicles. The roll cage designed is light weight, in accordance with the driver's safety and adds to the aesthetics of the car. The vehicle should a stable and a strong frame for which various materials are taken into consideration out of which chromoly (AISI 4130) is found to be suitable. The Finite Element Analysis (FEA) of model is done in Solid works 2018. Later, various loading conditions are applied on the design for static simulation to test all modes of failure. Based on the results obtained from these tests, the design is modified accordingly. The FEA of the roll cage needs to be done because of the failure of structural members which can lead to accidents. The roll cage should act as the central mounting structure for all the components. Conventionally, roll cage is a truss welded at joints. The chassis should withstand high energy transfer waves while acceleration and deceleration. This makes the chassis design more compatible and driver safe against accidental situations [1]. For designing a roll cage all the loads acting on it need to be considered. The main loads acting on an automobile's roll cage are given by C.D. Naiju, Anamalai K. et. al. in [2] as: (i) Longitudinal torsion (ii) Vertical bending (iii) Lateral bending (iv) Horizontal lozenging. Torsional load acts on the opposite corners of the roll cage. Suspension loads acts on the two ends of the car considering the frame to be a torsional spring. Torsion study has been demonstrated with experimental proofs by W.B. Riley and A. R. George in [3]. Deflections decreases with increase in stiffness when the back of the chassis was loaded according to a study on static analysis by A. Ryan in [4]. FEA has highly improved both the standards of engineering designs and the methodology of the design process in many industrial applications. Jason Denny et.al. in [5] researched on the development of a monocoque chassis through an iterative FEA process to develop a solar car. The methodology includes an iterative process considering various geometries and changes, with the aim of achieving a single sustainable frame. The suspension mounting points are also analyzed to ensure their rigidity under concentrated loading. A. K. Das et.al. [6] studied emerging growth of population and need to find and alternative source of energy in automobile industry. He worked on the design and fabrication of solar powered car and its performance evaluation. Nader A. Nader et. al. [7] also worked on design and building of Solar Car considering lightweight, aerodynamics and efficiency as major components while designing. P.S. Sreeragh et. al. [8] have worked on the designing of a chassis to fulfill rules and guidelines as indicated by ISIE for ESVC. Suraj Aru et. al. [9] have worked on the design of tubular space frame also called as the roll cage of a vehicle using FEA.

1.2. Problem Statement

The weight of roll cage is an important factor in designing. The overweight of the roll cage might result in reduced performance of the vehicle. Also, keeping in mind the rules as established by ISIE, the size of the vehicle has been reduced. Reduction in the size of the vehicle and procuring maximum power from the solar cells considering less weight of the roll cage along with driver's safety needs to be addressed. For reducing weight of the vehicle suitable material has been used without hampering the strength to weight ratio. For this purpose, AISI 4130 has been used with minimized dimensions. Also, to increase the power output from the solar cells optimization in the roll cage has been done by increasing the area for the solar cells. The frontal area has been increased by giving an inclination of 17° to the frontal surface of the roll cage with respect to the top plane.

2. Objective

- To design the roll cage structure considering the driver's safety.
- To design roll cage in accordance with the rulebook of ISIE ESVC 2019.
- To reduce weight of roll cage.
- To obtain maximum power output from the solar cells by optimizing the roll cage.
- To optimize the roll cage by selecting suitable material.
- To use the pipe with minimized dimensions.
- To design a vehicle for driver's ergonomics which considers driver's comfort, optimal placement of steering wheel, pedal positions, spacious cockpit.
- To avoid the failure in the front collision, rear collision, side collision, rolling and torsion in rough conditions.

3. Material selection

Various materials are taken into consideration for FEA of roll cage. The parameters considered while material selection are their weight, strength, reliability, availability and manufacturability. Initially, AISI 4130, AISI 6063, AISI 6061 and AISI 1020 are taken. Every material has different properties and advantages. Properties of various materials are detailed in table 1.

Table 1. Properties of materials.

Properties	AISI1020	AISI4130	AA6063 T6	AA6061 T6
Yield Strength	351.57 MPa	460MPa	214MPa	276 MPa
Ultimate Strength	420 MPa	670MPa	241 MPa	310 MPa
Density	7900kg/m3	7850kg/m3	2700kg/m3	2700kg/m3
Young's Modulus	205GPa	205GPa	68.9GPa	70GPa
Moment of Inertia	$2.027*10^{-8}m^4$	$2.027*10^{-8}m^4$	$1.263*10^{-7}m^4$	$1.263*10^{-7}m^4$

Frontal impact loading condition is considered while selecting the material. Various pipes with different properties are selected for analysis and based on the calculations and their results the final dimensions and material is finalized. According to the constraints in the rulebook the vehicle can be driven at a speed of not more than 60kmph. The following calculation are included by Abhinav Sharma et. al. in [10].

$$W_{net} = Net\ Work\ Done,\ F = Force,\ d = Distance\ Travelled$$

Now,

$$W_{net} = \frac{1}{2}mv_{final}^2 - \frac{1}{2}mv_{initial}^2 \tag{1}$$

$$W_{net} = | -\frac{1}{2}mv_{initial}^2 | \tag{2}$$

But,

$$W_{net} = Impact\ force \times d \tag{3}$$

For static analysis, vehicle is considered to come in rest 0.1 seconds next to the collision. If a vehicle moves at a speed of 60kmph, it will travel a distance of 1.66m after the impact by Oturkar Sania et.al. in [11]. From the above equations, it is found to be:

$$Impact\ force = \frac{1}{2}mv_{initial}^2 \times \frac{1}{d} \tag{4}$$

On using the values in eq. (4),

$$Impact\ force = \frac{1}{2} \times 235 \times (16.66)^2 \times \frac{1}{1.66}$$

$$Impact\ force = 19632.852\ N$$

Therefore, the maximum force to be applied is 19632.852 N. Assuming this force to be 20KN. Thus, material for roll cage should has max. yield point than the impact force. Frame has been analysed for rollover and collision. The design has been made in such a way that there are no redundant links in the frame for reduced weight without reducing the strength of the frame. Above discussed materials with all dimensions are taken for FEA and several results are obtained. Material is selected after analysing different materials with different dimensions for the values of FOS and various stresses acting on the frame. Below is the tabular representation of the results **(all results are based on static studies on SOLIDWORKS-2018)** obtained:

Table 2. Results of simulation done on AISI 4130.

O.D. \ Thickness	1mm	1.25mm	1.5mm
1 inch	FOS: 1.2 Stress: 1.2×10^{-2} N/m^2 Weight: 31.08 kg	FOS: 1.4 Stress: 1.23×10^{-2} N/m^2 Weight: 37.60 kg	FOS: 1.5 Stress: 6.32×10^{-2} N/m^2 Weight: 43.98 kg
1.5 inch	FOS: 2.6 Stress: 7.54×10^{-4} N/m^2 Weight: 44.71 kg	FOS: 3.2 Stress: 7.48×10^{-4} N/m^2 Weight: 54.65 kg	FOS: 3.8 Stress: 6.82×10^{-4} N/m^2 Weight: 54.45 kg
2 inch	FOS: 4.6 Stress:1.009×10^{8} N/m^2 Weight: 57.88 kg	FOS: 5.6 Stress: 8.17×10^{7} N/m^2 Weight: 71.11 kg	FOS: 7.3 Stress: 6.27×10^{7} N/m^2 Weight: 84.17 kg

Table 3. Results of simulation done on AISI 1020.

O.D. \ Thickness	1mm	1.25mm	1.5mm
1 inch	FOS: 0.88 Stress: 9.36×10^{-3} N/m^2 Weight:31.27 kg	FOS: 1.1 Stress: 1.65×10^{-3} N/m^2 Weight: 37.84 kg	FOS: 1.2 Stress: 1.01×10^{-3} N/m^2 Weight: 44.26 kg
1.5 inch	FOS: 2 Stress: 7.23×10^{-4} N/m^2 Weight:44.99 kg	FOS: 2.5 Stress: 8.56×10^{-4} N/m^2 Weight: 55 kg	FOS: 3 Stress: 4.88×10^{-4} N/m^2 Weight: 64.86 kg

2 inch	FOS: 3.4 Stress: 2.56×10^{-4} N/m^2 Weight: 58.25v kg	FOS: 4 Stress: 1.76×10^{-4} N/m^2 Weight: 71.56 kg	FOS: 5 Stress: 2.91×10^{-4} N/m^2 Weight: 84.70 kg

Table 4. Results of simulation done on AA 6063-T6.

O.D. \ Thickness	1mm	1.25mm	1.5mm
1 inch	FOS: 0.54 Stress: 1.27×10^{-2} N/m^2 Weight: 10.69 kg	FOS: 0.66 Stress: 1.22×10^{-2} N/m^2 Weight: 12.93 kg	FOS: 0.76 Stress: 7.51×10^{-3} N/m^2 Weight: 15.13 kg
1.5 inch	FOS: 1.2 Stress: 7.05×10^{-4} N/m^2 Weight: 15.38 kg	FOS: 1.5 Stress: 9.14×10^{-4} N/m^2 Weight: 18.80 kg	FOS: 1.8 Stress: 7.43×10^{-4} N/m^2 Weight: 22.17 kg
2 inch	FOS: 2.1 Stress: 2.56×10^{-4} N/m^2 Weight: 19.91 kg	FOS: 2.5 Stress: 2.6×10^{-4} N/m^2 Weight: 24.46 kg	FOS: 3 Stress: 1.51×10^{-4} N/m^2 Weight: 28.95 kg

Table 5. Results of simulation done on AA-6061-T6.

O.D. \ Thickness	1mm	1.25mm	1.5mm
1 inch	FOS: 0.69 Stress: 1.14×10^{-2} N/m^2 Weight: 10.69 kg	FOS: 0.85 Stress: 1.07×10^{-2} N/m^2 Weight: 12.93 kg	FOS: 0.98 Stress: 9.28×10^{-3} N/m^2 Weight: 15.13 kg
1.5 inch	FOS: 1.6 Stress: 8.28×10^{-4} N/m^2 Weight: 15.38 kg	FOS: 2 Stress: 1.11×10^{-3} N/m^2 Weight: 18.80 kg	FOS: 2.3 Stress: 4.24×10^{-4} N/m^2 Weight: 22.17 kg
2 inch	FOS: 2.7 Stress: 2.95×10^{-4} N/m^2 Weight: 19.91 kg	FOS: 3.2 Stress: 2.44×10^{-4} N/m^2 Weight: 24.46 kg	FOS: 3.9 Stress: 1.84×10^{-4} N/m^2 Weight: 28.95 kg

From the above inferences, it can be said that a vehicle with a maximum moving velocity of 60kmph can withstand an impact load of 20KN when material is selected appropriately considering the above results, AISI 4130 with dimensions of 1inch outer diameter and 1.25mm thickness is best suited for designing the roll cage.

4. Design Procedure

The design procedure is combined process of synthesizing older designs and developing new ideas. Initially, a basic prototype is considered for FEA and further development on the model is done based on the results of FEA. The structure is then tested to obtain some useful evidence to finalize the frame after synthesis. The members in the roll cage are used to distribute load across the frame under any impact and to transfer energy from the source. Also, the cage provides drivers safety in case of roll over. Roll Cage is modelled using Solid Works-2018 software (tab.6). The model is then subjected to various loading conditions by FEA. Based on the results derived from FEA, the roll cage is finalized. Appropriate material is selected to epitomize the roll cage. Various data regarding the material properties is entered.

Table 6. Different views of rollcage.

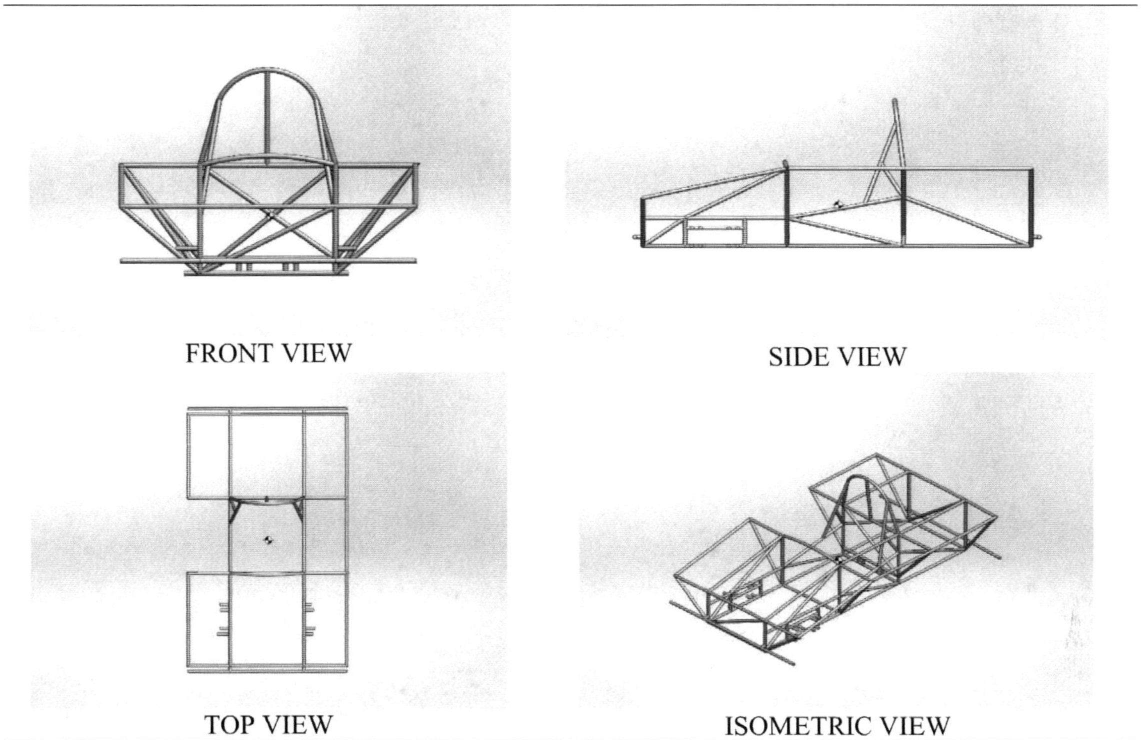

5. Finite Element Analysis

Using solid modelling and FEA in SolidWorks 2018, an optimized structure with maximum strength, maximum solar output and minimized weight is designed. Finite element of the cage is subjected to analysis which is of two main types:

5.1. Static Analysis

Under this analysis, chassis is subjected to various loads and impact tests i.e. front impact test, rear impact test, roll over test, torsional tests etc. for static conditions with suitable constraints as per different conditions. In static analysis, vehicle is subjected to maximum impact loads under worst conditions.

5.2. Dynamic Analysis

Under this analysis, vehicle is analysed under moving conditions. It is carried out for head-on collision as well as front impact, rear impact, side impact and roll over. Vehicle is tested for maximum moving velocity.

6. Calculation of various impact forces under static analysis

6.1. Front Impact

For front impact analysis, it is assumed that the car hits the wall at a moving velocity of 60kmph and comes to rest in 0.1 seconds by Oturkar Sania et.al. in [11] and Bharat Kumar Sati et. al. [12].
Using Work-Energy Theorem:
From eq. (1) and (2),

$$W_{net} = \frac{1}{2} m v_{initial}^2 \quad (5)$$

$$W_{net} = \frac{1}{2} \times 235 \times (16.66)^2$$

$$W_{net} = 32612.783\ N.m$$

Now,

$$W_{net} = F_{front} \times d \quad (6)$$

Also,

$$d = impact\ time \times v_{initial} \quad (7)$$
$$d = 0.1 \times 16.66$$
$$d = 1.66\ m$$
$$F_{front} = \frac{W_{net}}{d} \quad (8)$$
$$F_{front} = 19575.49\ N$$

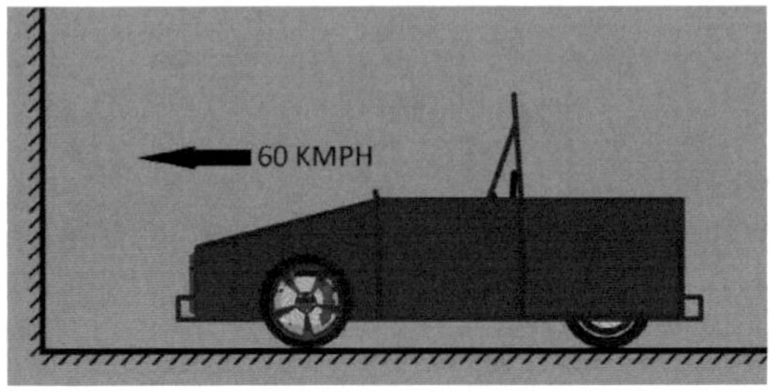

Figure 1. Car collision shown under frontal impact

6.2. Rear Impact

For rear impact analysis, it is assumed that a solar vehicle is hit by another moving solar vehicle. The vehicle comes to rest after 0.2 seconds by Bharat Kumar Sati et. al. [7]. Using Work-Energy Theorem in eq. (1) and (2),

$$W_{net} = \frac{1}{2} m v_{initial}^2$$
$$W_{net} = \frac{1}{2} \times 235 \times (16.66)^2$$
$$W_{net} = 32612.783\ N.m$$

Now,

$$W_{net} = F_{rear} \times d \quad (9)$$

Also,
From eq. (7),

$$d = 0.2 \times 16.66$$
$$d = 3.332\ m$$
$$F_{rear} = \frac{W_{net}}{d} \quad (10)$$
$$F_{rear} = 9787.75\ N$$

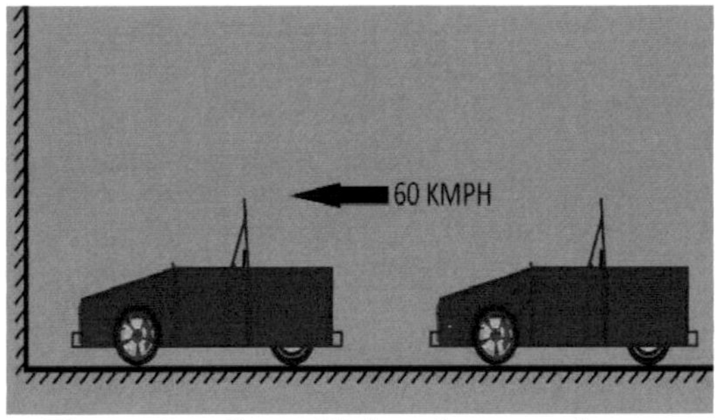

Figure 2. Car collision shown under rear impact.

6.3. Side Impact

For side impact analysis, it is assumed that another vehicle hits the vehicle from sides. The vehicle comes to rest after 0.2 seconds by Bharat Kumar Sati et. al. [8].
From Work-Energy Theorem:
From eq. (5),

$$W_{net} = \frac{1}{2} \times 235 \times (16.66)^2$$
$$W_{net} = 32612.783\ N.m$$

Now,

$$W_{net} = F_{side} \times d \quad (11)$$

Also,
From eq. (7),

$$d = 0.2 \times 16.66$$
$$d = 3.333\ m$$
$$F_{side} = \frac{W_{net}}{d} \quad (12)$$
$$F_{side} = 19575.49\ N$$

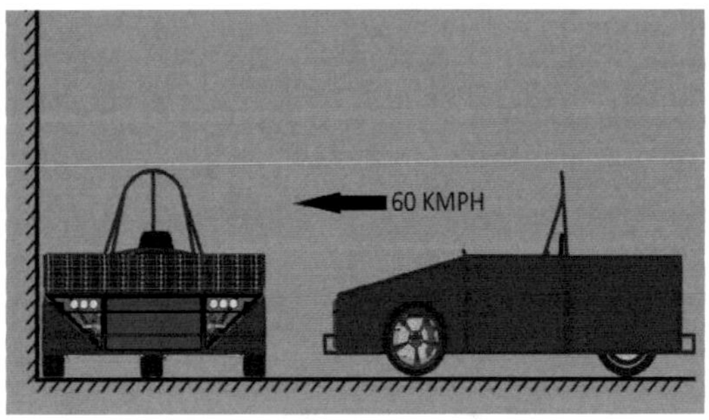

Figure 3. Car collision shown under Side Impact

6.4. Roll Over

For roll over analysis, car is considered to hit to the ground from a height of 3 meters. The roll over is tested for the strength of roll hoop. The vehicle comes to rest in 0.1 seconds by Bharat Kumar Sati et.al.[8].
During fall,

Potential Energy of the vehicle = Kinetic Energy of the vehicle

$$\frac{1}{2}mv^2 = m.g.h \tag{13}$$

On solving above equation,

$$v = \sqrt{2.g.h} \tag{14}$$

$$v = \sqrt{2 \times 9.81 \times 3}$$

$$v = 7.67\ m/sec$$

$$d = t \times v(15)$$

$$d = 0.1 \times 7.67$$

$$d = 0.76\ m$$

Now,
By Work-Energy Theorem,
From eq. (3)

$$W_{net} = \frac{1}{2} \times 235 \times (7.67)^2$$

$$W_{net} = 6912.32\ N.m$$

$$F_{roll-over} = \frac{W_{net}}{d} \tag{15}$$

$$F_{roll-over} = 9095.157\ N$$

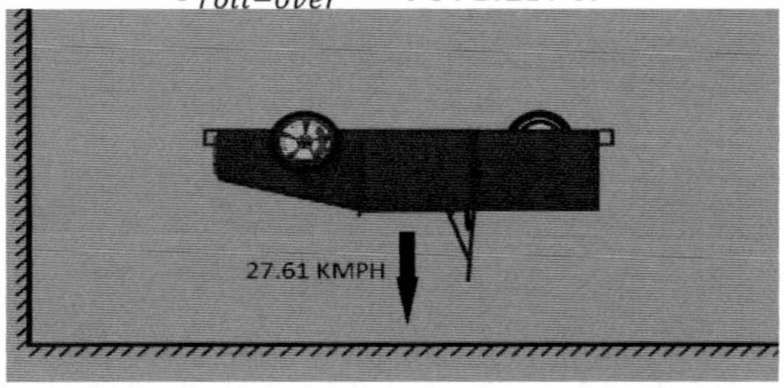

Figure 4. Car Collision shown under Roll Over

6.5. Force under torsion

For torsional analysis, acceleration is taken to be 2 times of the gravitational force (a=2.g).
Using Impulse-Momentum Theorem;

$$F_{torsional} = m.a \tag{16}$$

$$F_{torsional} = 235 \times (2 \times 9.81)$$

$$F_{torsional} = 4610.70\ N$$

7. Analysis of Chassis

7.1. Front Impact Analysis

For frontal impact we have fixed the geometry at the A-arm and swing arm mounting points. Load of 20KN is applied at the frontal crash zone of the vehicle. All the analysis done in SOLIDWORKS have been shown below (fig. 5-7):

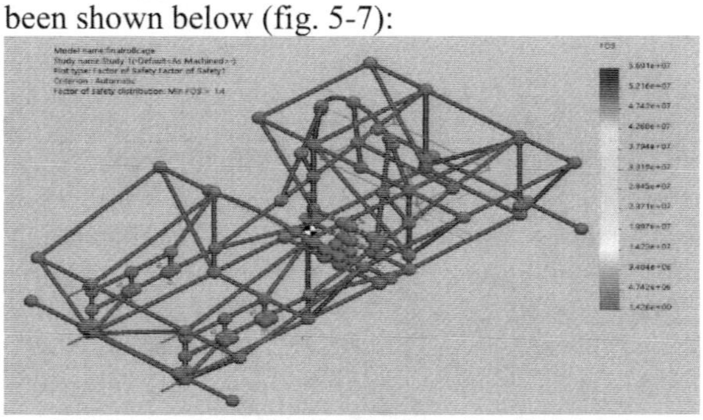

Figure 5. FOS (Front Impact Analysis).

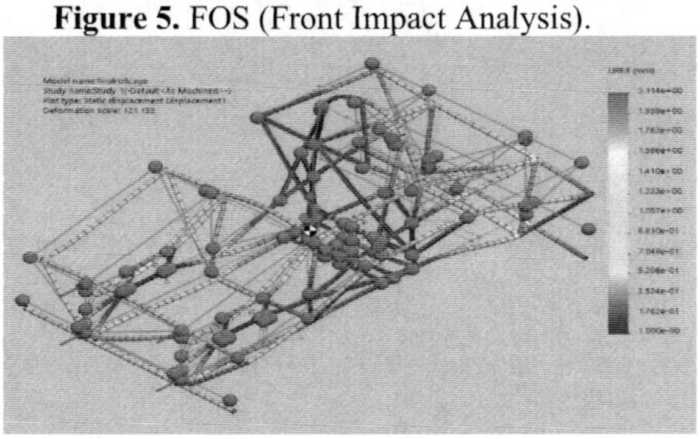

Figure 6. Displacement (Front Impact Analysis).

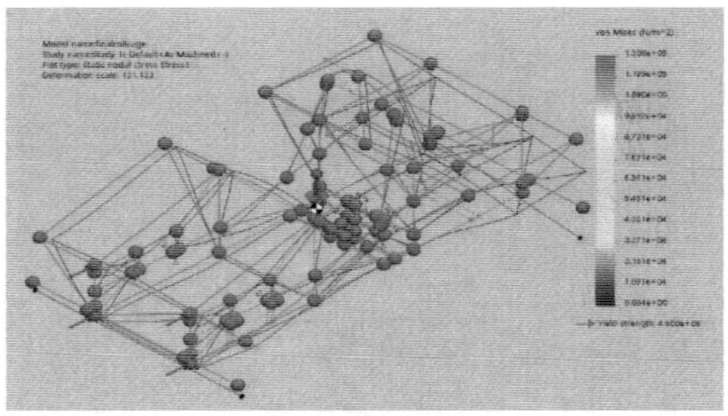

Figure 7. Stress (Front Impact Analysis).

The simulation shows below mention results:

Results for Frontal Impact Test

Min. FOS	1.4
Von Mises Stress	1.38×10^5 N/mm^2
Max. Displacement	2.14mm

7.2. Side Impact Analysis

For side impact we have constrained the opposite side of crash zone. Load is applied and the simulation done on SOLIDWORKS are shown as under (fig. 8-10):

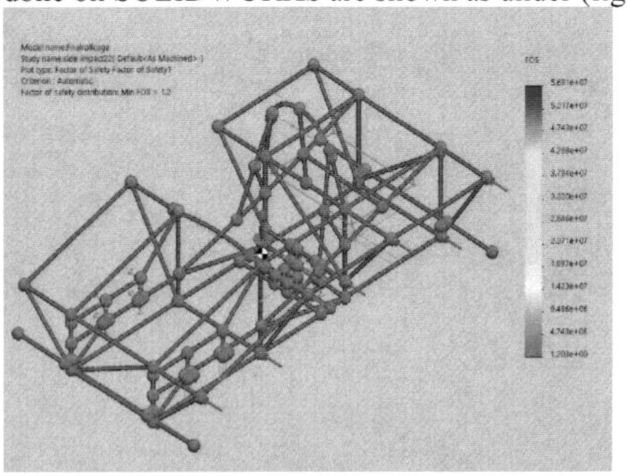

Figure 8. FOS (Side Impact Analysis).

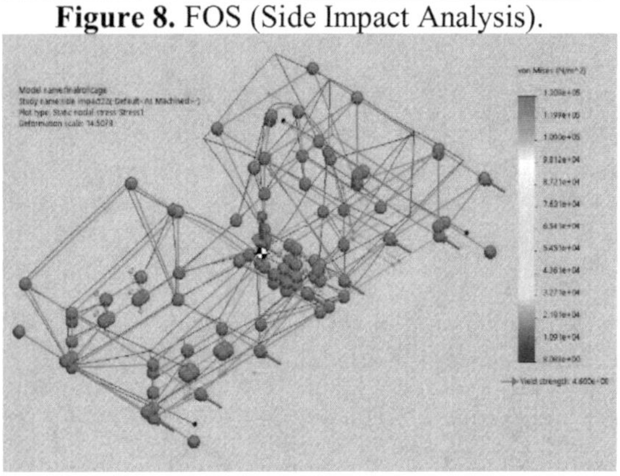

Figure 9. Stress (Side Impact Analysis).

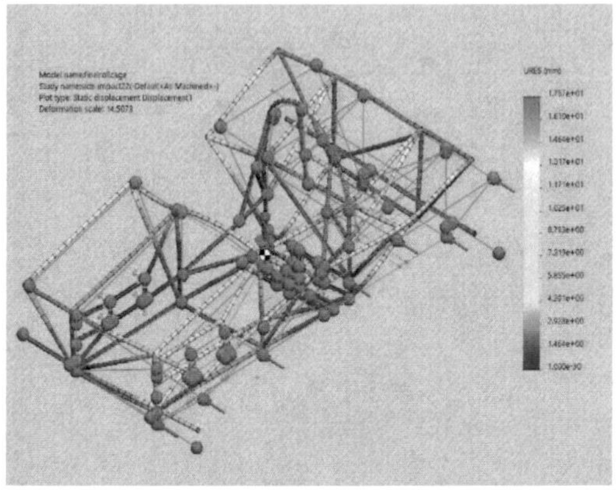

Figure 10. Displacement (Side Impact Analysis).

The results of simulation are shown as under:

Results for Side Impact Test	
Min. FOS	1.2
Von Mises Stress	1.308×105 N/mm2
Max. Displacement	mm

7.3. Roll Over Analysis

In this case the vehicle is assumed to be front rolling and all mounting points are fixed. Also, after rolling the vehicle tends to slide therefore a frictional force acting on the point of contact of the chassis and the ground is considered. Simulation done on SOLIDWORKS are shown as under (fig. 11-13):

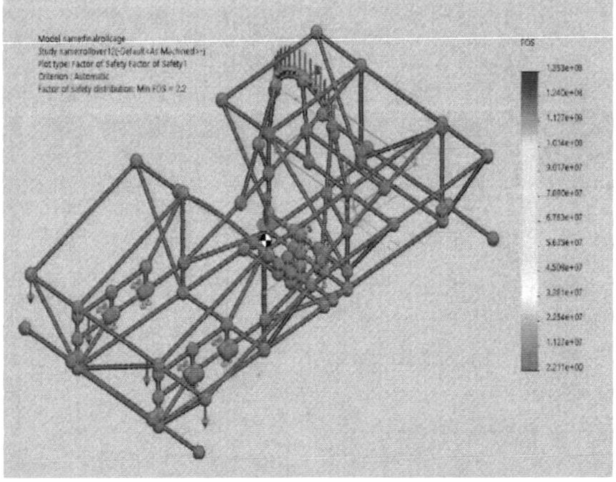

Figure 11. FOS (Roll Over Analysis).

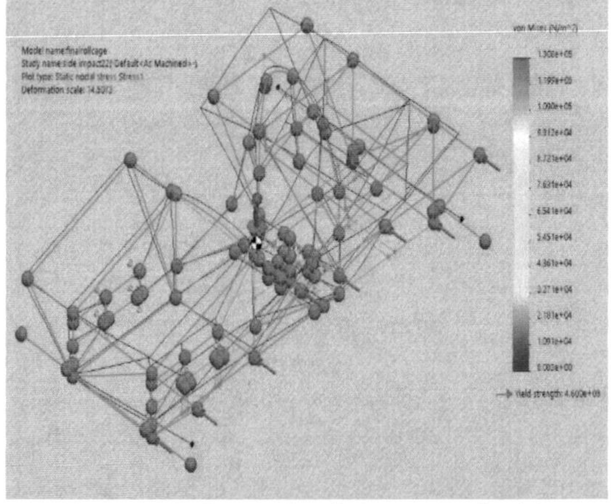

Figure 12. Stress (Roll Over Analysis).

IOP Conf. Series: Materials Science and Engineering **748** (2020) 012021 doi:10.1088/1757-899X/748/1/012021

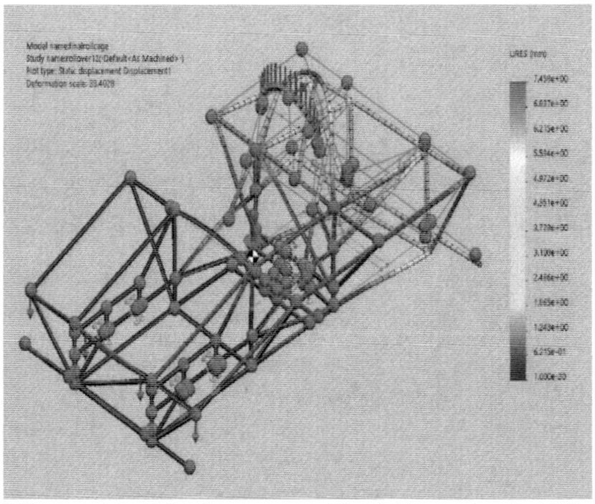

Figure 13. Displacement (Roll Over Analysis).

The results of simulation are shown as under:

Results for Roll Over Test

Min. FOS	2.2
Von Mises Stress	1.602×103 N/mm2
Max. Displacement	7.45 mm

7.4. Torsional Analysis

For torsional analysis, rear part of the chassis is fixed and load is applied conjugate on either side. The simulation done on SOLIDWORKS are shown as under:

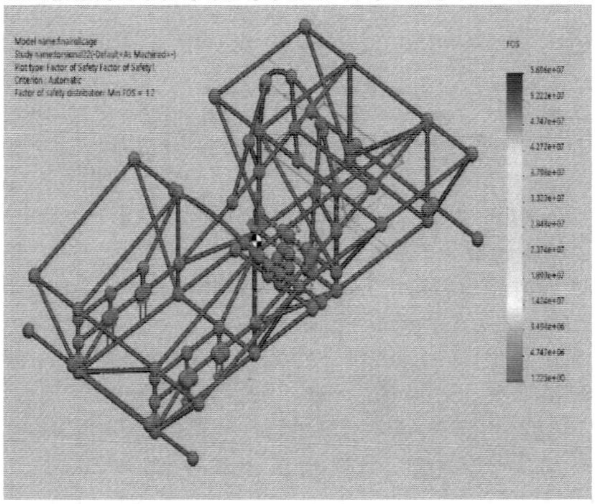

Figure 14. FOS (Torsional Analysis).

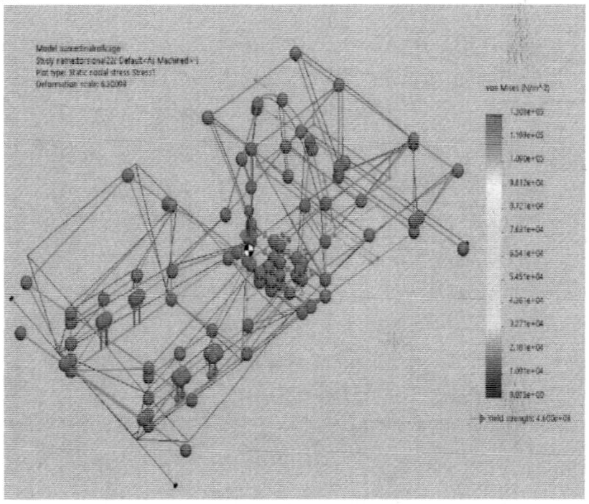

Figure 16. Stress (Torsional Analysis).

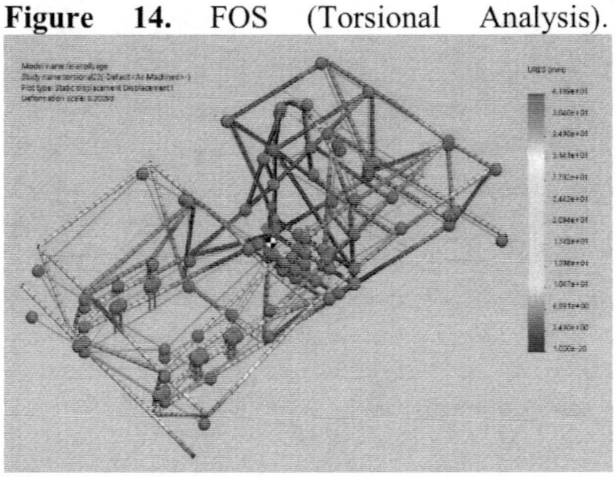

Figure 15. Displacement (Torsional Analysis)

The results of the simulation are shown as under:

Results for Torsional Test

Min. FOS	1.2
Von Mises Stress	1.308×105 N/mm2
Max. Displacement	mm

.

8. Results and Conclusion

After analysing the results, following inferences are derived:

- FOS is an important factor while deciding material for designing the roll cage.

- The pipe for manufacturing cage should be such that it can bear load under critical condition.
- The dimensions should be chosen appropriately such that it should not increase weight of the frame.

Considering FOS and pipe dimensions, we have concluded the use of AISI 4130 (chromoly) for manufacturing chassis with optimized dimensions with maximum strength and considerable weight. After simulation the FOS for the wireframe is observed to be 1.5 which is nominal and optimized. The dimension of the pipe is finalized with an iterative process where the chassis is analyzed under various pipe dimensions out of which pipe with 1inch outer diameter and 1.25 mm thickness is selected. Also, the cost of chromoly is less as compared to other considered materials.

9. References

[1] Brinkworth E, Jaggard D et al 2000 *Application of FEA Techniques to a Hybrid Racing Car Chassis Design SAE-2000-01-3538* Proc. of the 2000 Motorsports Engineering Conf. and Exposition, Dearborn, Michigan p-361.

[2] Naiju C D, Annamalai K, Nikhil Prakash et al 2012 Analysis of a Roll Cage Design against Various Impact Load and Longitudinal Torsion for Safety *Applied Mechanics and Materials (Trans Tech Publications Switzerland) Vol. 232* pp 819-822

[3] Riley W B, George A R Design, *Analysis and Testing of a Formula SAE Car Chassis SAE-2002-01-3300* Proc. of the 2002 Motorsports Engineering Conf. and Exposition, Indianapolis, Indiana p-382

[4] Ryan A 2008 *Formula SAE Race Car Analysis: Simulation and Testing of the Engine as a Structural Member* F2008 SC-005.

[5] Denny J, Veale K, Adali S, Leverone F 2018 Conceptual Design and Numerical Validation of a Composite Monocoque Solar Passenger Vehicle Chassis *Engineering Science and Technology* p 1067–1077.

[6] Das A K, Prusty P and Ojha M M S 2017 Design and Fabrication of a Solar Powered Vehicle and its Performance Evaluation *IOSR-JMCE Volume 14 Issue 5 Version I* p 65-68

[7] Nader N A, Ghoneim M and Alsayed R S 2015 Design and Building of Solar Race Car for the World Solar Challenge *Universal Journal of Mechanical Engineering Vol. 3(4)* p 122 - 130

[8] Sreeragh P S, Sowkath M, Bhuvanesh G, Mani P K 2018 Design and Fabrication of Car Chassis Frame and Hub Motor Driving System *International Journal of Pure and Applied Mathematics, Volume 119 No. 15* p 1757-1765

[9] Aru S, Jadhav P, Jadhav V, Kumar A, Angane P, 2014 Design, Analysis and Optimization of Multi Tubular Spaceframe *IJMPERD, Vol. 4, Issue 4* p 37-38.

[10] Sharma A, Singh J and Kumar A 2015 Optimum Design and Material Selection of Baja Vehicle *International Journal of Current Engineering and Technology Vol.5 No.3* E-ISSN p2277-4106, P-ISSN p 2347-5161

[11] S Oturkar, Gujarathi K 2013 An Introduction to Computational Frontal Static Stress Analysis of a Baja Car *International Journal of Scientific and Research Publications Vol. 3* p 1-5

[12] Sati B K, Upreti P, Tripathi A and Batra S 2016 Static and Dynamic Analysis of the Roll Cage for an All-Terrain Vehicle *Imperial Journal of Interdisciplinary Research (IJIR) Vol. 2 Issue 6*

IOP Conf. Series: Materials Science and Engineering **748** (2020) 012022 doi:10.1088/1757-899X/748/1/012022

Achieving Lean and Improving Sustainability through Value Stream Mapping for Complex Manufacturing

S Samant*[1] and R Prakash[2]

[1]Department of Mechanical Engineering, Amity School of Engineering and Technology, Noida, India Email: ssamant1@amity.edu *Corresponding author
[2]Ravi Prakash Amity Institute of Technology, Noida, Uttar Pradesh, India

Abstract
Lean manufacturing has helped companies to achieve reduced lean wastage, productivity, quality, and resource optimization. Value stream mapping (VSM) is an important tool in achieving lean orientation. It helps companies in enhancing the competitiveness of their production systems. Although, VSM is a very useful tool to achieve lean orientation, it suffers from several drawbacks. It is static pencil-and-paper tool. It cannot handle complex value streams. It does take into account monetary measures e.g. profitability of the value stream. Also, it does not include environmental impact of the value stream. This research paper overcomes the drawbacks of the VSM tool through an innovative framework involving the usage of discrete event simulation, box score, and life cycle assessment (lca).

1. Introduction

The manufacturing companies are today facing tremendous global competition. Lean manufacturing helps companies in enhancing their competitiveness by reducing costs and waste minimization,[1] The activities that are wasteful in a production system have to be minimized systematically.[2] There are many tools to achieve lean orientation e.g. standardized work, poka-yoke, 5S, VSM. VSM is an important tool to achieve lean orientation. VSM, however, suffers from several limitations:

- VSM is a static tool. It provides only a static snapshot of the production system.
- VSM cannot easily handle complex production scenarios. VSM can easily handle simple single product-based manufacturing situations. However, complex production scenarios involving multi-level bills of materials (BOM), or multi-product product situations cannot be easily handled.
- VSM does not take into account monetary measures within the value stream e.g. costs, profitability etc.
- VSM does not take into account environmental impact associated with the value stream.

This research paper overcomes the limitations of the VSM through some of the following counter-measures (Refer to the table 1):

Table 1: Limitations of VSM

Limitation of the VSM	Counter-measure employed in this research paper
VSM is a static tool.	• VSM is a static tool which only provides a static snapshot of the manufacturing situation. This limitation is overcome through the use of discrete event simulation software Extendsim.
VSM cannot easily handle complex production scenarios.	• VSM cannot easily handle complex production scenarios e.g. multi-level BOM, multi-product

	manufacturing situations. This limitation is overcome through the use of discrete event simulation software Extendsim.
VSM does not take into account monetary measures with the value stream	• The accounting framework involving the use of monetary measures such as profitability has been included through the application of lean box score.
VSM does not take into account environmental impact of the value stream.	• VSM does not include the environmental impact of the value stream.

The table 2 given below compares the traditional VSM process with the proposed framework.

Table 2: Comparison of conventional VSM with the proposed methodology

Requirements	**Conventional VSM**	**Proposed Approach**
Static Approach	●	●
Application to complex production streams	○	●
Dynamic Approach	○	●
Finding optimal level of resources (man and machine resources)	○	●
Insight into financial performance.	○	●
Prioritization of kaizen continuous improvement Initiatives	○	●
Ability to perform feasibility check	○	●
Achieve global optimization	○	●
Environmental impact of the value stream	○	●

One of major handicaps of the VSM process is that it does not include environmental impact of the value stream. The proposed framework uses a discrete event simulation model, combined with lean box score, and lca. The proposed methodology is demonstrated for a lamp post manufacturer case.

2. Literature review

Lean manufacturing emerged from Toyota Motor Corporation. This philosophy was called Toyota Production System. [3,4] The production system is organized and oriented in such a manner that eight lean wastes (Muda) are minimized. VSM is one of the most important tools for achieving lean orientation. VSM as a tool was introduced for achieving lean orientation.[5] VSM is very powerful tool for achieving lean manufacturing. However, VSM has certain limitations (refer to table 1, and table 2). Researchers have suggested many approaches for overcome these limitations. The approaches include heuristics, discrete event simulation *etc.* Some of the selected research works have been summarized in the table 3.

Table 3: Literature Survey

Author(s)	**Contribution**
Khaswala and Irani [6]	Suggested Value network mapping (VNM) approach for complex flows.
Braglia, Carmignani, and Zammori [7]	Suggested Improved VSM approach (IVSM) approach for Complex production streams.

Author(s)	Contribution
McDonald et al. [8]	Using Discrete event simulation (DES) for lean for Engineering-to-order (ETO) products.
Lian and Landeghem [9]	Proposed DES for lean.
Abdulmalek and Rajgopal [10]	Proposed VSM combined simulation (for steel mill – process industry))
Seth et al. [11]	Proposed, simplifications, and approximation in VSM. .
Alperen Bal, Cemil Ceylan & Caner Taçoğlu [12]	Combined VSM with DES for a emergency unit.
Braglia, Frosolini, Zammori [13]	Suggested fuzzy algebra with VSM
Parthanadee and Buddhakulsomsiri [14]	Demonstrated achieving lean through DES.
Samant, Mittal and Prakash [15]	For a chassis production system, suggested the use of VSM combined with DES.
Tapping et al [16]	Step by step method for achieving lean using value stream map (VSM).
Faulkner and Badurdeen [17]	Proposed metrics for environmental performance. Suggested methodology for sustainable value stream mapping.
Vinodh et. al. [18]	Proposed a methodology for integrating VSM with LCA.
Torres and Gati [19]	Proposed a managerial tool Environmental VSM (EVSM) to Address environmental and economic aspects of a production system.
Paju et. al. [20]	Proposed an assessment based on VSM. The framework includes VSM, DES, LCA etc.

Research Gap and Aim of Research:
From table 3, it is clear that achieving green and lean simultaneously through VSM is a research topic that has not been explored completely. There is a room for an alternate framework for achieving lean orientation for the value stream. It is difficult to achieve lean orientation, especially for complex value streams using traditional VSM approach.

3. Methodology

To overcome these limitations of VSM mentioned in section 1, this research paper proposes a framework which integrates LCA with VSM. It also integrates lean box score and environmental metrics into the framework. The main focus of this research is to study the effect of environmental impact of the value stream. This research also helps in our endeavor to have a lean and green production system.

3.1 Case Study

The framework has been explained using a case study. This case study pertains to a manufacturer of lamp post (refer figure 1) This product is currently in design/prototyping stage. The idea here is that it may be prudent to take up lean manufacturing at the design stage, because most of capex decisions are taken early in the design cycle. This is in line with concurrent design principles. Most of the available literature on lean manufacturing studies focusses on cases where full scale production is taking place. This has limited usefulness as all the capex decisions have already been made.

Figure 1: lamp post base

Any organizational improvement effort should begin by jotting down organizational priorities. Organizational Priorities are given below:

- Meet the production target of 2000 pieces per week
- Meet the quality target: First time right 99%
- Timely delivery: meet the on-time-delivery target of 99%
- Minimize Production Cost
- Meet the revenue target: $32,000 per week
- Minimize environmental impact of the value stream.

At this point, the process flow diagram is created detailing the sequential representation of the manufacturing processes used for manufacturing the product. The process flow diagram is shown in figure 2.

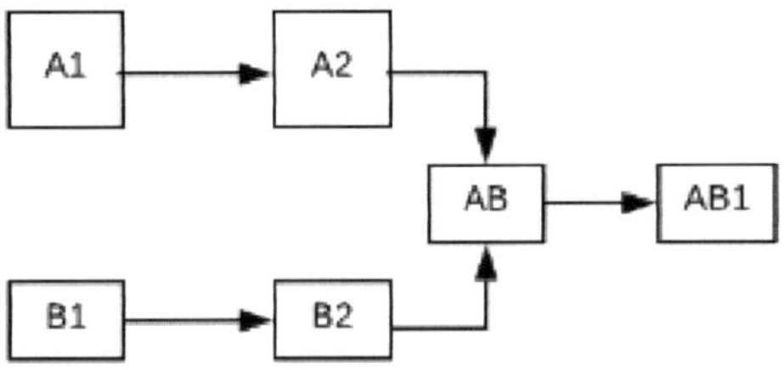

Figure2: Process flow diagram

The process parameters for the processes shown in the process flow diagram is shown in table 4.

The current process parameters (estimated) during initial stage (current value stream map) are given in table 4.

Table 4: Process Parameters

Process Name	Process Code (refer figure 2)	Cycle Time (minutes)	Set-up Time (min)	Transfer Batch Size (initially assumed value)	No of Machines	No of Operators	Weekly Depreciation Cost (USD)	Operating Cost (per operation) (USD)
Welding	AB	NORMAL (2, 0.2)	6	5	9		11.5	0.068
Painting	AB1	NORMAL (30, 4.5)	7	5	7		5.8	0.036
Deburring	B2	NORMAL (10, 1.5)	2	5	6		1.9	0
Drilling	A2	NORMAL (30, 4.5)	4	5	11	40*	11.5	0
Blanking	A1	NORMAL (2, 0.2)	4	5	5		96.2	0
Tube Shearing	B1	NORMAL (2, 0.2)	6	5	5		2.2	0

*Shared between operations; Cost of operator $99 (per week); Number of operators 25; Number of trolleys 25

The process parameters will be used for achieving lean orientation. This has been detailed in Annexure – A.

At this time, we must try to carry out life cycle assessment (LCA) of the products constituting the assembly. Life cycle assessment (LCA) is multi-step process for computing the environmental burden of a product/service. Typically, it includes the following stages:
a) goal and scope
b) inventory analysis
c) impact assessment
d) interpretation.

Goal and Scope: Some of key parameters that are defined during the goal and scope stage are discussed below:

- Functional unit: The functional unit for the LCA is a single product.
- Goal and Scope: Goal: The goal of LCA is to carry out environmental impact assessment of all the products constituting the assembly in all stages of the product life cycle (cradle to grave).
- System boundary: The system boundary includes all the upstream and downstream processes. The system includes the technosphere (economic system). Activities in the technosphere cause undesirable effects in the ecosphere.

Life Cycle Inventory (LCI): The life cycle inventory represents quantified flows (both mass and energy flows) representing unit processes. The flows (mass flows) and the processes used are given in table 5.
Life Cycle Impact Assessment (LCIA): In this step the environmental impact analysis due to LCI is carried out in order to find the Life Cycle Environmental Impact (LCIA). LCIA results are shown in table 6. The tool used here is lcacalculator.com.

Table 5: Life Cycle Inventory

Stage	Inventory List (for various processes)	Quality	Units
Raw Material Extraction	**Raw Material Extraction**		
	Plate	10	Kg
	Tube	16	Kg
Transport of Raw Materials	**Transport of Raw Materials**		
	Transport Freight Lorry (3.5 - 7.5) T(Plate and Tube)(Kg)	20	Km
Manufacturing	**Plate**		
	Stamping / Laser cutting		
	Machining (25%)		
	Welding	*	
	Tube		
	Hot Rolling		
	Machining (25%)	*	
Transport of Finished Goods	**Transport of Finished Goods (T8)**		
	Transport Freight Lorry (3.5 - 7.5) T(Assembly Weight Kg)	150	Km
Disposal	**Disposal (T9)**		
	Lamppost Assembly		

IOP Conf. Series: Materials Science and Engineering **748** (2020) 012022 doi:10.1088/1757-899X/748/1/012022

(* The software works the quantities for the unit processes internally)

Table 6: Life Cycle Impact Assessment (Global Warming Kg CO_2)

Stage	Inventory List (for various processes)	Quality	Units
Raw Material Extraction	**Raw Material Extraction**		
	Plate	24	Kg
	Tube	35	Kg
Transport of Raw Materials	**Transport of Raw Materials**		
	Transport Freight Lorry (3.5 - 7.5) Tonne (Plate and Tube) (Kg)	0.25	Kg
Manufacturing	**Plate**		
	Stamping / Laser cutting	4.2	Kg
	Machining (25%)	15	Kg
	Welding	0.6	Kg
	Tube		
	Hot Rolling	7.6	Kg
	Machining (25%)	24	Kg
Transport of Finished Goods	**Transport of Finished Goods (T8)**		
	Transport Freight Lorry (3.5 - 7.5) T(Assembly Weight Kg)	1.9	Kg
Disposal	**Disposal (T9)**		
	Lamppost Assembly	0.1	Kg

A cumulative assessment of the degree of lean orientation, as well as the environmental impact can to be carried out. This research paper proposes the value stream map in conjunction with lean and green box score (as the accounting framework for this cumulative assessment) for a comprehensive evaluation of the lean orientation, as well as the environmental impact of the value stream. The value stream map is shown in figure 3. This VSM differs from the traditional VSM in a sense that environmental impact trend lines have been added at the bottom of the VSM. The box score provides insights into the value stream from the financial, resource capacity, operational standpoint.

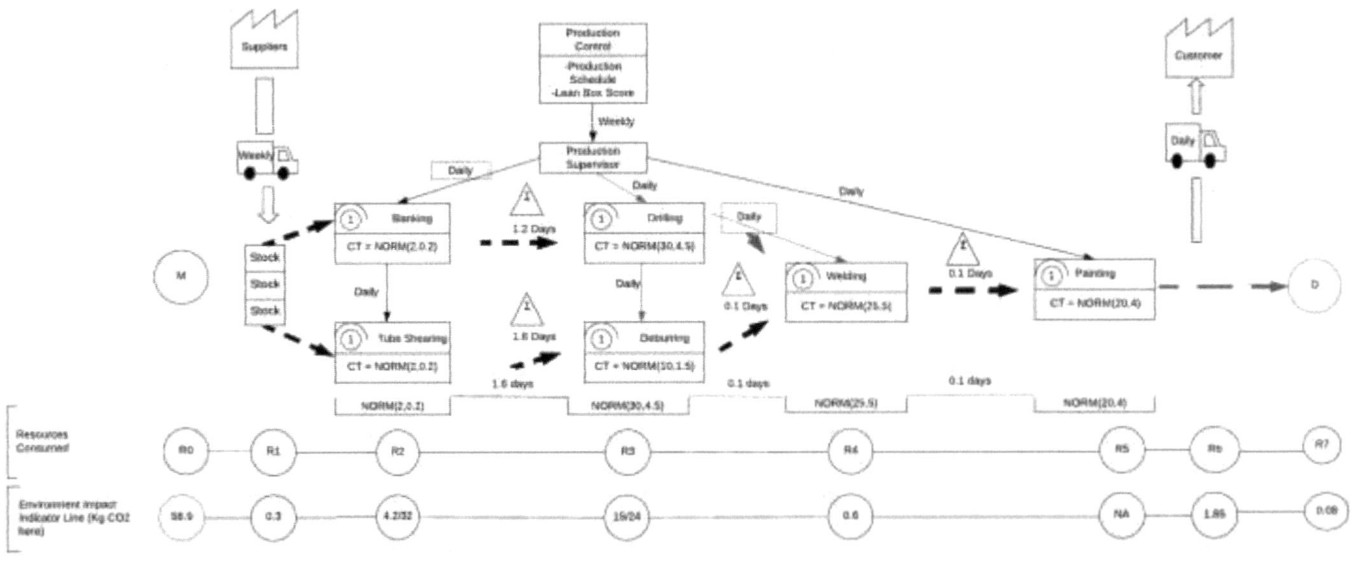

Figure 3: Value stream map

4. Results and Discussions

The lean box score prior to implementation of the kaizens is given in table 7.

Table 7: The lean and green box score

Lean Box Score		
Operational	Dock-to-Dock Days	4.25 days
	First Time Through (FTT)	90%
	On Time Delivery/Service Level	100%
	Floor Space	-
Utilization	Productive	[0.6, 0.6, 0.4, 0.6, 0.1, 0.1]
	Non-productive	[0.03, 0.04, 0.02, 0.02, 0.04, 0.04]
	Available	[0.4,0.3,0.6,0.4,0.9,0.9]
Financial	Sales per person (USD)	320
	Inventory Value	412
	Revenue	8000
	Material Costs	5000
	Conversion Costs (USD)	3308
	Value Stream Profit (USD)	-720
	Sales per Product (USD)	16
Environmental	Climate Change (KgCO2)	113

The evaluation provides key insights into the value stream, and helps in coming up with the kaizen continuous improvement ideas. The following kaizen improvement ideas have been identified: The continuous improvement efforts (kaizens) can be towards achieving lean orientation (lean kaizens), or towards reducing environmental impact (environmental kaizens). The lean and the environmental kaizens have been summarized in table 8.

Table 8: The proposed Kaizen improvement initiatives

Lean Kaizens		
Kaizen No	Description of Kaizen	Implementation
1	Reduce Conversion Cost	- Resource Optimization: it can be seen that the conversion cost contributes maximum towards the overall cost. So, the reduction of conversion cost deserves maximum attention. Conversion cost can be reduced through resource optimization. This includes both man and machine resources. Resource optimization can be carried out using the Discrete Event Simulation software. Extendsim (evolutionary optimizer) has been used here. A more detailed discussion of resource optimization has been discussed in Appendix A.
	Reduce environmental burden associated	- There is an environmental burden associated with the value stream. The environmental burden associated with the value stream is listed in table 6. The environmental burden has been calculated using online tool (lcacalculator.com (based on Ecoinvent database)). Table 6 gives a comparison of environmental impacts during various stages

2	with the value stream	of the product life cycle. This environmental impact can be reduced using the following steps: o From figure 4 it can be seen that raw material extraction is the major contributor to the environmental impact. This can be reduced considerably by increasing the recycled content in the raw materials. o From figure 5 it can be seen that of all the manufacturing processes, machining process is the next biggest contributor to environmental burden, Machining process adds considerably to the carbon emissions. From figure 5 it can be seen that by reducing machining can reduce carbon emissions considerably. The product designers should try to focus on reducing the machining process as much as possible.

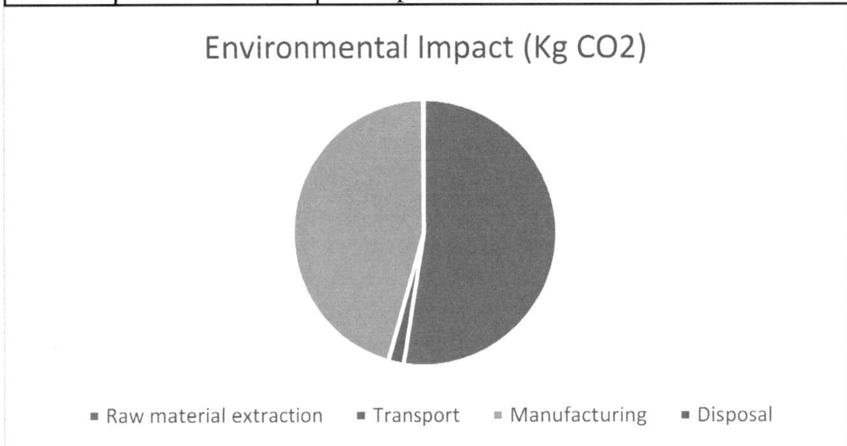

Figure 4: Breakup of Environmental Impact due various life cycle stages of product

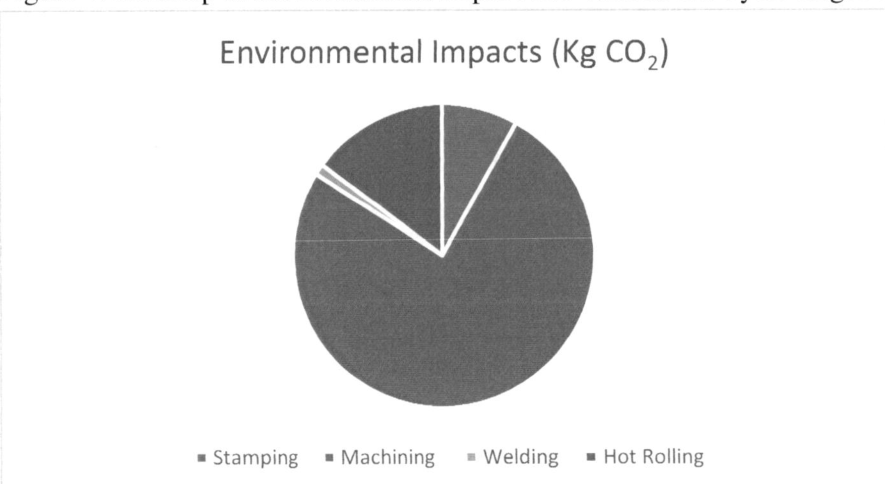

Figure 5: Breakup of Environmental Impact due various manufacturing processes

The lean box score results achieved through evolutionary optimizer are given below:

Table 9: Lean box score (future state)

Lean Box Score

Operational	Dock-to-Dock Days	5
	First Time Through (FTT)	90%
	On Time Delivery/Service Level	100%
	Floor Space	-
Utilization	Productive	[0.9, 0.8, 0.7, 0.9, 0.4, 0.4]
	Non-productive	[0.04, 0.06, 0.03, 0.02 0.17, 0.20]
	Available	[0.1,0.1, 0.3, 0.1, 0.4, 0.4]
Financial	Sales per person (USD)	320
	Inventory Value	317
	Revenue	8000
	Material Costs	5000
	Conversion Costs (USD)	1631
	Value Stream Profit (USD)	1052
	Sales per Product (USD)	16
	Cost Per Product (USD)	14
Environmental Impact		40% reduction in Kg CO2 emissions can be targeted (approximately)

The proposed methodology is very useful for achieving lean orientation for the complex value stream. It helps to optimize on the relevant parameters of the complex VSM. It helps to focus and prioritize the kaizen improvement ideas. It helps to check feasibility of the future state VSM through the optimization of lean box score.[21] Some of the unique advantages of this innovative approach:

a) Ability to achieve lean orientation for a complex value stream
b) Unique plug and play and approach in modeling the value stream made up of standardized components
c) Ability to prioritize kaizen improvement ideas
d) Ability to perform feasibility checks, i.e, whether a certain kaizen improvement idea is feasible. The parts can be produced with the available resources at the required performance level.

5. References

[1] Hines P and Taylor D, 2000 Going lean *Cardiff, UK: Lean Enterprise Research Centre Cardiff Business School* pp3-43

[2] Womack J P and Jones D T 1996 Beyond Toyota: how to root out waste and pursue perfection *Harvard business review* 74(5) pp140-158

[3] Ōhno Taiichi 1988 Toyota production system: beyond large-scale production. Boca Raton, FL: CRC Press, Taylor & Francis Group

[4] Monden Y 2015 Toyota Production System: An Integrated Approach to Just-In-Time Boca Raton: CRC Press

[5] Rother M and Shook J 2003 Learning to see: value stream mapping to add value and eliminate muda *Cambridge MA: Lean Enterprise Institute*

[6] Khaswala Z N and Irani S A 2001 Value network mapping (VNM): visualization and analysis of multiple flows in value stream maps *In Proceedings of the Lean Management Solutions Conference* (pp. 1-18)

[7] Braglia M, Carmignani G and Zammori F 2006 A new value stream mapping approach for complex production systems *International Journal of Production Research* **44** (18-19) 3929–3952

[8] Mcdonald T, Aken E M V and Rentes A F 2002 Utilising Simulation to Enhance Value Stream Mapping: A Manufacturing Case Application *International Journal of Logistics Research and Applications* **5** (2) 213–232

[9] Lian Y H and Van Landeghem H 2002 An application of simulation and value stream mapping in lean manufacturing *In Proceedings 14th European Simulation Symposium* (pp 1-8) c) SCS Europe BVBA

[10] Abdulmalek F A and Rajgopal J 2007 Analyzing the benefits of lean manufacturing and value stream mapping via simulation: A process sector case study *International Journal of Production Economics* **107** (1) 223–236

[11] Seth D, Seth N and Dhariwal P 2017 Application of value stream mapping (VSM) for lean and cycle time reduction in complex production environments: a case study *Production Planning & Control* **28** (5) 398–419

[12] Bal A, Ceylan C and Taçoğlu C 2017 Using value stream mapping and discrete event simulation to improve efficiency of emergency departments *International Journal of Healthcare Management*, **10**(3) pp 196-206

[13] Braglia M, Frosolini M and Zammori F 2009 Uncertainty in value stream mapping analysis. *International Journal of Logistics Research and Applications* **12** (6) 435–453

[14] Parthanadee P and Buddhakulsomsiri J 2012 Production efficiency improvement in batch production system using value stream mapping and simulation: a case study of the roasted and ground coffee industry *Production Planning & Control* 25 (5) 425–446

[15] Samant S, Prakash R and Mittal V K 2018. Resource optimisation for an automobile chassis manufacturer through value stream mapping enhanced with simulation technique and constraint programming *International Journal of Industrial and Systems Engineering* **28** (3) 379

[16] Tapping D, Luyster T and Shuker T Value Stream Management: Eight Steps to Planning, Mapping and Sustaining Lean Improvements CRC Press

[17] Faulkner W and Badurdeen F 2014 Sustainable Value Stream Mapping (Sus-VSM): methodology to visualize and assess manufacturing sustainability performance *Journal of Cleaner Production* **85** 8–18

[18] Vinodh S, Ruben R B and Asokan P 2015 Life cycle assessment integrated value stream mapping framework to ensure sustainable manufacturing: a case study *Clean Technologies and Environmental Policy* **18** (1) 279–295

[19] Torres Jr AS and Gati AM 2009 Environmental value stream mapping (EVSM) as sustainability management tool *In PICMET'09-2009 Portland International Conference on Management of Engineering & Technology* (pp 1689-1698) IEEE

[20] Paju M, Heilala J, Hentula M, Heikkila A, Johansson B, Leong S and Lyons K 2010 Framework and indicators for a Sustainable Manufacturing Mapping methodology *Proceedings of the 2010 Winter Simulation Conference*

[21] Maskell B H, Baggaley B and Grasso L 2011 Practical lean accounting: a proven system for measuring and managing the lean enterprise Boca Raton, FL: CRC Press

IOP Conf. Series: Materials Science and Engineering **748** (2020) 012022 doi:10.1088/1757-899X/748/1/012022

6. Annexure - A

The resource optimization is carried out by Extendsim optimizer. The extendsim optimizer is an evolutionary optimizer. This optimizer is an integral part of the software Extendsim (discrete event simulation software). The optimizer is based on evolutionary algorithm. It mimics the common biological evolutionary processes such as mutation, recombination and selection.

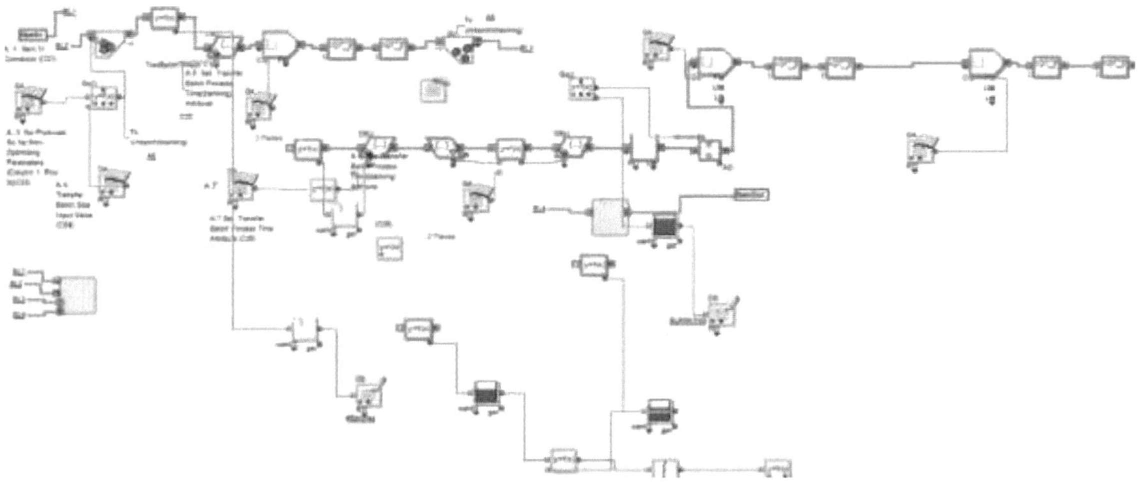

Figure 6: Extendsim simulation model (for a single process)

Selected parameters from the optimization model shown in figure 8 are linked to optimizer. The optimizer seeks the desired goal. The optimizer block within Extendsim is presented in Figure 8. Top part of the optimizer window stores the parameters to be optimized. These parameters are derived from simulation model. These parameters are linked to the parameters within the simulation model. The objective function is stored in the bottom part of the window. The simulation model is run multiple times with different values of the parameters. The evolutionary optimizer tries to come up with an optimal solution with favorable values of the lean box score (objective function). Here the lean box score parameter: value stream profit is the objective function. The decision variables are the resource (man and machine) whose optimal values are being sought.

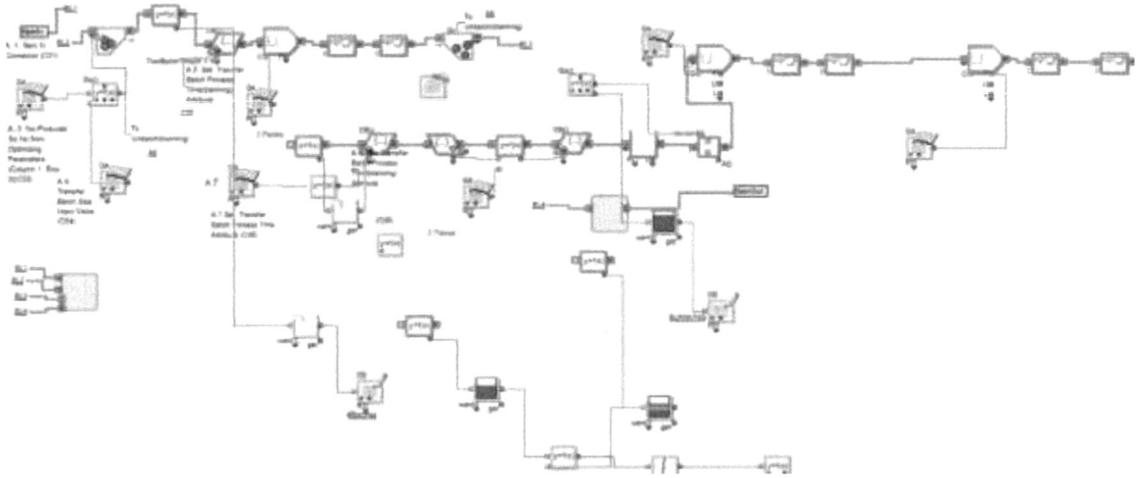

Figure 7: Extendsim simulation model (for a single process)

Selected parameters from the optimization model shown in figure 8 are linked to optimizer. The optimizer seeks the desired goal. The optimizer block within Extendsim is presented in Figure 8. Top part of the optimizer window stores the parameters to be optimized. These parameters are derived from simulation model. These parameters are linked to the parameters within the simulation model. The objective function is stored in the bottom part of the window. The simulation model is run multiple times with different values of the parameters. The evolutionary optimizer tries to come up with an optimal solution with favourable values of the lean box score (objective function). Here the lean box score parameter: value stream profit is the objective function. The decision variables are the resource (man and machine) whose optimal values are being sought.

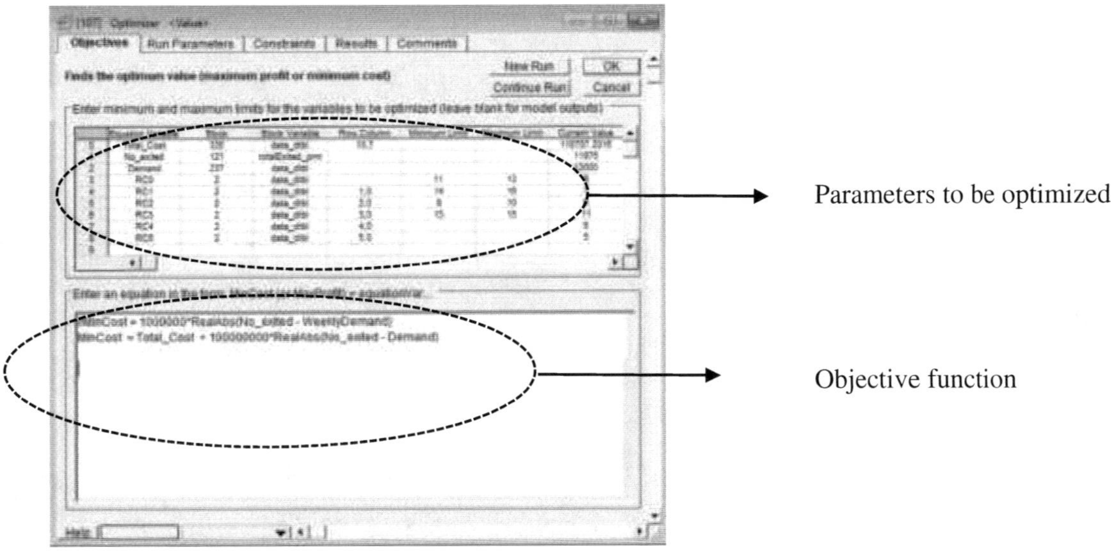

Figure 8: Extendsim evolutionary optimizer

Optimization model:

$$\text{Value Stream Profit} = \left(\textstyle\sum_1^n(\text{MiSi}) \;-\; \left(\sum_1^k(\text{Hm Rm}) + \sum_1^t(\text{Ho Oo})\right)\right);$$

Mi = Number of pieces of product sold of type i
Si = Price of product i
Hm = Number of hours resources a resource of type m employed
Rm = Resource rate of type m
Ho = Time of operations of type o
Oo = Operation cost of type o

After the optimization, we are one step closer to lean orientation. The lean box score are shown in table 9. The resource optimization leads to line balancing as well.

IOP Conf. Series: Materials Science and Engineering **748** (2020) 012023 doi:10.1088/1757-899X/748/1/012023

Effect of microcapsules addition on impact properties of carbon fiber reinforced polymer and glass fiber reinforced polymer composites

Kanu Priya Jhanji[1] , Asokan R[2], Amit Kumar R[3], Shivanshu Sarkar[4], Mahathi Guptha K[5] and Sathish Kumar[6]

[1,2,3,4,5,6] School of Aeronautical Sciences, Hindustan Institute of Technology and Science, 603103, Chennai, India

Email: *kanujhanji@gmail.com

Abstract: Fiber reinforced composites are ubiquitous in the aerospace and automobile industries due to their significant superior characteristics such as strength to weight ratio, corrosion resistance, fatigue resistance, specific strength and specific modulus. Impact is common kind of damage found in aircraft and automobile structures. The cause of damage may be sudden strike from a bird, falling of tool during servicing or striking of debris during transportation. Hence, predicting the behavior of composites under impact is necessary and also some mechanism to repair these damages should be found. So, this work aims at determining the effect of inclusion of microcapsules containing healing agents on impact behavior of Fiber Reinforced Polymer (FRP) composite. The resin stuffed and hardener stuffed microcapsules were prepared by in-situ polymerization. These capsules of diameter ranging from 0.19 μm to 0.30 μm were mixed in epoxy resin system during the manufacturing of FRP laminates. Experimental investigation was performed under ambient conditions to observe the variation in impact properties of Carbon Fiber Reinforced Polymer (CFRP) and Glass Fiber Reinforced Polymer (GFRP) composites. It was observed that value of peak load was boosted with addition of capsules and more energy was absorbed during elastic deformation and less during plastic deformation. Also, partial healing was observed due to addition of capsules at 2.5 wt. %.

Keywords: self-healing, low velocity impact, composites, healing agents, microcapsules.

1. Introduction

FRP composites are applied in structures of many components that require use of advanced materials to satiate the demand of superiority in multiple properties through a single material. Such components are from industries like marine, civil, aerospace etc. The FRP composite materials are formed by combination of polymer matrix and fiber reinforcements. The matrix transfers the load to reinforcements and reinforcements aligned in various directions carry load in corresponding directions [1-5]. The structures made from such materials are anisotropic in nature, so damage detection, its prediction and its repair are highly complicated. It not only requires skilled professionals and advanced machineries but also demands a lot of time and effort.

Damage occurring due to impact by any foreign matter in case of composite structures can lead to their failure that might be or might not be visible [6-8]. Parts with detected damages can be repaired or replaced. But if the damage remains undetected, it can result in catastrophic failure of structure and that can be a threat to the personnel working around the structure. So, the behavior of composites under impact has been investigated by many researchers. Impact energy and specimen thickness plays a critical role in composites subjected to impact. Composite undergoes lower damage at lower impact energy and damage increases with the increase in impact energy. Plates with higher thickness have higher resistance to impact and they also redistribute the loads better than plates with lower thickness [9]. Many factors like fiber volume fraction, exposure temperature and exposure time can also influence impact resistance. Due to variation in fiber volume fraction and exposure temperature, the increase or decrease in impact strength of composite mainly depends on the source of failure i.e. fiber or matrix [10]. Many researchers included additives like CNTs, CNF and MCWNTs in required amount in the matrix system to improve the impact resistance of the composites [11, 12, 13]. Addition of nanofillers in the resin system requires proper dispersion to provide better properties. Methods like ultrasonication can be used for obtaining such fine dispersion [14]. The inclusion of Shape Memory Alloy wires in basalt fiber reinforced composite improved the impact resistance due to hybridization [15].

Most of the research is focused on improving impact resistance. But once the damage has occurred, providing permanent solutions is necessary. Traditional way of detecting damages and repairing them requires a lot of time and skill and is also a costly affair. So, many researchers focused on finding new solutions like implementation of self-healing technology in composites. Such materials can restore their original properties by healing themselves. Self-healing can be included in FRP composites through various techniques. These techniques can be subdivided into two groups i.e. intrinsic and extrinsic healing. Intrinsic self-healing completes by reversibility of cross-linking of bonds whereas extrinsic self-healing uses some capsules or vascular network that breaks by damage and as a result, releases the encapsulated healing agents [16-18]. Inclusion of such capsules can modify the properties of both epoxy system as well as composite materials. Such capsules containing healing agent and curing agent when dispersed into the resin system to produce glass fiber epoxy composite subjected to damage caused by indentation resulted in 80% recovery of the indented zone [19]. This work has been done to understand the change in behavior of CFRP and GFRP laminate under impact load upon inclusion of microcapsules in them during their fabrication. Laminates without microcapsules were also prepared for easy comparison to notice the change. The damage behavior of all the laminates was observed visually by capturing the images after damage.

2. Experimental work details

2.1 Materials Used and Methods of formation

Plain weave fabrics made up of Toray T300 Carbon fibers and E-glass fibers, having fibers woven in two directions 0° and 90° were used along with DGEBA epoxy (LY 556) and amine hardener (HY951) for this study. Materials applied in microcapsule synthesis were melamine, ethylene maleic anhydride, formaldehyde, ammonium chloride, styrene, citric acid, sodium hydroxide and resorcinol. Fabrication of composite laminates with microcapsule infusion to it was procured in the following three steps.

2.1.1 Formation of resin and hardener filled microcapsules

Both resin filled capsules and hardener filled capsules were prepared by in-situ polymerization in a similar way as described in (17) with minute modifications. Borosilicate glass beaker of 2000 ml capacity was used for synthesis of these capsules. Firstly, 31.25 g of melamine and 67.75 g of formaldehyde were poured in a 200 ml beaker and kept in an hot air oven where melamine and formaldehyde mixture was heated at a constant temperature of 70 °C for half an hour in order to form the cross links between melamine and formaldehyde. The pH of the mixture was kept between 9-10 to maintain the basic state while heating, which was achieved by addition of triethanolamine. After 30 minutes the mixture was removed from hot air oven and cooled at atmospheric temperature of 32 °C. For encapsulation of DGEBA based epoxy, 600 ml distilled water was poured in 2000 ml beaker and 200 g of DGEBA based epoxy, 1.5

IOP Conf. Series: Materials Science and Engineering **748** (2020) 012023 doi:10.1088/1757-899X/748/1/012023

g of ammonium chloride, 1.5 g of resorcinol and 6 g of styrene were added to it. The mixture was stirred with low shear propeller for 5 minutes at 500 rpm. Melamine formaldehyde blend was added to it and mixture was continuously stirred in ambient conditions for one hour at 800 rpm. The formed capsules were washed, filtered and dried at 60 °C in hot air oven. In the similar way, hardener was filled in melamine formaldehyde capsules. The hardener filled capsules were dipped in catalyst overnight so that catalyst permeates through shell walls. Hardener capsules were rinsed and dried at room temperature. Both the capsules were dried for 48 hours. Both the resin and hardener filled capsules formed were white in colour and are shown in Figure 1.

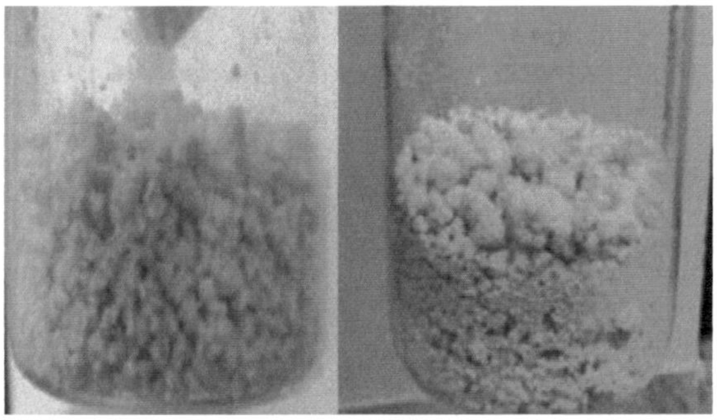

(a) (b)

Figure 1. (a) Resin Microcapsules (b) Hardener microcapsules

2.1.2 Preparation of resin system

After successful drying of capsules, the capsules were mixed in weight percentage of 2.5% in epoxy resin though continuous stirring at room temperature with mechanical stirring equipment at 400 rpm until the capsules were mixed in the resin properly. Both resin filled capsules and hardener filled capsules were added in ratio of 50:50.

2.1.3 Fabrication of Carbon fiber/epoxy and Glass fiber/epoxy Composite Laminate

Both Carbon fiber (Toray T300) reinforced epoxy laminate and E-glass fiber reinforced epoxy laminate were created by Hand Lay-up Process. Fabrication was done on a granite table and surface was cleaned with a paint thinner solution to remove any dirt or grease present to ensure that the surface was clean and even. A square area of 400 mm * 400 mm was marked on the surface. Two 100 micron Mylar sheets were cut into same dimension (400 mm * 400 mm) and kept aside. The 5000 mm * 5000 mm plain weave T300 carbon fabric was cut into dimension of 350 mm*350 mm to obtain 12 layers of same dimension from it. One Mylar sheet was placed on the surface. The surface of the Mylar sheet was cleaned and wax was applied. Wax acted as surface release agent that led to easy removal of the laminate after curing and also it provided good surface finish. The LY556 resin was mixed with HY 951 hardener in the ratio 10:1 in a 1000 ml beaker. Once the wax was dried, first layer of resin/hardener mixture was laid on the surface followed by laying of T300 carbon fabrics. The second layer of resin was applied on the surface and a metal roller was used to wet the surface of fibers with resin. In a similar fashion, all 12 layers were applied. On the top layer, a Mylar sheet with waxed surface was attached and rolled out for proper sticking and smooth surface finish. The surface was pressed with heavy metal plates. The laminate was allowed to cure for 24 hours. In the similar way, four laminates were created i.e. Carbon Fiber Reinforced Polymer with Microcapsules (CFRPWC), Carbon Fiber Reinforced Polymer with no Microcapsules

(CFRPNC), Glass Fiber Reinforced Polymer with Microcapsules (GFRPWC), and Glass Fiber Reinforced Polymer with no Microcapsules (GFRPNC).

2.1.4 Preparation of specimens

The specimens of required dimensions (150mm *100mm) were cut from laminates by using water jet cutting. Water jet cutting with suitable velocity is preferred over conventional methods like cutting by hand saw and advanced methods like cutting by laser to avoid micro-cracks and damage to surface due to dissipation of heat while cutting. Water jet cutting also provides good surface finish.

3. Details of Specimen Testing

3.1 Scanning Electron Microscopy (SEM)

Scanning Electron Microscopy was used to determine the size of the capsule formed. Scanning electron microscope imaged the capsules by using focused beam of electrons of short wavelength, which in turn gave high resolution. The SEM test setup is shown in Figure 2.

Figure 2. SEM test set-up.

Scanning Electron Microscope images of both resin filled capsules and hardener filled capsules are shown in Figure 3. It was observed from the Figure 3 that size of resin filled microcapsules and hardener filled microcapsules ranged from 0.19μm to 0.30μm and 0.21μm to 0.28μm respectively.

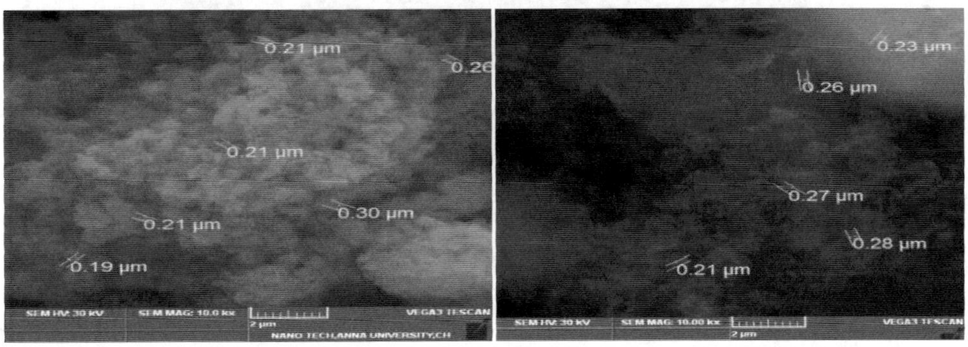

(a) (b)

Figure 3. SEM image of (a) resin filled microcapsules and (b) hardener filled microcapsules

3.2 Drop weight impact test

As drop weight impact tester can impact the specimens at higher energy than pendulum and can also rupture specimens of different shape and materials, it was used to test fiber reinforced composite laminate specimens. Firstly, the specimen was fixed between the clamps to avoid its movement during the test. Then the parameters like impact velocity, impactor height, impactor mass; test type and testing standard were given as input and values of parameters are shown in Table 1. The impactor was set at the vertical distance of 1593.21 mm from the specimen, so that when it hits the specimen by free fall under gravity, the kinetic energy will be 30 J. The impactor was set at the center of the clamps to get a central impact on the specimen. All the specimens were tested in a similar way. Drop weight impact tester is shown in Figure 4. Impact tests were conducted in accordance with ASTM D5628.

Table 1. Input parameters in the software for impact test

S.No.	Name of parameter	Value
1.	Impact Velocity	5.59 m/s
2.	Impact Height	1593.21mm
3.	Carriage Mass	1.3 Kg
4.	Total Mass	1.926 Kg
5.	Impactor shape	Hemispherical
6.	Impactor diameter	12.7 mm
7.	Test type	Biaxial

Figure 4. Drop weight impact test set-up

4. Results and Discussion

This section deals with the results of drop weight impact test of Carbon Fiber Reinforced Polymers and Glass Fiber Reinforced Polymers. The sub-sections scrutinize the impact force response, energy response and deformation response of the CFRP and GFRP laminates. In total, 12 specimens were tested, that were divided into the following four categories (a) CFRPNC-Carbon Fiber Reinforced Polymer with no Capsules (b) CFRPWC- Carbon Fiber Reinforced Polymer with Capsules (c) GFRPNC-

IOP Conf. Series: Materials Science and Engineering **748** (2020) 012023 doi:10.1088/1757-899X/748/1/012023

Glass Fiber Reinforced Polymer with no Capsules (d) GFRPWC- Glass Fiber Reinforced Polymer with Capsules. Three specimens from each category were tested under the impact loading at room temperature.

4.1 Impact force response of CFRP and GFRP laminates

Impact force response of carbon fiber reinforced laminates with no capsules (CFRPNC) and with capsules (CFRPWC) is shown in Figure 5. Impact force response of glass fiber reinforced laminates with no capsules (GFRPNC) and with capsules (GFRPWC) is shown in Figure 6. As all the specimens in the corresponding category shown the similar curves, from each category impact force response for one specimen is shown to understand the load Vs time curve.

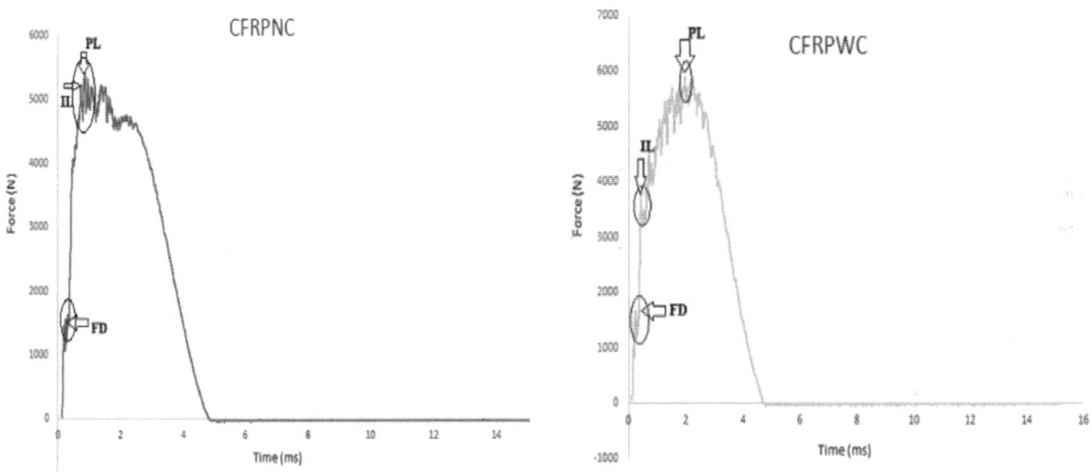

(a) (b)

Figure 5. Load Vs Time Response of (a) CFRPNC specimen and (b) CFRPWC specimen

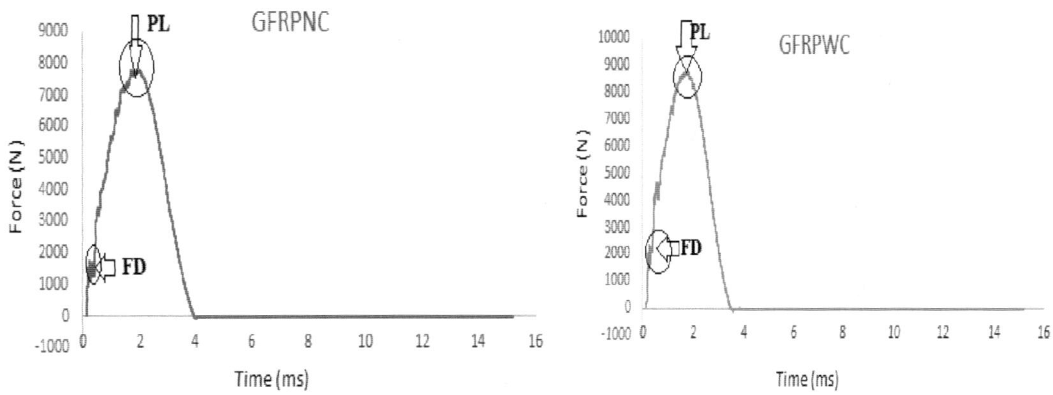

(a) (b)

Figure 6. Load Vs Time Response of (a) GFRPNC specimen (b) GFRPWC specimen

Load Vs time curves of all carbon fiber reinforced epoxy laminates show three important load points i.e. First damage (FD), Incipient load (IL) and Peak load (PL). The first damage (FD) corresponds to the first drop in load. It is the point where the first damage occurred, but the laminate was able to take more loads and its effect was negligible on load bearing capacity of laminates. Incipient load (IL) is load corresponds to the point where damage appears majorly after which damage propagates very fast, in curves it can be implied by first intense drop in load which is followed by oscillations in the curve. Peak load (PL) is load corresponds to the point where material will not be able to take any more load, in curves it is represented by highest point. The curve did not witness significant oscillations between the points FD and IL. The oscillations were witnessed corresponding to progression of damage in all the specimens following which specimens failed. The incipient load, peak load and load corresponds to first damage are displayed in Table 2.

Table 2. Details of first damage load, incipient load and peak load obtained from impact test

Specimen Details	Incipient load (N)	Peak load (N)	First damage load (N)
CFRPNC 1	5215.89	5410.66	1463.83
CFRPNC 2	4528.69	4972.13	1201.68
CFRPNC 3	4306.97	5032.14	1194.33
CFRPWC 1	3771.66	5808.78	1515.28
CFRPWC 2	3891.70	5898.20	1287.44
CFRPWC 3	3825.56	5879.82	1409.93
GFRPNC 1	-	7794.44	1509.15
GFRPNC 2	-	7382.85	1255.18
GFRPNC 3	-	7638.87	1277.63
GFRPWC 1	-	8617.62	1602.25
GFRPWC 2	-	8740.11	1821.52
GFRPWC 3	-	8560.04	1640.23

It was observed from the obtained values that addition of microcapsules increased the peak load and load where first damage occurred whereas incipient load decreased. Both categories of glass fiber reinforced epoxy laminates behaved differently from the carbon fiber reinforced epoxy laminates. The load Vs time curve shows two important points i.e. first damage (FD) and peak load (PL). The load Vs time curves of GFRPNC and GFRPWC category did not show incipient load point and there were no significant oscillations which proved that the damage growth happened in a stable way. It is clear from the Table 2 that specimens of both GFRPNC and GFRPWC groups have shown higher peak loads as compared to CFRPNC and CFRPWC group of specimens. Like carbon fiber epoxy laminates, peak load and first damage load values of glass fiber epoxy laminates boosted with addition of epoxy and hardener filled microcapsules.

4.2 Load Vs displacement response of CFRP and GFRP laminates.

The load Vs displacement curves obtained during impact test are shown in Figures 7 (a) and 7 (b). As the curves of all the specimens of a category showed the similar characteristics, load Vs displacement curve of one specimen from each category is displayed. Figure 7(a) indicates the load Vs displacement curve for CFRPNC and CFRPWC specimens and Figure 7 (b) depicts the load Vs displacement curve for GFRPNC and GFRPWC specimens. From Figure 7 (a), it is clear that both the curves contain two phases i.e. loading phase and unloading phase. Both the curves of “CFRPNC” and “CFRPWC” are linear till first damage (FD), at first damage there is minor fluctuation of load and also few load fluctuations were observed between FD and IL. This behavior shows that, the laminate was deformed continuously with very less damage progression. Once the peak load point was reached, the curve showed higher intensity fluctuations due to severe progression of damage, followed by failure. The unloading phase is different from the loading phase due to dissipation of energy during the impact. It was also observed that CFRPWC specimen has shown initially less displacement as compared to CFRPNC specimen. Due to addition of microcapsules, after the first damage, some capsules broke and delivered healing agents to fill the damage, which in turn reduced the overall damage and corresponding displacement. It can be seen from the Figure 7 (b) that the area between loading and unloading of curve GFRPWC is smaller than curve GFRPNC. This reduction in area corresponds to healing and less displacement. Also, curve GFRPWC shows more smoother profile till peak load than curve GFRPNC and after peak load both the curves show a smooth unloading phase. Like CFRP specimens there were no fluctuations witnessed after peak load. Few fluctuations correspond to stable damage growth.

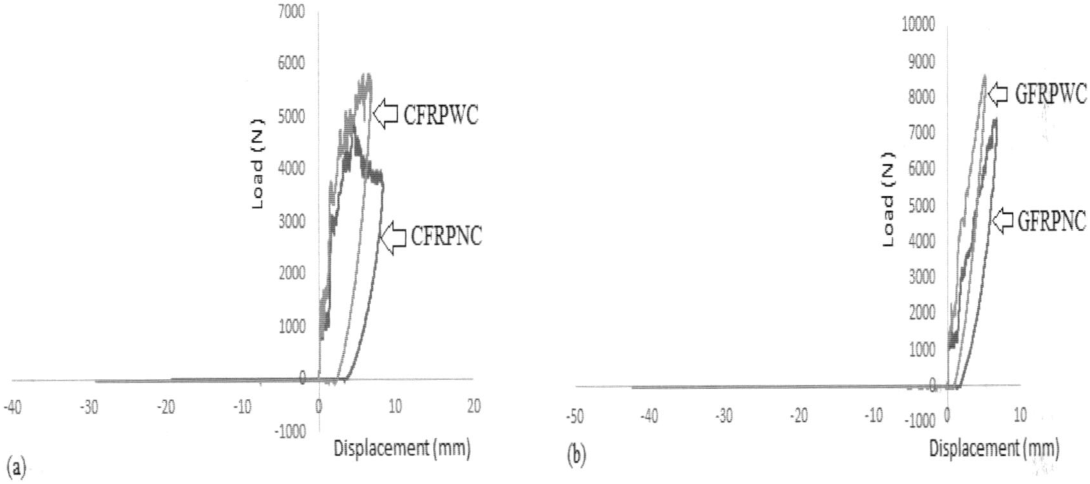

Figure 7. Load Vs Displacement curve for (a) CFRPNC and CFRPWC specimens (b) GFRPNC and GFRPWC specimens

4.3 Energy Vs time response of CFRP and GFRP laminates

Energy Vs Time response of CFRPNC and CFRPWC is shown in Figure 8 (a) and Energy Vs Time response of GFRPNC and GFRPWC specimens is shown in Figure 8 (b). Two types of energy is marked in the curves i.e. Energy absorbed elastically (EEL) and Energy absorbed plastically (EAB). EEL represents elastic energy, elastic energy is the energy absorbed by the specimen during elastic deformation which results from hitting of specimen with impactor. Once the peak value of deformation is reached, this energy is released. When energy applied through impact is higher than EEL, the extra amount of energy is absorbed during plastic deformation and damage occurrence. This extra amount of energy absorbed is represented by EAB. From Figure 8, it is clearly depicted that CFRPWC and GFRPWC specimens shown higher values for EEL and lower values of EAB than CFRPNC and GFRPNC specimens. This shows that by addition of capsules, the energy absorption by elastic deformation is increased and energy absorbed during plastic phase and damage is decreased. The damage and plasticity is reduced by addition of capsules. The detail of EEL and EAB of all the specimens is shown in Table 3.

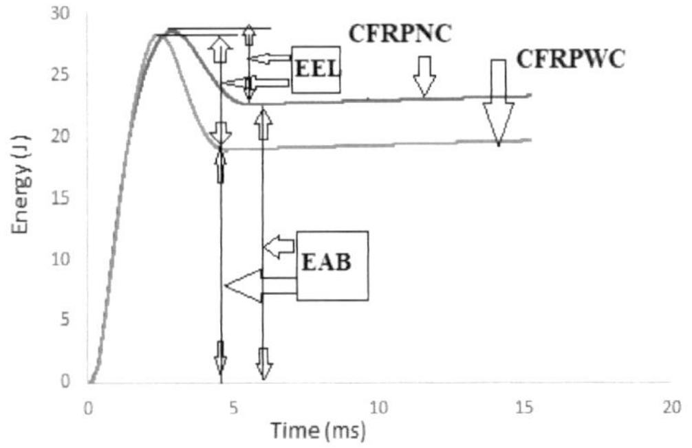

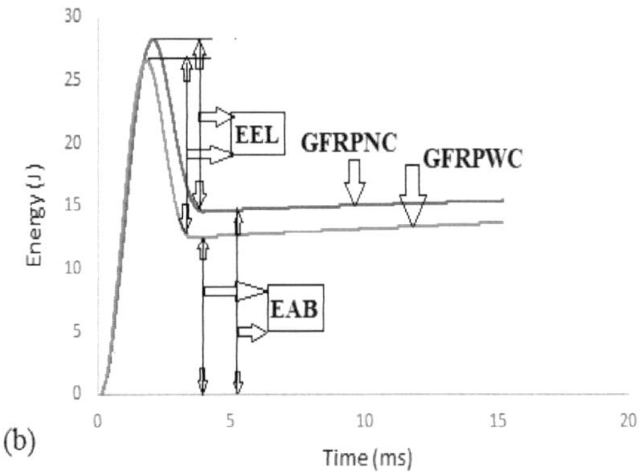

Figure 8. Energy Vs time curve for (a) CFRPNC and CFRPWC specimens (b) GFRPNC and GFRPWC specimens.

Table 3. Details of energy absorbed elastically (EEL) and energy absorbed plastically (EAB)

Specimen Details	EEL (J)	EAB (J)
CFRPNC 1	6.83	21.37
CFRPNC 2	5.89	22.66
CFRPNC 3	5.90	22.65
CFRPWC 1	9.16	19.25
CFRPWC 2	9.41	18.91
CFRPWC 3	9.47	18.82
GFRPNC 1	13.26	14.71
GFRPNC 2	12.96	15.26
GFRPNC 3	12.81	15.67
GFRPWC 1	14.07	12.47
GFRPWC 2	14.10	12.50
GFRPWC 3	13.97	12.38

4.4 Damage and Healing Assessment

The damaged area of the specimens was inspected visually. The region in the vicinity of the damage of CFRPNC and CFRPWC specimens and of GFRPNC and GFRPWC specimens is shown in Figure 9 and Figure 10 respectively. Figure 9 (a), (b) shows the front face of the CFRPNC and CFRPWC specimens, this face came into direct contact with the impactor. It can be seen from the photograph that CFRPNC specimen has shown an indentation along with cracks at the point of impact whereas the CFRPWC specimen has shown four cracks without any indentation. Figure 9 (c) and (d) shows the back face of CFRPNC specimen indicating a pyramidal damage and CFRPWC specimens indicating smooth surface with a single visible vertical crack respectively. This shows that the damage was healed partially. Figure 10 (a) and (b) shows the front face of GFRPNC and GFRPWC specimens. GFRPNC specimen shows less damage on the front face than GFRPWC specimens. Delamination and indentation is more on front face of GFRPWC specimen as compared to GFRPNC specimen. Figure 10 (c) and (d) shows back face of the GFRPNC and GFRPWC specimens. It is clear from the GFRPNC specimen that a pyramidal damage occurred whereas GFRPWC showed a smooth surface with delamination inside the layer (white in color).

As the amount of healing agents delivered at the damage site was less than required, so only partial healing happened.

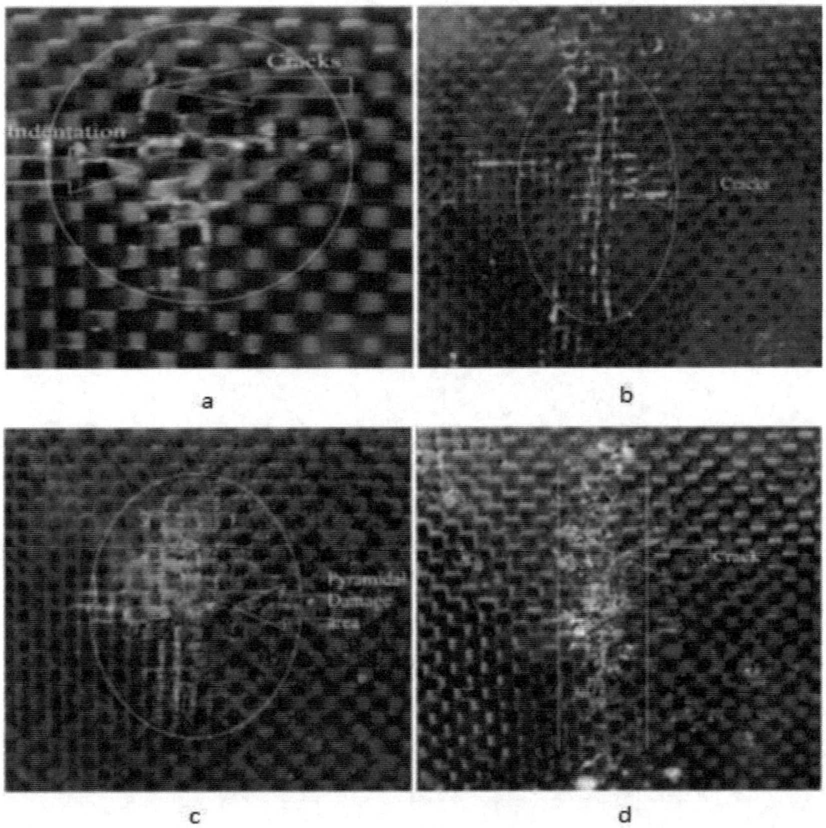

Figure 9. Damage area of (a) front face of CFRPNC (b) front face of CFRPWC (c) back face of CFRPNC (d) back face of CFRPWC

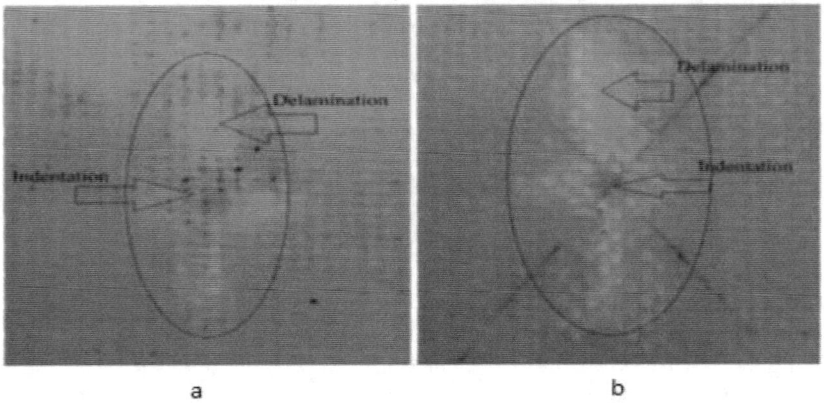

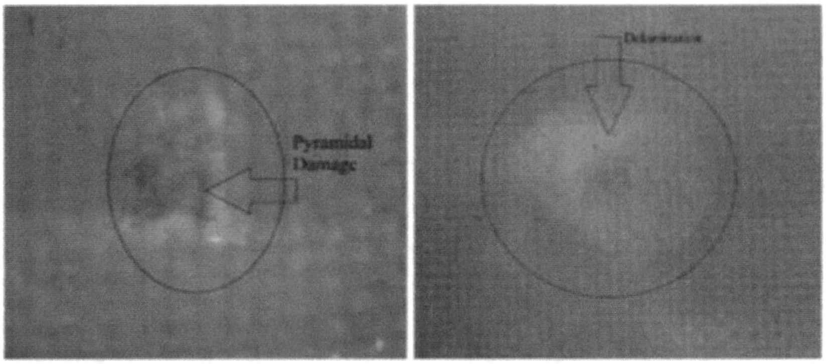

c d

Figure 10. Damage area of (a) front face of GFRPNC (b) front face of GFRPWC (c) back face of GFRPNC (d) back face of GFRPWC

5. Conclusion

Drop weight impact test was executed on Carbon Fiber Reinforced Epoxy Laminates and Glass Fiber Reinforced Epoxy Laminates. Four types of laminates namely CFRPNC, CFRPWC, GFRPNC and GFRPWC were tested. CFRPNC and GFRPNC did not contain any capsules and CFRPWC and GFRPWC contain capsules filled with resin and hardener. From the test results the following points were determined:

- Inclusion of capsules increased peak load in both CFRP and GFRP laminates.
- Inclusion of capsules reduced the energy absorbed during plastic deformation and damage progression whereas energy absorbed during elastic deformation was increased.
- Addition of 2.5 weight % capsules is not enough to heal the damage completely, only partial healing was achieved.

References

[1] Mangal Giri P D 1999 Composite Materials for Aerospace Applications *Bull. Mater. Sci.* **22** 657-664

[2] Mahajan G V and Aher V S 2012 Composite Material: A review over Current Development and Automotive Application *International Journal of Scientific and Research Publications* **2** 1-5

[3] Nikhil Nayak V 2014 Composite Materials in Aerospace Applications *International Journal of Scientific and Research Publications* **4** 1-10

[4] Sanjay Kumar Sardiwal , Md. Abdul Sami , Sai Anoop B V , Gudipudi Susmita , Lahari Vooturi and Syed Arshad Uddin 2014 Advanced Composite Materials in Typical Aerospace Applications *Global Journal of Researches in Engineering: D Chemical Engineering* **14** 5-10

[5] Autar K. Kaw 2006 Mechanics of Composite Materials Second Edition CRC press *Taylor and Francis*

[6] Sultan M T H, Hodzic A, Staszewski W J and Worden K 2010 A SEM- Based Study of Structural Impact Damage *Applied Mechanics and Materials* **24** 233-238.

[7] Gholizadeh S 2018 A Review of Impact Behavior in Composite Materials, Proceedings of Academics era 38th *International Conference London* UK

[8] Sandeep Agrawal, Kalyan Kumar Singh and Sarkar P K 2014 Impact Damage on fibre-reinforced polymer matrix composite-A review *Journal of Composite Materials* **48** 317-332

[9] Bibi I S M and Aizat A A R 2017 Study of Impact Damage Behavior in Woven Carbon Fiber Plates *Procedia Engineering* **170** 47-54

[10] Amal A M B 2012 Impact behavior of glass fibers reinforced composite laminates at different temperatures *Ain Shams Engineering Journal* **3** 105-111

[11] Prashant Rawat and Kalyan K S 2017 Damage Tolerance of Carbon Fiber Woven Composite Doped with MWCNTs under Low-Velocity Impact *Procedia Engineering* **173** 440-446

[12] Polyxeni Dimoka, Spyridon P, Christine K and Vassilis K 2019 Assessing the Damage Tolerance of Out of Autoclave Manufactured Carbon Fibre Reinforced Polymers Modified with Multi-Walled Carbon Nanotubes *Materials* **12** 1080

[13] Hamed Saghafi , Mohamad Fotouhi and Giangiacomo Minak 2018 Improvement of the Impact Properties of Composite Laminates by Means of Nano-Modification of the Matrix—A Review *Applied Sciences* **8** 2406.

[14] Amit kumar R, Asokan R, Kanu Priya Jhanji, Debajit Das and Vishnu Sai Nalla 2019 Investigation of Tensile Properties of Carbon / Epoxy Composite Joints with and without Carbon Nano-Tubes *International Journal of Vehicle Structures and Systems* **11** 209-213

[15] Hao Li, Jingbiao Liu, Zhenqing Wang, Zhengwei Yu, Yanfei Liu and Min Sun 2018 The Low Velocity Impact Response of Shape Memory Alloy Hybrid Polymer Composites *Polymers* **10** 10-26

[16] Stefan Zechel, Robert Geitner, Marcus Abend, Michael Siegmann, Marcel Enke, Natascha Kuhl, Moritz Klein, Jürgen Vitz, Stefanie Gräfe, Benjamin Dietzek, Michael Schmitt, Jürgen P, Ulrich S Schubert and Martin D Hager 2017 Intrinsic self-healing polymers with a high E-modulus based on dynamic reversible urea bonds NPG Asia *Materials* **9** 420

[17] Yongjing Wang, Duc Truong Pham and Chunqian Ji 2015 Self-healing composites: A review *Cogent Engineering* **2** 1075686

[18] Lee J, Bhattacharyya D, Zhang M Q and Yuan Y C 2015 Mechanical properties of a self-healing fibre reinforced epoxy composites *Composites Part B* **78** 515-519

[19] Kanu Priya Jhanji, Asokan R, Amit Kumar R, Sharanyaa Pandima Dhevi S Mahathi Guptha K and Probal Chowdary 2019 Effect of Healing Agent Filled Microcapsules on Tensile Strength of Glass / Epoxy Composite Laminate *International Journal of Vehicle Structures and Systems* **11** 250-254

[20] Pulkit Garg, Pallav Gupta, Devendra Kumar and Om Parkash 2016 Structural and Mechanical Properties of Graphene reinforced Aluminum Matrix Composites *Journal of Materials and Environmental Science* **7** 1461-1473

Response of adhesively bonded single lap joint doped with MWCNT and Fullerene C_{60} in the adhesive to the tensile strength of the joints

Amit Kumar R[1], Asokan R[2], Kanu Priya Jhanji[3], Sahil Sarkar[4] and Debajit Das[5]

[1, 2, 3, 4, 5] School of Aeronautical Sciences, Hindustan Institute of Technology and Science, 603103, Chennai, India

Email: *mail2amit.17@gmail.com

Abstract: The basis for this research is to create joints that have greater joint strength and capability of carrying greater damage load. The structural damage behaviour of adhesively bonded single lap joints (ABSLJ's) was investigated. Laminates formed by carbon fibre woven in plane weave pattern and bonded by epoxy were used as adherend. Since a variety of nano materials are currently in use and are said to improve the damage strength of the composite when dispersed in them during the fabrication, two such nano materials namely multi walled carbon nano tube (MWCNT) and Fullerene C_{60} in different weight percentages of 0%, 1%, 3%, 5% and 7% were introduced by the method of ultrasonication to disperse them in the adhesive. A two part epoxy ROTEX EP - 416 resin and ROTEX EH – 153 hardener mixtures were used as adhesive for creating the joints. Results stipulate that the addition of MWCNT nanofillers in 1 wt % and 3 wt % in the adhesive has lead to a considerable increase in the tensile damage load of the ABSLJ's by about 93% and 199% respectively. While the addition of 1 wt % Fullerene C_{60} in the adhesive increased the failure load but there was not much significant improvement when compared with the same percent of MWCNT.

Keywords: MWCNT, Fullerene C_{60}, ABSLJ, Ultrasonication.

1. Introduction

The employment of nano materials have shown an increase in recent years due to their ability to generate better material properties like increased strength and assistance in creating a tougher and stiffer component. These nano composites are known to be multiphase solid in nature that have one of the phase dimension of the material structure less than 100 nm and the nanoscale fillers are suppose to be zero-dimensional (embedded particles), one-dimensional (nanotubes and nanorods), two-dimensional (nanocrystalline multilayers), nanostructured materials [1]. Implementation of adhesive bonding technique along with reinforced nano particles for fastening composites is advantageous in terms of cost, mechanical properties, environmental condition resistance and greater strength [2, 3].

Mechanical stirring with the use of automatic speed controlled stirrer can be used to disperse the nanofillers in the resin. For proper dispersion a suitable rpm (speed of motor) and mixing time is required to be maintained so that there are no lumps formed during the dispersion process and the nanofillers gets completely dispersed. Care should be taken to avoid usage of stirrer having sharp edges as the sharp edges may sometime damage the nanofillers if the rpm is too high [4]. Adding nanofillers such as fullerene in the adhesive increases the adhesive strength by a greater percentage. Involvement of temperature during hardening cannot be undone when considering the adhesive strength. At lower temperature, the strength decreases when compared to the material hardened at higher temperature regime [5]. Further there are joints which find various applications in sectors of automobiles, aerospace, marine etc. These joints are basically single lap joint created by different techniques like by using rivets, nuts and bolts, adhesives etc. Among these adhesive bonded joints are noteworthy and are swiftly in use. Along with the use of conventional nano particles, carbon based particles have gained importance in improving the joint strength when mixed with adhesives to create the joint. Flexural strength increase of 20.7 % was observed with the addition of nanofillers like MWCNT at 1% by weight in the epoxy to blend a glass fibre composite thereby showing an improvement in the damping properties [6]. Also an increment of 47% and 86% was observed in the bonding strength of the single lap joints created with the use of 0.1 wt % and 0.3 wt % of MWCNT respectively [7].

Likewise a growth of 26.4% and 24% respectively in the flexural strength, fracture toughness in case of a quasi isotropic laminate was observed when alumina nano particles were added to the epoxy. While for a unidirectional laminate, the addition of alumina showed negligible improvement in flexural properties [8]. When Al_2O_3, SiC and MWCNT particles were added to Epocast 50 – Al/946 epoxy to be used as an adhesive to create an adhesive bonded joint, the dispersion of Al_2O_3, SiC nano particles was simple due to their lower aspect ratio for greater weight percentage of nano particles. The MWCNT is slightly different due to its high aspect ratio and dispersion gets burdensome even with lower weight percentage of nano particles. The advancement attained in the tensile strength and shear strength for the bonded joints having 0.5 wt % MWCNT, 1.5 wt % SiC and 1.5wt % Al_2O_3 are 7.5%, 4%, 0.5 % and 5.5%, 4.9%, 6.3% respectively [9].

The use of different nano fillers at different weight percentages help in determining the properties of the joints but as the weight percentage keeps on increasing, a subsequent decrease in properties is observed and the dispersion becomes difficult [10, 11]. The ability of CNT's to show resistance towards water showed that the epoxy specimen with CNT added to it showed greater water resistance compared to a normal epoxy specimen [11]. Mixing MWCNT along with the adhesive improves the bonding strength and resists the propagation of crack due to strong interfacial bond strength [12]. Addition of graphene nano particles in the adhesive with 10 layers of carbon fibre as an adherend leads to an increase in the peak load of approximately 21 % at 1 wt. % and 57 % at 2 wt. % [13]. Incorporating CNT's in the adhesives improved the adhesion efficiency and provided a tailored galvanic behaviour of nano composites [14]. The addition of 1 Vol % of Zirconia nano particles lead to an increase in shear strength of about 60%. Additionally the joints containing Zirconia showed decreased contact angle with water [15]. Two different failure modes are considered in the adhesive bonded joints, one is adhesive failure where failure materializes at the interface between the adhesive and adherend due to weak adhesion of the boundary layer while other is cohesive failure where due to fracture in the adherend, the adherend fails first before the adhesive does [16].

2. Experimental Work

2.1 *Materials*

A plain weave carbon fibre fabric as shown in Figure 1a, T300 of 200 gsm manufactured by Toray Composite Materials, was used to create a laminate to be used as the adherend. The required thickness was obtained by using 11 layers of the carbon fibre. For preparation of the laminate, bonding of each layer of carbon fibre fabric was done by using a two-part epoxy resin namely ROTEX EP – 306 and hardener ROTEX EH – 758. For joint creation a separate bonding material was used which is also a two-part epoxy adhesive ROTEX EP- 416 and the hardener was ROTEX EH-153 as the epoxy adhesives are considered as the strongest adhesives.

Table 1. Properties of Carbon/Epoxy Laminate

Properties	**Values**
Longitudinal Young's Modulus (E_1)	70 GPa
Transverse Young's Modulus (E_2)	70 GPa
In-Plane Shear Modulus (G_{12})	5 GPa
Major Poisson Ratio (υ_{12})	0.10

MWCNT of 99% purity, 10 – 15 nm average diameter, 5μm average length and 400 m^2/g and Fullerene C60 of 99% purity with 1600g/cm^3 of relative density was used as the filler material to be added in the adhesive for creating the joints. Properties of the adherend and adhesives (one for creating the laminate and other for creating joint) are mentioned in Table 1 and 2.

Table 2. Properties of Adhesive System

Properties	**Values**
FOR RESIN	
Viscosity cps	30000-50000
Specific Gravity	1.15-1.20
FOR HARDENER	
Viscosity cps	25000-40000
Specific Gravity	0.90-0.99

2.2 *Laminate Preparation*
CFRP laminate shown in Figure 1b was prepared by using hand layup technique. Carbon fibre fabrics of dimensions 500 mm * 500 mm were used for preparing the laminate. A mylar sheet of required dimensions were cut and cleaned by using thinner and then wax was applied for easy removal of the laminate and also for removal of air bubbles. ROTEX EP – 306 resin along with ROTEX EH – 758 hardener was mixed in the ratio of 10:1 by weight and applied over the mylar sheet. A layer of carbon fibre fabric was laid over the resin hardener mixture and a roller was used to remove the excess resin and the air bubbles. The same process was repeated for the next 11 layers.

Finally a cleaned and waxed mylar sheet was laid over the final layer of carbon fibre and then required amount of weight was put in order to apply pressure. The prepared laminate was then cured at room temperature for 24 hours following which curing was carried out in an oven for 2 hours at 60° C. Two such laminates were prepared.

Figure 1. a. Plane weave Carbon fibre fabric b. Carbon fibre/epoxy laminate

2.3. *Specimen Preparation*

2.3.1 *Preparation of Specimen and Preparation of Specimen Surface*

The laminate was cut by using abrasive water jet cutting technique to produce the required specimen. The dimension of the specimens are 101.6 mm * 25.4 mm as per ASTM D-5868-01 standard as mentioned in Figure 2. This method of cutting was chosen over the other methods due to its advantage of producing smooth edges and avoids delamination at the edges. A high speed jet of water mixed with sand particles which act as abrasive was used as a medium for cutting the specimens. The average thickness of the cut specimens was measured to be 2.98 mm. Further the surface of specimens was prepared for making the joints. The surface preparation was necessary in order to create a strong and long lasting bond between the adhesive and adherend. Thus a part of the specimen was rubbed with 40 grit sand papers to make the surface rough where joint was supposed to be made. The specimens were then cleaned with acetone to remove any dust or loose particles formed after rubbing with sand paper.

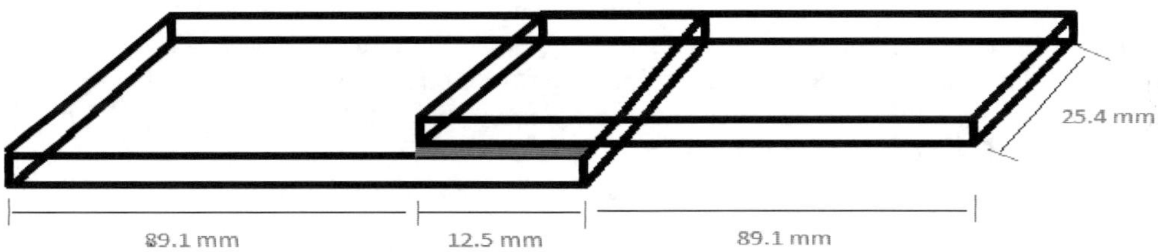

Figure 2. ASTM D 5868 – 01 standard

2.3.2 *Preparation of Adhesive System*

This process is of prime importance, as it requires proper and homogeneous dispersion of nanofillers in the adhesive. Dispersion of nanofillers done mechanically provides even distribution thereby enhancing its properties, while the dispersion done by high speed mixing creates bubbles that cannot be easily removable, though both the methods can be used for dispersion [17]. Thus in the current study, mechanical/ultrasonication was carried out to disperse MWCNT and Fullerene C_{60} into ROTEX EP - 416 resin as shown in Figure 3. During the process, firstly 15 g of resin was taken and then 1 wt %, 3 wt %, 5 wt % and 7 wt % of MWCNT and Fullerene C_{60} was added respectively and kept in ultrasonicator at a frequency of 30 Hz for 45 minutes to obtain proper dispersion. During the process the temperature was maintained at 60 °C. Further to the mixture, an equal amount of ROTEX EH – 153 hardener was added.

During the sonication process, it was observed that epoxy containing 1 wt % and 3 wt % of MWCNT and Fullerene C_{60} respectively got easily and completely dispersed but as the nanofiller percentage was increased to 5 wt % and then to 7 wt %, the nanofillers even though partially dispersed, began to coagulate resulting in uneven and dense dispersion. An unreinforced mixture of adhesive was also prepared by hand stirring of resin and hardener.

Figure 3. Ultrasonication of nanofillers (MWCNT and Fullerene C_{60}) in the Adhesive

2.3.3 *Preparation of Joints*

To prepare the sample of ABSLJs unreinforced adhesive and adhesive reinforced with nanofillers were applied to one of the rough and cleaned surface of the adherend covering the overlap distance. Metal shims were placed to provide a uniform thickness to the adhesive. On top of the adhesive layer, the rough surface of another adherend was placed covering the overlap and making a complete single lap joint. The joint was then cured by clamping for 8 hours at room temperature. The metal shims were then removed and further the empty space on the sides was filled up with the respective adhesive mixture. Care was taken during removal of the metal shims so as to not disturb the alignment and bonding. Once the joint was filled with adhesive after removal of the shims, they were clamped again and left to cure for 24 hours at room temperature and further curing was done by placing the ABSLJs in oven at 60 °C for 6 hours. 3 samples were made for each joint type taking the total to 27 sets. Once the curing was completed, ABSLJs were obtained as shown in Figure 4 and the average thickness of the joint was found to be 3.23 mm.

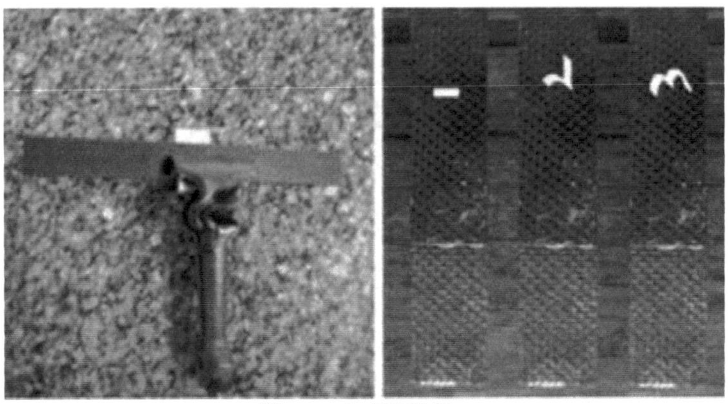

Figure 4. ABSLJ's

2.4 *Testing Procedure*

The single lap joint test was carried out using a computer controlled universal testing machine at a temperature of 32 °C. Specimens were categorised into 5 main groups with each group containing 3 joint samples each as given in Table 3. Specimens with different configuration of joints were subjected to tensile loading as shown in Figure 5 at a rate of 1 mm/min. Overall thickness of adhesive layer was measured for all the joint configuration before the test and the average thickness was found to be 0.25 mm. Once the test was concluded, the joints were visualised for adhesive and cohesive failure.

Table 3. Designation and Configuration of various samples

Joint Designation	Adhesive/Reinforcement
C/EP416+EH153	C/EP416+EH153/Non reinforced
C/EP416+EH153/CNT/I	C/EP416+EH153/MWCNT/1%
C/EP416+EH153/F/I	C/EP416+EH153/Fullerene/1%
C/EP416+EH153/CNT/II	C/EP416+EH153/MWCNT/3%
C/EP416+EH153/F/II	C/EP416+EH153/Fullerene/3%
C/EP416+EH153/CNT/III	C/EP416+EH153/MWCNT/5%
C/EP416+EH153/F/III	C/EP416+EH153/Fullerene/5%
C/EP416+EH153/CNT/IV	C/EP416+EH153/MWCNT/7%
C/EP416+EH153/F/IV	C/EP416+EH153/Fullerene/7%

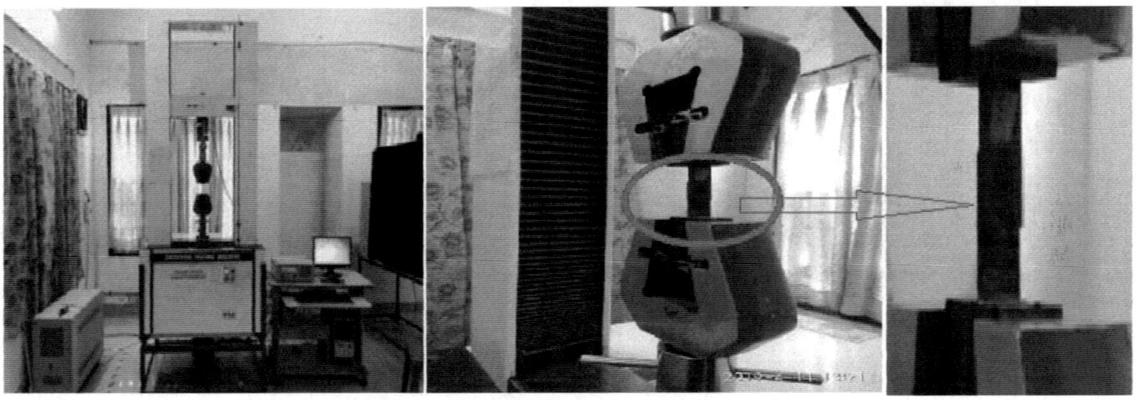

Figure 5. Tensile testing of ABSLJ's.

3. Results and Discussions

Three samples each from the above mentioned configuration given in Table 3 were tested to determine the maximum load at which the joints failed. The average value was taken to determine the average failure load given in Table 4 and Figure 6. The addition of nanofillers to the adhesive increases the failure/damage load but only till certain weight percentage of nanofillers. With addition of 1 wt % of CNT as a reinforcement to the adhesive in comparison with unreinforced adhesive, the average damage load increases by 93%. This indicates that the addition of MWCNT nanofillers substantially improves the bonding strength when compared to the bonds formed by the unreinforced adhesive. Increasing the weight percentage of MWCNT to 3% further increases the damage load to 199%. This increase in the damage load is due to homogeneous dispersion of MWCNT nanofillers in the adhesive. Also the failure load had increase because of better bonding between the adhesive and adherend. However the addition of Fullerene

C_{60} to the adhesive in 1 wt % and 3 wt %, the average damage failure load increased by 56% and 35% but the increase in considerably lower by 37% and 164% when compared with the similar weight percentage of MWCNT.

Further when the percentage of MWCNT was increased to 5 wt % and 7 wt % respectively, the average load decreased by 14% and further by 32% respectively when taken in comparison with the peak damage load obtained by using 3 wt % of MWCNT. Also providing an increase in the percentage of Fullerene C_{60} by 5 wt % and 7 wt % resulted in decrease in the average damage load in comparison with similar weight percent of MWCNT.

This decrease is observed due to the increase in amount of MWCNT and Fullerene C_{60} added to the adhesive which lead to improper dispersion thereby making it impossible to create a proper bond between the former and the later.

Table 4. Failure loads and average failure loads for each category of ABSLJ's

Joint Designation	Failure Load (N)	Average Failure Load (N)
C/EP416+EH153	1875 1842 1914	1877
C/EP416+EH153/CNT/I	3548 3641 3684	3625
C/EP416+EH153/F/I	2967 2892 2934	2931
C/EP416+EH153/CNT/II	5679 5583 5618	5626
C/EP416+EH153/F/II	2542 2586 2495	2541
C/EP416+EH153/CNT/III	4684 4819 4972	4825
C/EP416+EH153/F/III	1872 1951 1898	1907
C/EP416+EH153/CNT/IV	3890 3762 3847	3833
C/EP416+EH153/F/IV	1539 1598 1573	1570

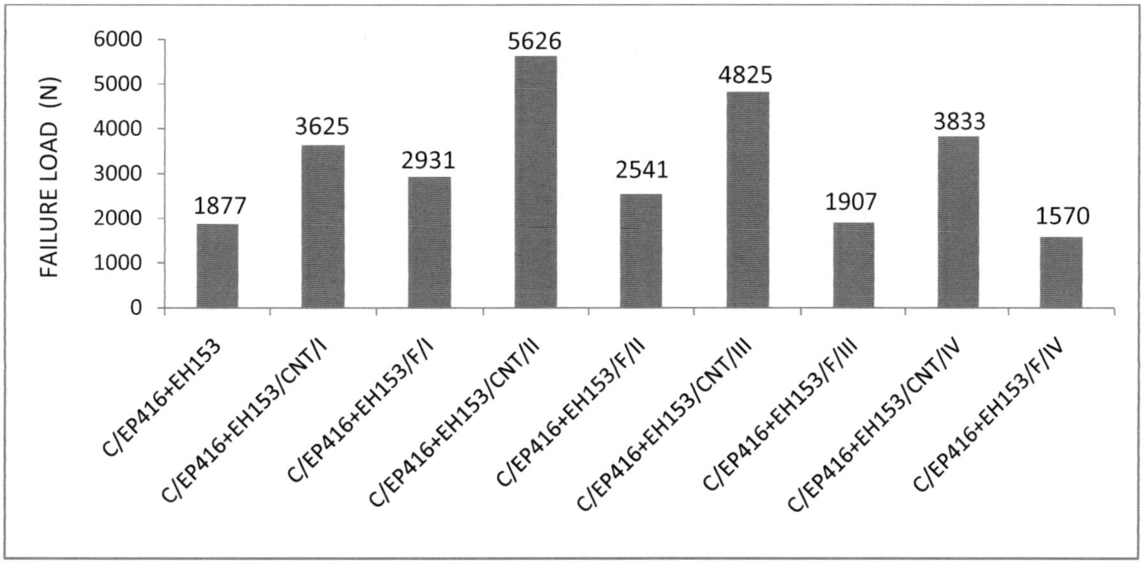

Figure 6. Average Failure load of different joint configuration

When we consider the load v/s displacement curve as shown in Figure 7 and Figure 8, it is evident that the addition of MWCNT and Fullerene C_{60} nanofillers to the adhesive improves the displacement capacity of the joints which gives an indication that the adhesively bonded joints doped with nanofillers are capable of taking more failure load. Similarly the load v/s displacement curve also saw a decreasing trend wherein the displacement capacity decreased with increase in MWCNT and Fullerene C_{60} percent but the displacement obtained by doping the adhesive with MWCNT is higher for all categories of joints when compared with the joints created by addition of Fullerene C_{60}.

It was also noted that, majority of the ABSLJ's suffered adhesive failure, where the failure initiated at edges and moved to the center, even though the adhesive thickness was maintained the same throughout, due to the weak bonding between the adhesive and the adherend.

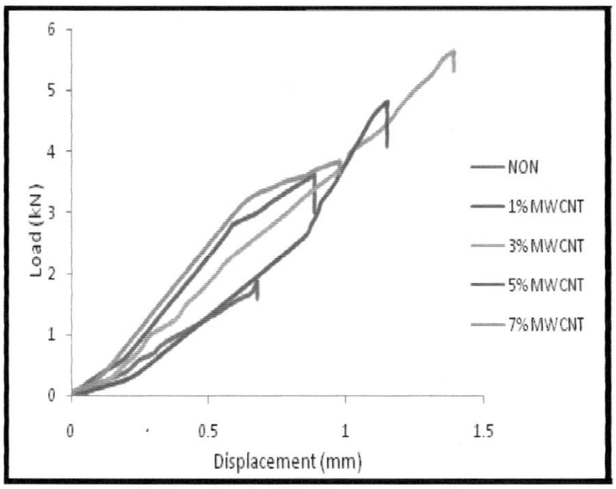

Figure 7. Load v/s Displacement Curve for MWCNT

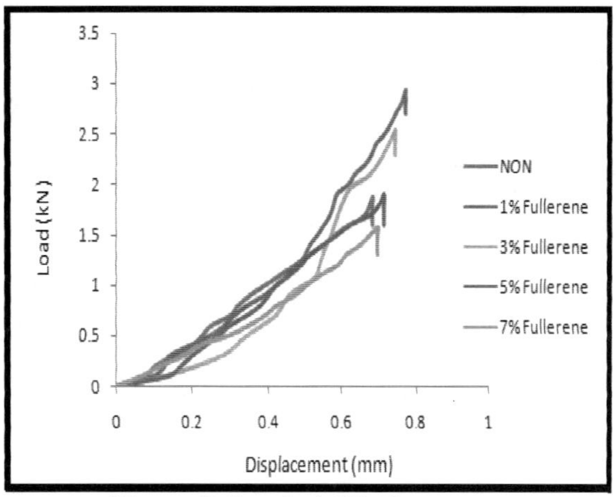

Figure 8. Load v/s Displacement curve for Fullerene C_{60}

4. Conclusion

The current work involved finding the effect of addition of various wt % of MWCNT and Fullerene C_{60} to the adhesive used to create ABSLJ's on the tensile failure load. From the results it is concluded that:

- The epoxy adhesive used here was capable of producing strong bond there and thereby can be classified as a rigid type of adhesive.
- The tensile damage load depends on dispersion of various percentages of nanofillers in the adhesive. Homogenous dispersion was obtained when the nanofillers were added in optimum weight percent thereby increasing the strength at the interface. While with the increase in percent of nanofillers, complete dispersion was not obtained resulting in decrease in the damage load.
- Addition of MWCNT nanofillers to the adhesive in 1 wt % and 3 wt % increases the average failure load by 93% and 199 % respectively but with further increase in percentage of MWCNT to 5% and 7%, the average failure load decreased. This is because of limit in the dispersion of nanofillers and also due to inadequate bonding between the adhesive and adherend.
- Adding Fullerene C_{60} to the adhesive in 1 wt % and 3 wt % led to increase in the average failure load but with further increase in percent of Fullerene C_{60}, the average failure loads decreased. It can be said that with the addition of Fullerene in the adhesive, there was no significant increment or decrement in the load carrying capacity of the joints.
- The displacement capacity of the joints with MWCNT was found to be greater when compared with non reinforced joints and joints with Fullerene C_{60} added in the adhesive.
- Adhesive failure was observed in the ABSLJ's due to insufficient adhesion between the adhesive and the adherend.

Reference

[1] Ashby M F, Ferreira P J and Schodek D 2009 Nanomaterials, nanotechnologies and design : An introduction for engineers and architects *Elsevier/Butterworth Heinemann* ISBN: 9780750681490.
[2] Hsiao K`T, Alms J and Advani S G 2003 Use of epoxy/multiwalled carbon nanotubes as adhesives to join graphite fibre reinforced polymer composites *Nanotechnology* **14** 791-793
[3] Dorigato A, Pegoretti , Bondioli F and Messori M 2010 Improving Epoxy Adhesives with Zirconia Nanoparticles *Composite Interfaces* **17** 873–892
[4] Kanu Priya Jhanji, Asokan R, Amit Kumar R, Sharanyaa Pandima Dhevi S and Mahathi Guptha K 2019 Effect of Healing Agent Filled Microcapsules on Tensile Strength of Glass/ Epoxy Composite Laminate *International Journal of Vehicle Structures & Systems* **11** 250-254
[5] Muradyan V E, Arbuzov A A, and Smirnov Y, 2009 Adhesive Strength of Fullerene-Doped Epoxy amine compositions *Russian Journal of General Chemistry* **79** 797–799
[6] Alnefaie K A, Aldousari S A and Khashaba U A 2013 New development of self-damping MWCNT composites *Composites Part A* **52** 1–11
[7] Amit Kumar R, Asokan R, Kanu Priya Jhanji, Debajit Das and Vishnu Sai Nalla 2019 Investigation of Tensile Properties of Carbon / Epoxy Composite Joints with and without Carbon Nano-Tubes *International Journal of Vehicle Structures & Systems* **11** 209-213
[8] Khashaba U A 2014 Development and characterization of high performance nano - hybrid GFRE composites for structural applications *Journal of Composite Structure* **116** 523–534
[9] Khashaba U A, Aljinaidi A A aand Hamed A A 2014 Nanofillers modification of Epocast 50A1/946 epoxy for bonded joints *Chinese Journal of Aeronautics* **27** 1288–1300
[10] Meguid Sun 2004 On the tensile and shear strength of nano-reinforced composite interfaces

Materials and Design **25** 289-296

[11] Yu S, Tong M N, Critchlow G 2009 Wedge test of carbon – nanotube - reinforced epoxy adhesive joints *Journal of Applied Polymer Science* **111** 2957-2962

[12] Srivastava V K 2011 Effect of carbon nanotubes on the strength of adhesive lap joints of C / C and C / C–SiC ceramic fibre composites *International Journal of Adhesion & Adhesives* **31** 486–489

[13] Neto AS, Lopes da Cruza DT, Avila AF 2013 Nano-modified adhesive by graphene: the single lap-joint case *Material Research* **16** 592-596.

[14] Gkikas G, Sioulas D, Lekatou L, Barkoula N M and Paipetis A S 2012 Enhanced bonded aircraft repair using nano-modified adhesives, *Materials and Design* **41** 394–402.

[15] Dorigato 2010 Improving Epoxy Adhesives with Zirconia Nanoparticles *Composite Interfaces* 17 873-892.

[16] Messler R W 1993 Joining of Advanced Materials (*Troy, New York: Butterworth Heinemann*), 1st Edition, eBook ISBN: 9781483292144.

[17] Ayo S A , Olawumi O S and Michael O D 2016 Effect of dispersion method and CNT loading on the quality and performance of nanocomposite soy protein/CNTs adhesive for wood application, *Advances in Natural Sciences: Nanosciences and Nanotechnology* **7** 035005

Model Prediction of Microcantilever using DOE for Stress and Eigen Frequency Analysis for Force Measurement

Monica Lamba[a]*, Meetu Nag[b], Himanshu Chaudhary[a], Kulwant Singh[a]

[a] Department of Electronics and Communication Engineering, Manipal University Jaipur, Rajasthan, 303007, India

[b] Department of Mechatronics Engineering, Manipal University Jaipur, Rajasthan, 303007, India

E-mail: lambamonica346@gmail.com

Abstract. In this paper, stress and eigen frequencies are analyzed through prediction of microcantilever model using design of experiment approach. The important factors affecting displacement sensitivity, stress and eigen frequencies were identified such as geometrical dimensions of microcantilever (length, breadth and thickness), modulus of elasticity of materials, applied force. Based on the selection of factors and levels, experiments were designed using L9 orthogonal array. The designed experiments were simulated with finite element method in COMSOL Multiphysics software. Further, optimum values of dimensions of microcantilever and material property are used for stress and eigen frequencies analysis. The evaluated frequencies under eigen frequency analysis revealed that designed and simulated Gallium Arsenide (GaAs) microcantilever is prone to vibrate at these frequencies, whereas stress analysis illustrates the response of structure and material properties towards reliability.

Keyword: - Microcantilever, Force, Design of Experiment, Eigen frequency, stress analysis

1. Introduction

Recently, Micro-electro-mechanical systems (MEMS) technologies have gained a lot of attention due to its extra ordinary features and potential in field of microbotics, automobiles, aerospace, medical sciences, applications [1-4]. MEMS devices are fabricated, by taking advantage of miniaturization, low cost, less power, easy integration, highly resistant of heat, vibration and shock [5]. MEMS technology is mainly categorized into two classes i.e. Sensors and actuators [6]. There are various types of sensing and actuation mechanisms available in the industry. The selection of sensors and actuators depends upon, type application where it is implemented. This paper mainly focused on the sensing part of the MEMS technology. As per current market survey of MEMS sensor market is projected to reach $12.3 billion by 2024, at a CAGR (Compound annual growth rate) of 7% from 2017 to 2024 [7].Different types of mechanism are available for converting one form of energy into another, for force measurement microcantilever is one of the flexible, versatility, low cost and highly sensitive mechanical structure that can be utilized for vast range of applications including chemical, physical and biological [8-11]. For achieving this various input factors of microcantilever are identified which are affecting displacement, stress sensitivity of microcantilever in correspondence with the applied force. It is analytically proven from the famous Stoney's equation that the displacement of the microcantilever depends on dimension of microcantilever and material properties under consideration [12]. Loui et al. reported that microcantilever with length greater than width are optimal for point-loading applications such as microscopy and force measurements [13]. Lim et al.; Nachippan et al. mentioned that by increasing the length reducing the thickness and material with lower modulus of elasticity improves the sensitivity [14-15]. Neethu et al. analyzed from their study that sensitivity of the microcantilever increases by increase in length and reducing thickness [16]. Zhang et al. reported that minimizing the effective mass near the microcantilever's free end and the clamping width at the fixed end significantly increases the overall sensitivity of microcantilever [17]. From the above mentioned literature different factors affecting the sensitivity of the microcantilever is identified and taken into consideration for investigation in this paper.

Regardless of the above mentioned research work in the field of force measurement with microcantilever, still there exist certain gap towards optimization of microcantilever dimension with appropriate material identification from selected. To conquer this gap, in this paper initially factors affecting displacement and stress of microcantilever are identified, from the literature and analytical proven equation. Further identified factors are utilized for designing an experiment using Taguchi optimization to see the cumulative of various input factors on output. Taguchi approach is utilized as it successfully reduce the design time without compromising on the quality of the products as mentioned by Ghani et al.; Rosa et al. [18-19]. Designed experiments were simulated using COMSOL Multiphysics 5.3a software and the optimized model of microcantilever with respect of its dimension and material is obtained with higher displacement sensitivity. For this investigation materials under consideration are Silicon, Germanium and Gallium Arsenide. At high frequencies Gallium Arsenide (GaAs) microcantilever are less noisy and gives better results than silicon. Due to optical characteristics gallium arsenide material is most widely used in infrared light-emitting diodes, solar cells, laser diodes etc.

2. Design Methodology

2.1. Simulation Consideration

The purpose of this work is to utilize the mechanical properties of microcantilever for the designing of force sensor, which is able to operate in nanoNewton(nN) range. Microcantilever is fixed from one end and free from another end. Force is applied at the moving part of the cantilever and indicates output in form of displacement. As per the investigations conducted by [20-21] rectangular shape of moving part is more suitable for force measurement as compared to other shapes. The specification of the microcantilever is summarized in Table 1, indicates the operating range of sensor along with dimension of microcantilever. Dimension of the microcantilever is to be optimized using Taguchi approach. Mitra et al.; Taguchi et al. reported that Taguchi approach uses a specially design of orthogonal array to study the effect of various parameters with fewer number of experiments [22-23].

Table 1. Design parameter consideration of Piezoresistive MEMS Force sensor.

Parameter Considered	Value
Force range	0-100 nanoNewton(nN)
Dimension of Microcantilever	L μm × B μm× H μm (To be optimized by considering different values)
Simulated Platform	COMSOL Multiphysics 5.3a

The procedure of Microcantilever optimization using Taguchi approach is illustrated in the Figure 1.

IOP Conf. Series: Materials Science and Engineering **748** (2020) 012025 doi:10.1088/1757-899X/748/1/012025

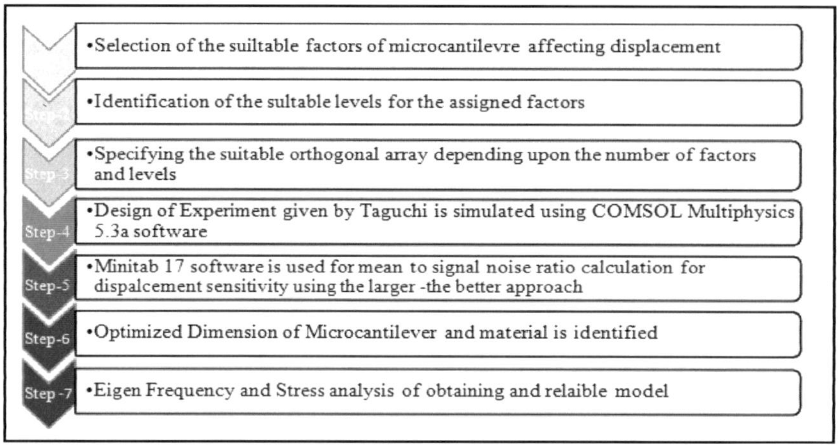

Figure 1. Optimization Methodology using Taguchi

2.1.1. Selection of factors for Microcantilever optimization: - Applied force at the tip of the microcantilever is responded in form of displacement. The deflection or displacement of the microcantilever is given by the relation shown below in equation (1) [24]

$$\Delta = \frac{4FL^3}{EWH^3} \tag{1}$$

Where is the deflection or displacement of the microcantilever, F is the applied force, E is young's modulus of elasticity, L, W and H is the length, width and height of the cantilever respectively. This equation indicates that Force (F) and deflection (Δ) are input and output quantities respectively. Further L, W, H is associated with microcantilever dimension optimization whereas E is material dependent parameter. So, the selected factors for optimization are L, W, H and E of the microcantilever. For optimizing the microcantilever length, breadth and thickness parametric sweep is applied in (50- 300) μm, (5- 30) μm and (1-5) μm ranges respectively in COMSOL Multiphysics 5.3a. The levels are selected on the bases of observed major variations in simulated results as illustrated in Table 2.

Table 2. Factors and levels selected for Microcantilever Optimization.

Factors	Level 1	Level 2	Level 3
Length(μm) [A]	100μm	150 μm	200 μm
Breadth(μm) [B]	10 μm	15 μm	20 μm
Thickness(μm) [C]	1 μm	2 μm	3 μm
Material [D]	GaAs (Gallium Arsenide)	Ge (Germanium)	Si(Silicon)

Material properties are also one of the important factors that need to be considered for simulation for designing of microcantilever. For this investigation the in-built properties of Silicon, Germanim, Gallium Arsenide in COMSOL Multiphysics software is tabulated in Table 3, indicates the young's modulus of elasticity, density and poison's ratio.

Table 3. Materials and its Properties for Simulation in COMSOL Multiphysics 5.3a.

Material Properties	Material Name		
	Silicon(Si)	Germanium(Ge)	Gallium Arsenide(GaAs)
Young's Modulus of elasticity(GPa)	170	103	85.5
Density [kg/m^3]	2329	5323	5320
Poisson's Ratio	0.28	0.275	0.31

2.2. Numerical modeling

For stress and eigen frequency analysis type of study selected in COMSOL Multiphysics 5.3a software is stationary and eigen frequency respectively. The device is modeled using solid mechanics module, as it is based on solving problem using Navier's equation and compute results in form of stress, displacement and strain. For boundary conditions, all boundaries are participating for the designed model. Boundary load is applied to boundary 4 whereas fixed constrain is boundary 1.As force is a vector quantity the direction of applied force is also taken into consideration. Force is applied in x-direction on the boundary 4.Free tetrahedral fine meshing shown in Figure 2, is selected with minimum and maximum element size of 0.04 and 4 requited for finite element analysis.

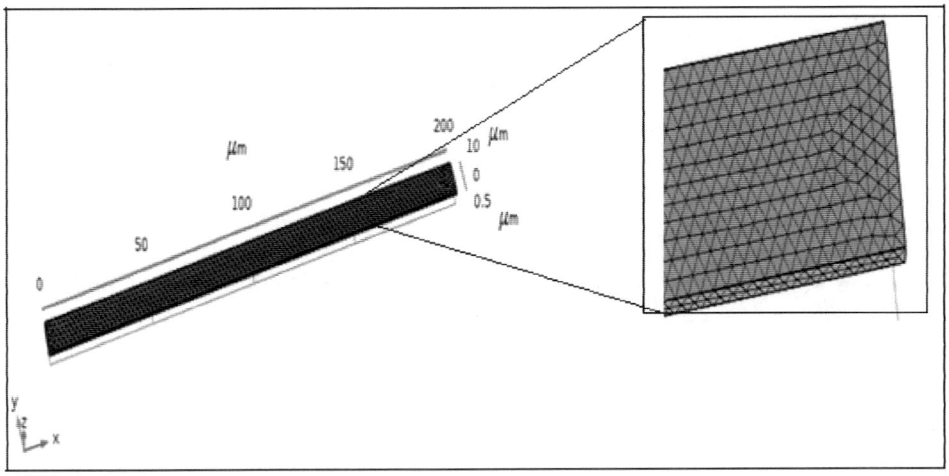

Figure 2. Free tetrahedral meshing of designed and simulated microcantilever

3. Design of Experiments Using Taguchi Approach

This work has adopted Taguchi approach for the design of experiment. It utilizes a special design of orthogonal array to study various input factors with fewer number of experiments. Three different types of signal to noise ratio methods are used to study the effect of various input factors on output i.e. Larger-the Better, Smaller-the Better, Nominal-the-better. In this paper larger-the-better approach has been used, as in accordance with the applied force displacement required to be highest. Depending upon the number of factors and levels orthogonal array using Taguchi approach is selected. For this investigation with four different factors at three different levels are there and Taguchi approach reduces the number of experiments from 3^4=81 to 9 experiments. Taguchi design of experiment methodology reduces the number of experiment from 81 to 9 without compromising the quality of output [21-22]. By using different combination of length, breadth, thickness and material the designed L9 orthogonal is shown in Table 4.

Table 4. Taguchi L9 (34) orthogonal array for selected factors and levels

Experiment No.	[A] (µm)	[B] (µm)	[C] (µm)	[D] (Material)
1.	100	10	1	GaAs
2.	100	15	2	Ge
3.	100	20	3	Si
4.	150	10	2	Si
5.	150	15	3	GaAs
6.	150	20	1	Ge
7.	200	10	3	Ge
8.	200	15	1	Si
9.	200	20	2	GaAs

The above tabulated nine experiments are designed and simulated using COMSOL Multiphysics 5.3a software using structural mechanism module followed by assigning boundary conditions and fine tetrahedral meshing. COMSOL software indicates the output in form of displacement corresponding with each designed experiment. The mean SN ratio of simulated outputs is calculated using Minitab 17 software by considering larger-the-better approach. The simulation output is used to evaluate sensitivity of force sensor design.

4. Results and Discussion

From the results it can be revealed that whenever there is need of conversion of one form of physical quantity (force) in another form (displacement) microcantilver is one of the simplest and cost effective mechanisms to do so. From this investigation, dimension of microcantilever is optimized using larger the better mean SN (Signal to noise) ratio. Combination of 200µm length, 10µm breadth, 1µm thickness and Gallium arsenide (GaAs) material developes with maximum displacement as compare to other selected levels and factors in force range from 1nN-100 nN as indicated in Figure 3.

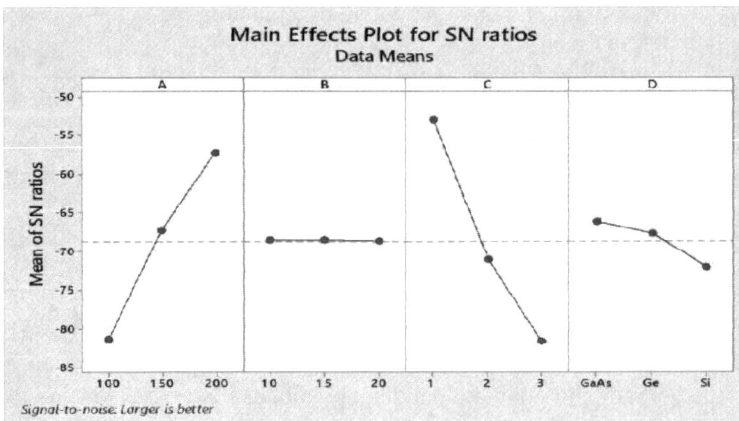

Figure 3. The Larger –the-better SN ratio for Microcantilever Dimension Optimization

Above mentioned dimensions and material both helpful in designing an optimized model of microcantilever for force measurement. From this optimized model two different types of analysis observed are indicated as follows:

4.1. Eigen Frequency analysis

Eigen frequencies are analyzed to identifies the exact value of frequencies at which mirocantilever is prone to vibrate[22]. When an object vibrates at a frequency equal to its Natural Frequency, its vibration amplitude increases appreciably this could lead to irreversible damage. The expression of the eigen frequency is given by the relation shown in equation (2)

$$f_n = \frac{1}{2\pi}\sqrt{\frac{Et}{\rho l^2}} \quad (2)$$

Where is natural frequency, E is young's modulus of elasticity, ρ is mass density, t is thickness and is the length of microcantilever. By simulating optimized Gallium arsenide microcantilever having dimensions 200μm length, 10μm breath, 1μm thickness first six eigen frequencies all in (KiloHertz) are identified and Tabulated in Table 5, indicates both the theoretical and the simulating frequencies. Theoretical frequency values are calculated using equation (2).

Table 5. Theoretical and simulated natural frequencies of optimized model of Microcantilever

Natural Frequency	Theoretical Frequency(KHz)	Simulated Frequency(KHz)
1.	15.60	16.293
2.	101.26	102.09
3.	161.15	162.17
4.	284.92	285.87
5.	559.11	560.28
6.	606.10	607.57

The simulated six natural frequencies all in Killohertz(KHz) is shown below in Figure 4, indicating the affect different frequencies on microcantilever.

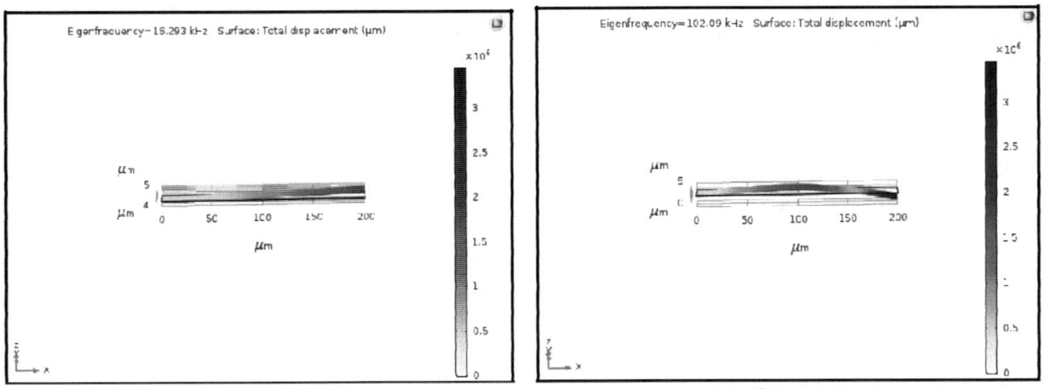

(a) (b)

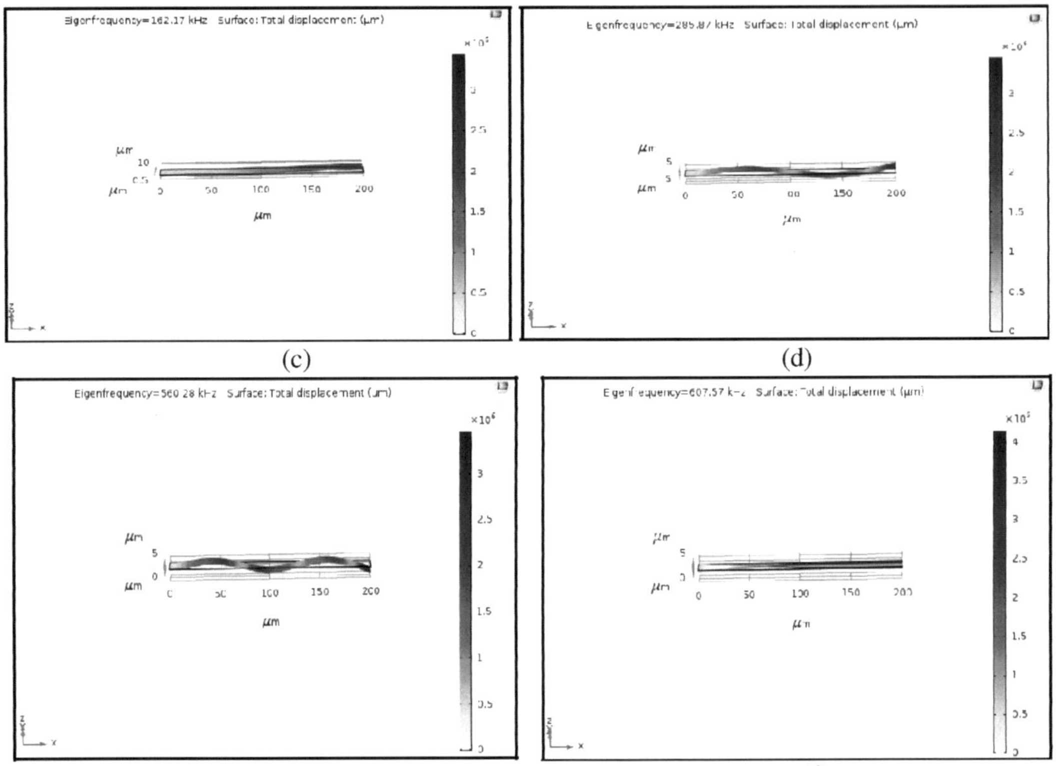

Figure 4. Simulated Eigen frequencies of GaAs optimized Microcantilever (a) 1 (b) 2(c) 3(d) 4(e) 5(f) 6

4.2. Stress Analysis

Stress causes deformation in the structure, can be observed in form of displacement shown in Figure 5,indicates that maximum stress is present at the joint of fixed and moving edges and as we move away from the joint stress reduces can easily be seen from the stress diagram with the variation in colour. Stress analysis is helpful in identifying the cause of the structure failure which is very important before specifying for an application where it can be implemented.

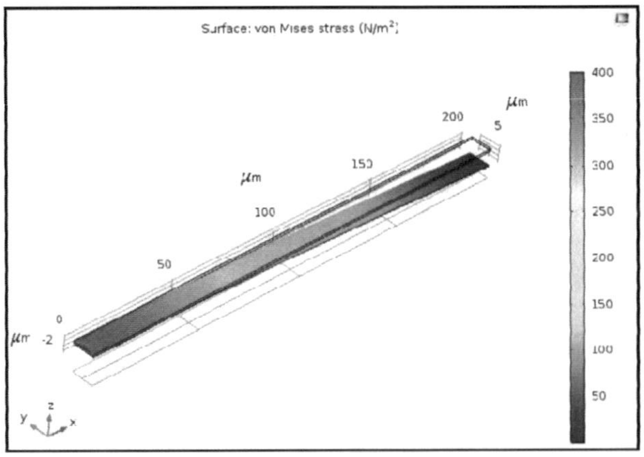

Figure 5. Optimized GaAsMicrocantilever Von Mises stress (N/ ^{2})

IOP Conf. Series: Materials Science and Engineering **748** (2020) 012025 doi:10.1088/1757-899X/748/1/012025

The range of force for which optimized model is operated is from 1nN-100nN. Figure 6(a) and (b) shows the linear fit curve in increasing order of force and their corresponding residual *v*/s independent plot. Similarly Figure 6(c) and (d) shows the plots for above mentioned parameters in decreasing order of force and their corresponding residual *v*/s independent plot.The relation between force(operating range) and displacement for the optimized gallium arsenide microcantilever model shows the linear relationship. As far as convergence study is concern the designed optimized models takes 11 itrations to solve the simulated model with a time period of 15mintues and 0.0016% of linear error,The linear error lies in the acceptable range.

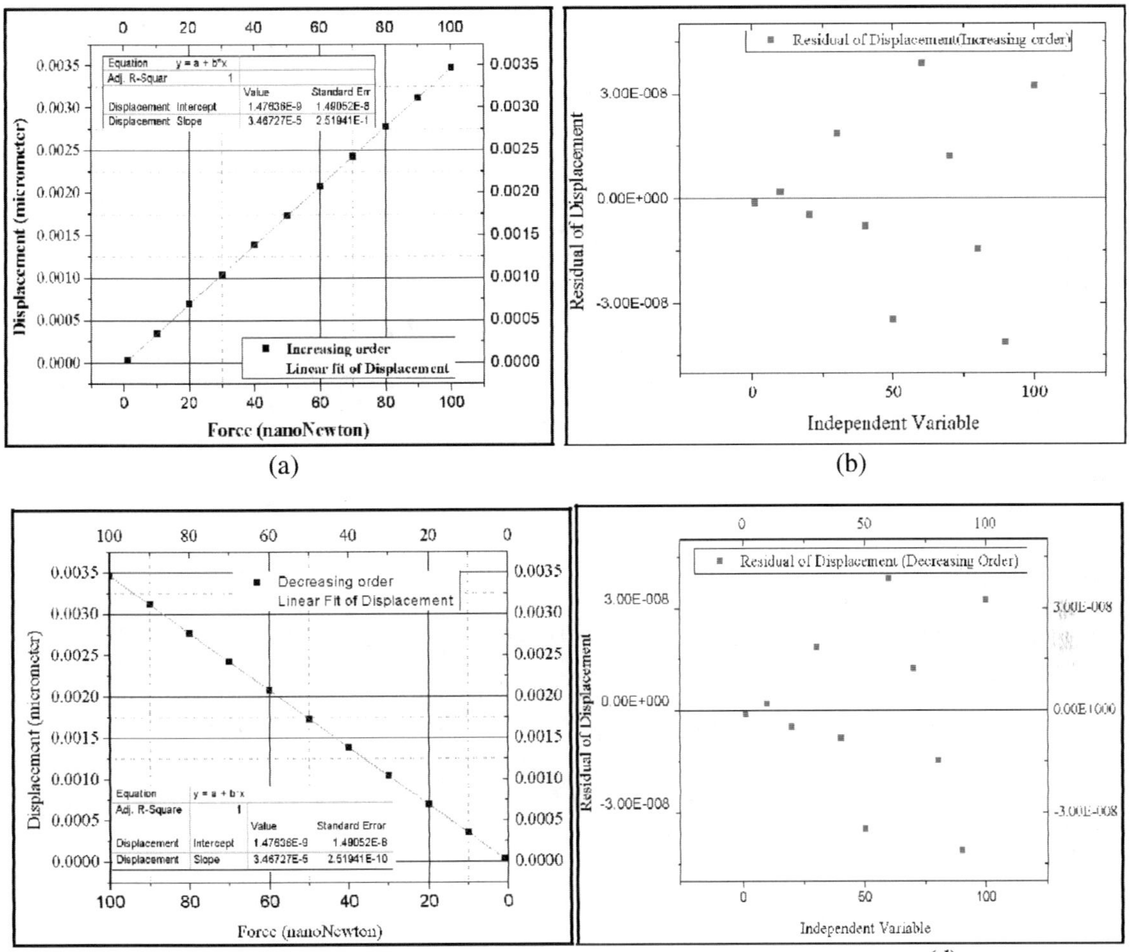

(c) (d)

Figure 6. (a) Linear fit curve in increasing order (b) Residual of sensor output in increasing order (c) Linear fit curve in decreasing order (d) Residual of sensor output in decreasing order

5. Conclusion

From this investigation it can be concluded that, Gallium Arsenide microcantilever with optimized dimensions of 200μm length, 10μm breadth, 1μm thickness responds to nanoNewton range forces with high sensitivity.The displacement sensitivity of 3.47E+04μm/nN is achieved with the optimized model. It is one of the cost effective method of converting force into displacement.Eigen frequency analysis provided a range of frequency from 15-608 KHz at which optimized model is not operated. Above and below this frequency range the optimized model can be operated.Stress analysis illustrates the response of structure and material properties towards reliability. Due to optical properties of the Gallium Arsenide material microcantilever is well suited in optical applications i.e. X-rays detection, solar cells, solar cars,optical fibers.

References

[1] Beyeler F, Muntwyler S, Nagy Z, Moser M and Nelson B J 2007 *IEEE Int. Conf. Intell. Robot. Syst.* pp 3803–3808
[2] Vinay P, Vamsi C V S S, Hemanth M, Saiteja A, Ali M A and Kumar P A 2017 *International Journal of Mechanical Engineering and Technology* **8** pp 424–434
[3] Lynch J P, Partridge A, Law K H, Kenny T W, Kiremidjian A S and Carryer E 2003 *J. Aerosp. Eng.* **16** pp 108–114
[4] Fitzgerald A M 2010 *IEEE Eng. Med. Biol. Soc.*pp 1–38
[5] Lueke J and Moussa W A 2011 *Sensors* **11** pp 1433–1460
[6] Faraday P P and University L 2002 *Prime Faraday Technology Watch-Wolfson School of Mechanical and Manufacturing Engineering* pp 1-51
[7] Dirjish M 2018 Printed/Flexible Sensor Market In An Upward Spiral *Fierce Electronics*.
[8] Vashist S K 2017 *J. of Nanotechnology* **3** pp1-18
[9] Karhade O G, Chiluveru S S, and Apte P R 2004 *IEEE/SEMI Advanced Semiconductor Manufacturing Conference and Workshop (IEEE Cat. No. 04CH37530). IEEE* pp 409-412
[10] Yang S M and Yin T I 2007 *Sensors Actuators, B Chem.* **120** pp 736–744
[11] Thundat T, Oden P I and Warmack R J 1997 *MicroscaleThermophysical Engineering* **3** pp185-190
[12] Feng X, Huang Y, and Rosakis A J 2007 *J. Appl. Mech. Trans. ASME* **74(6)** pp 1276–1281
[13] Loui A, Goericke F T, Ratto T V, Lee J, Hart B R, and King W P 2008 *Sensors and Actuators A : Physical* **147** pp 516–521
[14] Lim Y C, Kouzani A Z, Duan W and Kaynak A 2010 *IEEE/ICME Int. Conf. Complex Med. Eng. C.* pp 177–181
[15] Nachippan N M, Balaji V, Reddy S V S and Logesh K 2015 *Int. J. ChemTech Res.* **8** pp 349–356.
[16] Neethu K and Suja K J 2016 *Procedia – Procedia Comput. Sci.* **93** pp 146–152
[17] Zhang G, Zhao L, Jiang Z and Yang S 2011 *Journal of Physics D: Applied Physics* **44(42)** pp 425402
[18] Ghani J A, Choudhury I A, and Hassan H H 2004 *J. Mater. Process. Technol.* **145(1)** pp 84–92
[19] Rosa J L, Robin A, Silva M B, Baldan C A and Peres M P 2009 *J. Mater. Process. Technol.* **209(3)** pp 1181–1188
[20] Murat G and Shimoyama I 2004 *J. Micromechanical Microengineering* **14** pp 423-428
[21] Peiner E, Doering Æ L and Balke Æ M 2008 *Microsystem Technologies* pp 441–451.
[22] Mitra A 2005 *2nd ed. Prentice-Hall of India, New Delhi*
[23] Taguchi W Y, Chowdhury S 2005 Taguchi's quality engineering handbook *Wiley, Oxford*
[24] Feng, Huang Y and Rosakis A J 2007 *J. Appl. Mech. Trans. ASME*, **74(6)** pp1276–1281
[25] Mekalke G C 2016 *Int. J. Eng. Technol. Manag. Appl. Sci.* **4(2)** pp 60–66

A New User Identity Based Authentication, Using Security and Distributed for Cloud Computing

Sanjay Kumar[1], Syed Akbar Abbas Jafri[2], NishitArun Nigam[3,]
Nakshatra Gupta[4], Gagan Gupta[5] , Dr. S K Singh[6]
Student[2,3,4], Department of Information Technology, GCET
Asst. Professor[1,5],Department of Information Technology, GCET
Professor[6],Department of Information Technology, GCET
Email: dhananjay0397@gmail.com, skhakhil@gmail.com

Abstract. The Cloud based Computing is a trending technology in present IT scenario in which data is stocked in a Scattered or distributed manner & resources are shared among people. The data centers are widely distributed across the globe and are accessed by anyone at any time. Data is fetched, uploaded and manipulated by several users simultaneously hence identification and authentication of users who are fetching the data need to be done mandatorily without any error. In this Research Paper we have presented a new advanced security architecture for user identification which includes two factor authentication, AES based file encryption and decryption of data uploaded on cloud, admin verification and locking of users, fetching IP details of users, distributed database storage i.e. data is stored in tiers which means user login details is stored in one database &encryption/decryption details such as (files uploaded, key) is stored on different database. Whole work deals on enhancing security for cloud computing.

Keywords:Cloud Computing, Security architecture, AES, onetime password, Distributed database,Encryption, Description.

1. Introduction

In today's era of information technology, cloud computing has been one of the major outbreaks for developers and users. Cloud computing works on the concept of distributed computing. Cloud computing has a wide dimension of scope ranging from solving complex problems, huge volume of data storage, lesser storage cost, anywhere and anytime access of user data. In cloud computing a user can store their data and fetch that data, whenever they require, what they need is just to sign up to cloud service.

Now, accessing their services require some authentication method, traditionally used method make use of static password mechanism, in which user choose a password while registering and then login using that password. This method of keeping a password that is used for further login is known as one-factor authentication. But guessing these password (one-factor authentication) is easy, as user does not keep strong password, since they are hard to remember and even if they keep strong password then also one factor authentication is prone to key-logging attack (key logging attack is a attack in which attackers captures the key strokes of the host and gets its details).

One factor authentication is prone to password guessing because people don't often change their password. Therefore, security of the cloud is essential in today's world. cloud computing has a very high potential to emerge as a most used technology because of its wide dimensional scope, but still not used due to its security concern like where the data is stocked, its integrity concern etc.... all these concerns steps back users from using cloud services. The things of cloud storage that disturbs clients are decentralized distributed storage and scalability i.e. user do not know where and how its data is stored. User does not have the knowledge from which location and how data is fetched, and where the backup of the data is kept. Therefore, there must be proper cloud awareness among people to believe in service. Now by looking the above belief factor we understood that why cloud computing face challenge in emerging as lead technology, apart from these factors cloud data is also vulnerable to following attacks-

1. Password Guessing Attack- It includes key logging and various other technique to obtain users password. It may be either by hook or crook method.
2. Intermediate Middle Attack- It is an act of hacking server and gaining the password of the user. Here the attackers try to create a breach in the propagation medium and gain access to the secure data.
3. Phishing Attack- This is used to guess the user's password by creating a fake login page and fetching details.
4. Masquerade Attack- Attackers try to deceive the host by creating fake user identity.

Hence securing the users identity is the only way to gain user belief on cloud computing. Identity and Access Control for the Cloud Services must be provided to the user in cloud storage and computing services, this can be done using Admin Verification/Lock of users by checking the logs and looking for any abnormality in the IP pattern. Identity and access control must provide identity management for the cloud users.In this analysis we have suggested a better way of security mechanism to secure the cloud for identity-based user authentication and a new security architecture is used, here we have used a concept of distributed data storage to safeguard the user details and the stored data from the attacks of intruders and hackers. It also includes two factor authentications,AES based file encryption and description of data uploaded on cloud, admin verify and lock controls for users, fetching IP details of users, distributed database storage i.e. data is stored in tiers, user login details in one database and encryption decryption on different database i.e. the separate data gets stored on distinguished database.

2. RELATEDWORK

The data storage mechanism of the cloud is similar to the data storage mechanism of the storage peripherals like hard disc and pen drive but the only difference is where the data is collected (in cloud data is stored in remote location). Several studies on the security of cloud-based computing has previously been presented and analyzed.

The major four aspect of information security in cloud computing are –

1. Validation

2. Privacy
3. Solidarity
4. Obtain ability

Validation is the throughout process for Verifying the identity of the user/client, now in the previously researched work user authentication is done by password and data confidentiality is maintained using encryption/decryption algorithm. Efficient encryption algorithm must be used in order to ensure good security, to enhance the upload and retrieval of data. Previously the secure login and registration is done using one factor authentication i.e. making use of static password mechanism in which user choose the password while registering and login using that password which is vulnerable to be hacked.One factor authentication is prone to password guessing because it does not have any other security check. Encryption algorithm used in those security architectures was DES (Data Encryption Standard).It is a 64-bitFiestal block cipher based on the conceptual formulation of Fiestal rounds rule having in total 16 rounds and a massive key size of 56 bit. DES was most widely used encryption scheme adapted in the late 20th Century by N.I.S.T. (National Institute of standard and technology). it follows Fiestal Structure.

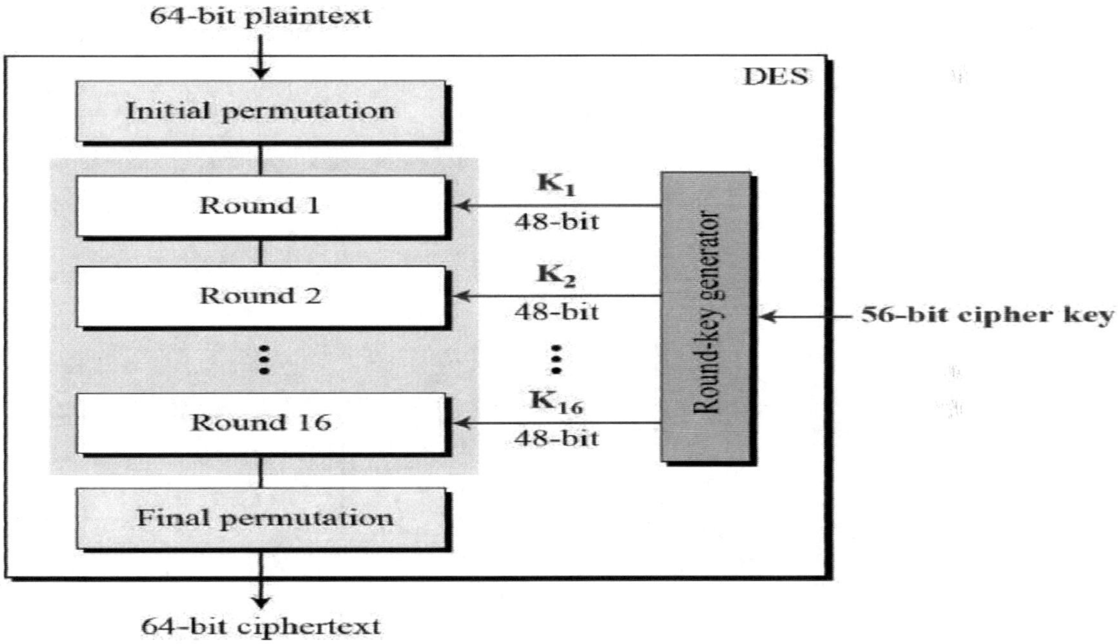

Figure 1DSS Algorithm Mechanism

Before the development & Implementation of DSS algorithm, the Simple Encipherment and Decipherment Technology was used to encode and decode the data stored in a computer based cloud related services .All the previously used techniques were brutally being cracked and misused as people have understood their loop holes and have identified their weak points which led to the low security in authenticating, validating and protecting user's data on cloud.

3. MODELPROPOSED

In our Research Modelwe have suggested work that deals with a new way of security mechanism including: -

- Admin Verification/Locking of User
- AES for Secured file encryption.
- Distributed data storage for user's ID details and Files Details
- One-timepasswordforauthenticationeach time user login.

All the algorithms and user authentication mechanism which have been proposed so far have some or more problem. the earliest cipher like (shift cipher etc.…) can be broken with ease on modern computation system. TheDES(Data Encryption Standard Algorithm) was patched/smashed in 1998 using a computer hardware that cost about $250000/- in 3 days. The triple DES was way too slow for quick and fastretrieval, updating &Manipulation. This problem can be solved if we Implement AES (AdvanceEncryption Standards) Algorithm for Encryption/DecryptionNFC is a new yet a very resourceful concept. The concern while developing such system was, would it also have practical application or would it just be limited to a theoretical term. But in recent times there has been an enormous growth in the number of NFC enabled devices.

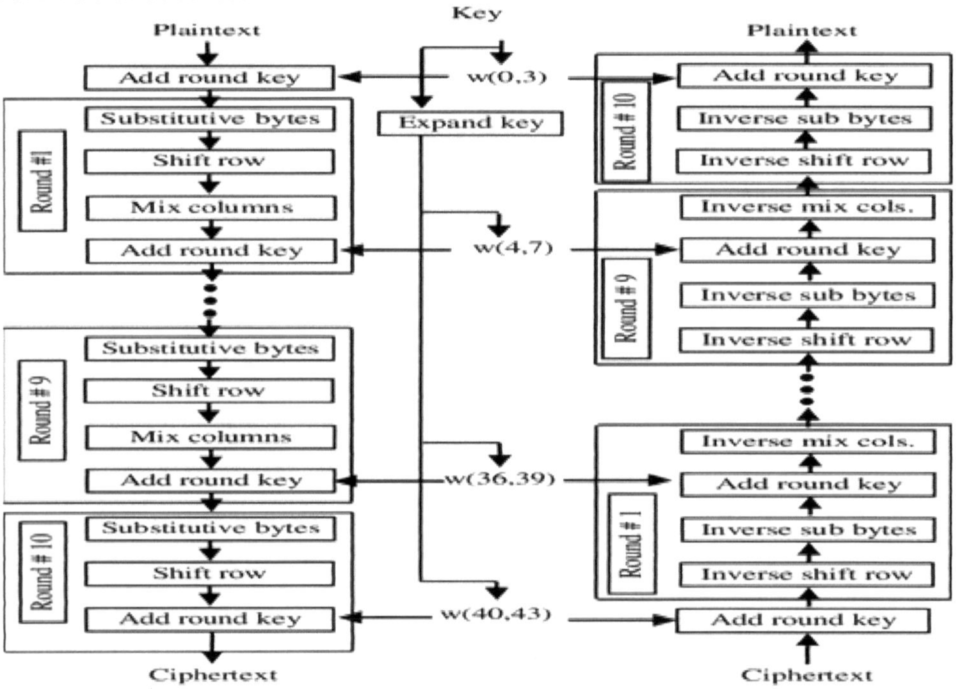

Figure 2. AES Algorithm Mechanism

Note that this AES algorithm mechanism wasn't developed by any organization; it is the work of the people who took part in alarge scalecompetition raised by N.I.S.T. to get an advancedencipherment and decipherment algorithm which follows a standard protocols and are organized in format. N.I.S.T. got several algorithm mechanisms on which several test were performed in which only 5 algorithms were considered for the final test in which the best algorithm found was AES (Advance Encryption Standard) Algorithm Security Proven architecture. AES is also a block based symmetry cipher system.

With several large companies investing in this technology, this technology looks to completely take over the contactless payment market in future. Google, Apple, Microsoft, Master Card are few of the many companies venturing and investing in NFC based payment systems.For the Security purposes it has a Registration page where the users first create a Strong Password for login purposes.

Figure 3. Login Page

After the user is registered, the Admin check for the authenticity of the user and accordingly verify the user or locks the user. User authenticity is checked by looking in the logs and if the user is found genuine then it is verified. After the verification of the user, user can easily access his/her account. To enhance the security TWO STEP VERIFICATION is done at every login i.e. user gets the OTP on their registered email id every time they login to cloud services.

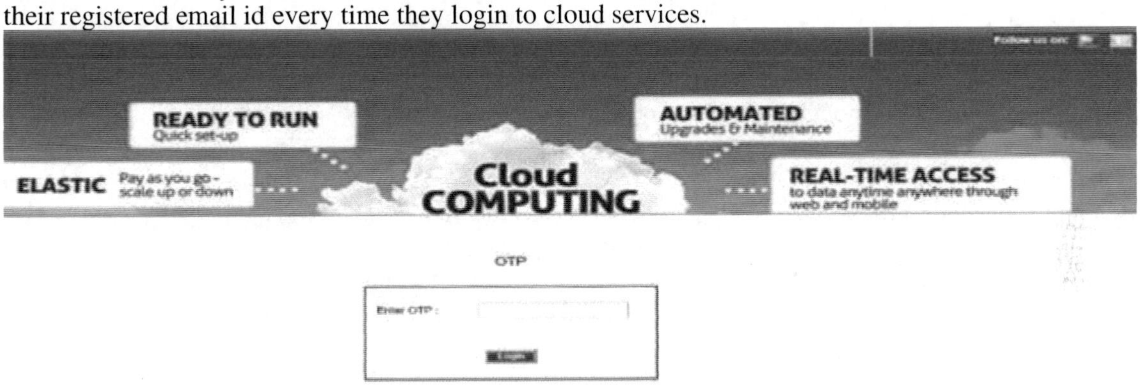

Figure4. OTP Match Popup

After the correct OTP matching, if the OTP is matched then user gets logged into the cloud service portal.Here the user can look into his/her Profile, Upload/Download Files On cloud using AES encryption/decryption Algorithm. Here the Database is distributed i.e. the database for the User Details , Encryption/Decryption & Key Storage is different so that if by any means hacker/attacker get into the User Detail Database , still he/she will not be able to access to the encryption and decryption algorithm key and therefore not be able to access uploaded data , since he/she will not have key so he will not be able to fetch or download the data .

IOP Conf. Series: Materials Science and Engineering **748** (2020) 012026 doi:10.1088/1757-899X/748/1/012026

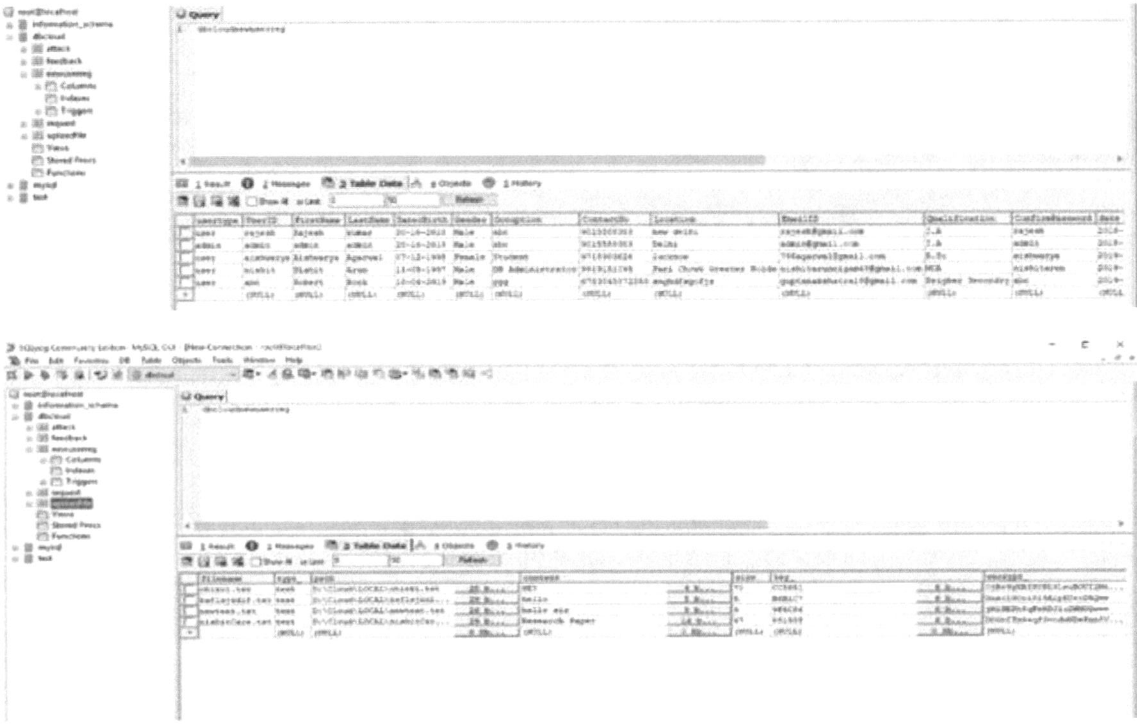

Figure5. Key Details

If by any means user gets logged into the system by stealing OTP or By – Passing OTP, then to fetch any details of the user he requires to enter the KEY, on failing which he gets a FAKE DATA of the information downloaded with garbagevalue.

This fake data will firstly deceive the hacker or the person who has gained un authorized access to the user's account and at the same moment of time it will protect the user's valuable amount of data to be stolen/copied/misused by attacker , hence we can suggest the this garbage data act as the bait to the attacker by which he feels he has succeeded hacking or creating a breach in the cloud architecture but unknowingly he has downloaded garbage content & triggered the alarm to the admin indicating that a malicious activity took place in the user's account .

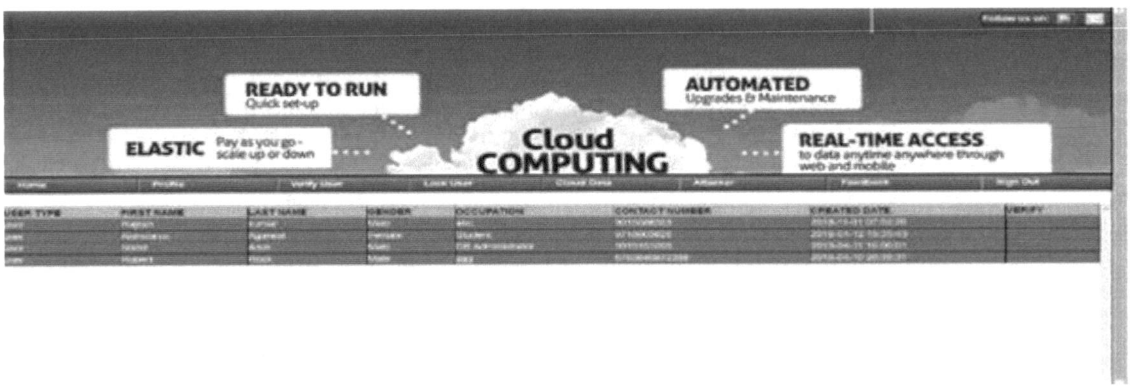

Figure6. Admin lock.

The admin can verify/lock the user by checking for the IPs,if VPN is connected then the IPs change every 5 minutes & hence by observing the malicious intentions of the user admin can lock the users account . After the account gets locked only the Admin is provided the right to Unlock the user using his best knowledge about the user.

Figure 7.AttackStocked IP Catch Details.

The Module work by grabbing the IP Address of the system through which the user has logged into the account, The IP Address of the system at that time is locked in the database and that IP Address are stored and stocked in the attack and hit database every 5 minutes.

The Admin can keep a check from which IP Address the Particular functionality is been performed like

1. From where the user has Logged In – LOGIN IP Address.
2. Date fetch and Upload Time with IP Address – IP Address recorded at manipulation time.
3. Date and Time the IP Address is being changed.

If the account is reported or any malicious activity is found then the Admin has a right to lock the user and perform necessary action to protect the data from being maliciously used.

A session window of 30 Minutes is kept, i.e. the user's gets automatically logged out from the account if he is kept inactive for that period/interval of tenure/time. This will help the user from the privileged access of their account from the user who try to take advantage from the user by gaining their faith and using their logged in account for un – ethical use and malicious intensions.

4 CONCLUSION

In this research paper we have suggested a big new/up-to-time security architecture for cloud based computing domain which includes two factor authentication , AES(Advance Encryption Standard) based file encipher and decryption of data uploaded on cloud , admin verification and locking of users, fetching IP details of users, distributed database storage i.e. data is stored in tiers which means user login details is stored in one database & encryption/decryption details such as (files uploaded , key) is stored on different database. Overall work deals on enhancing security for cloud computing, thisworkfortify security for completecloud-basedcomputing structure.

In our research based suggested working representation, ahacker/attacker cannot easily fetch information from the cloud databases and load their corrupted files as he require to getaccess all the database data centers, which is very tough and hard to breach or crack. All these attacks which could be performed on the previously implemented system could be reduced to zero enhancing the security of the cloud and at the same time identifying and servicing only the genuine clients.

References

[1] Yashpa lKadam, 5 October, 2011 "Security Issues in Cloud Computing A Transparent View", International Journal of Computer Science Emerging Technology, Vol-2 No ,316-322

[2] RohitBhadauria, RituparnaChaki, NabenduChaki, SugataSanyal, 2011 "A Survey on Security Issues in Cloud Computing".

[3] Mladen A. Vouk, 2008, "Cloud Computing – Issues, Research and Implementations", Journal of Computing and Information Technology CIT 16, 4,235–246

[4] Ye Hu, Johnny Wong, Gabriel Iszlai, Marin Litoiu, 2009 "Resource Provisioning for Cloud Computing", IBM Canada Ltd.

[5] Daniele Catteddu, Giles Hogben, 2009 "Cloud Computing:- Benefits, risks and recommendations for information security", November.

[6] Jiyi Wu, QianliShen, Tong Wang, Ji Zhu, Jianlin Zhang "Recent AdvancesinCloudSecurity",JOURNALOFCOMPUTERS,VOL.6,

[7] Trusted Computing Group, May2010"Solving the Data Security Dilemma with Self-Encrypting Drives".

[8] Hongwei Li, Yuanshun Dai, Ling Tian and Haomiao Yang, 2009 "Identity- Based Authentication for Cloud Computing", CloudCom, LNCS 5931, pp. 157–166,2009

[9] Sven Bugiel, Stefan Nurnberger, Ahmad-Reza Sadeghi, Thomas Schneider, 2011 "Twin Clouds: Secure Cloud Computing with LowLatency",CASED, Germany.

[10] Luis M. Vaquero, Luis Rodero-Merino, Daniel Morán, 2011 "Locking the sky: a survey on IaaS cloud security", Computing 91:93–118

[11] Yang Tang, Patrick P. C. Lee, John C. S. Lui, and Radia Perlman, 2010"FADE: Secure Overlay Cloud Storage with File Assured Deletion".

[12] Thuy D. Nguyen, Mark A. Gondree, David J. Shifflett, Jean Khosalim, Timothy E. Levin, Cynthia E. Irvine, 2010 "A Cloud-Oriented Cross-Domain Security Architecture", The Military CommunicationsConference, U.S. Govt.

[13] Kumar, Sanjay, AshishNegi, and J. N. Singh. , 2019 "Semantic Segmentation Using Deep Learning for Brain Tumour MRI via Fully Convolution Neural Networks." Information and Communication Technology for Intelligent Systems. Springer, Singapore, 11-19.

[14] Brian, A. LarsanAro, L. Arockiam, and P. D. S. K. Malarchelvi. 2014."An IOT based secured smart library system with NFC based book tracking." *International Journal of Emerging Technology in Computer Science & Electronics (IJETCSE)* 11, no. 5.

[15] Kumar, Sanjay, Gagan Gupta, and KunwarRajat Singh. 2015 "5G: Revolution of future communication technology." In *2015 International Conference on Green Computing and Internet of Things (ICGCIoT)*, pp. 143-147. IEEE.

[16] Monteiro, David M., Joel JPC Rodrigues, and Jaime Lloret. 2012 "A secure NFC application for credit transfer among mobile phones." In *2012 International Conference on Computer, Information and Telecommunication Systems (CITS)*, pp. 1-5. IEEE.

[17] Haselsteiner, Ernst, and KlemensBreitfuß. 2006"Security in near field communication (NFC)." In *Workshop on RFID security*, pp. 12-14.

[18] Dutot, Vincent. 2015 "Factors influencing near field communication (NFC) adoption: An extended TAM approach." *The Journal of High Technology Management Research* 26, no. 1: 45-57.

[19] Sharif, Rabia, Syed Waqas Ahmad, HirraAnjum, NaveedRamzan, and ShahidRaza Malik, 2014 "Effect of infusion time and temperature on decaffeination of tea using liquid–liquid extraction technique." *Journal of Food Process Engineering*37, no. 1: 46-52.

[20] Van der Aalst, Wil MP, and Akhil Kumar, 2003 "Xml–based schema definition for support of interorganizational workflow." *Information Systems Research* 14, no. 1: 23-46.

[21] Gavish, B. and Gerdes, J. 1998,"Anonymous Mechanisms in Group *Decision Support Systems C*ommunication. Decision Support Systems, 23(4), 297-328.

[22] Von Alan, R. Hevner, Salvatore T. March, Jinsoo Park, and Sudha Ram. "Design science in information systems research." *MIS quarterly* 28, no. 1 (2004): 75-105.

[23] Mulliner, C. 2009 " Vulnerability Analysis and Attacks on NFC-enabled MobilePhones". *In Proceedings of,* International Conference on Availability,Reliability and Security. 695-700.

[24] Strommer, Esko, JouniKaartinen, JuhaParkka, ArtoYlisaukko-oja, and IlkkaKorhonen, *2006* "Application of near field communication for health monitoring in daily life." In *International Conference of the IEEE Engineering in Medicine and Biology Society*, pp. 3246-3249. IEEE.

[25] Tiwari, Naveen Kumar, and Sanjay Khakhil. 2015 "Security System for DNS using Cryptography." *International Journal of Computer Applications* 120, no. 17.

[26] Gupta, Anil Kumar, ShashankShukla, SandeepSaxena, and Sanjay Khakhil,2015."A Journey towards Workflow Scheduling of Cloud Computing." *International Journal of Computer Applications* 123, no. 4

[27] Gupta, Anil Kumar, ShashankShukla, SandeepSaxena, and Sanjay Khakhil, 2016 "Bloom filter for elasticity in DBaaS." In *IEEE International Conference on Recent Trends in Electronics, Information & Communication Technology (RTEICT)*, pp. 672-677.

[28] Shukla, Shashank, Anil Kumar Gupta, SandeepSaxena, and Santosh Kumar, 2016 "An evolutionary study of multi-objective workflow scheduling in cloud computing." *International Journal of Computer Applications* 133, no. 14 14-18.

[29] Shukla, Shashank, Anil Kumar Gupta, Santosh Kumar Upadhyay, and Ashu Gupta. 2016"Wireless Sensor Data Collecting to MobileCloud Device." *International Journal of Computer Applications* 138, no. 11.

[30] AMAZON, MULTI-OBJECTIVE WORKFLOW SCHEDULING IN. "International Journal OF Engineering Sciences & Management Research."

Reducing soldering defects in mobile phone manufacturing company: A DMAIC approach

Anshula Garg[1], Kalhan Raina[2], Richa Sharma[3]

[1] Engineering Program/Project Management, Becton Dickinson and Company, Chandigarh, India-160101
[2] Process Engineering, Manufacturing Department, Lava International Limited, Noida, India-201301
[3] Mechanical Engineering Department, Amity University, Noida, India-201301
er.anshulagarg@gmail.com, raina.kalhan@gmail.com , richasahnivirgo@gmail.com

Abstract. A DMAIC (an effective tool) approach has been presented in this paper for reducing defects in soldering in an endeavor to approach six sigma quality level in a mobile phone manufacturing company. DMAIC helps expose and kill the root causes of defects in the process and thereby helps achieve higher customer satisfaction by means of improved quality. The case study was conducted in an electronics manufacturing industry engaged in an assembly of mobile phones. Post successful implementation of Six sigma - DMAIC in the assembly processes, ppm defectives dipped from 3800 to 200, which is a massive 94.74 percent reduction in ppm defectives.

Keywords: Six Sigma, Quality Control, DMAIC, Drag soldering, Defects, Mobile phone, Manufacturing, ppm

1. Introduction

Fierce competition these days, has made imperative for all the manufacturing industries to bring down their various types of waste. Minimizing wastes gives an industry unparalleled and distinctive edge over its rivals because it clearly shows how the former uses its diverse resources way more efficiently. Working on getting rid of these eight wastes of manufacturing viz. defects, over production, waiting time, non-utilized talent, transportation, inventory, motion and extra-processing are what turns a company into lean. Putting lean principles into action has its own rewards of making a company productive, profitable, better in terms of quality and hence achieve customer satisfaction.

Among these eight, one of the most unsought yet noteworthy waste materializes in the form of defects. This is where Six Sigma methodology comes to rescue and makes a company run lean by minimizing the ppm defectives. DMAIC stands for Define, Measure, Analyse, Improve and Control, and is a six-sigma tool used to define a problem onto a paper, scrutinize root causes of the defects and then control them by exercising manifold improvements in the process. In simpler words, DMAIC helps expose and kill the root causes of defects in the process and thereby helps achieve higher customer satisfaction by means of improved quality. Though it provides a sequence of well-oiled steps to check waste, a precise combination of approach, direction and tool is what you require for turning a project into success.

2. Literature Review

The DMAIC as an effective tool has attracted many researchers as an area of interest. The motive of this review is to highlight the significant improvements in the form of productivity and quality improvement. This section briefly discusses the implementations of six sigma – the DMAIC methodology and results achieved by various researchers.

Satyajit Mahato (2016) describes that Six Sigma is not limited to the manufacturing sector; it is also useful in a service industry. There was a reduction in default cases from 28.4% to 5.74%. Neha Gupta (2013) discussed quality improvement study applied to a yarn manufacturing firm based on six sigma methodology. The author specifically uses the DMAIC tool to streamline the processes and enhance productivity. Rahul (2014) investigated how the fast-changing economic conditions, such as global competition, product variety, customer demand for high quality product etc., had a major impact on manufacturing industries. To achieve these needs, the author used various strategies, such as ISO 9000, TQM, Six Sigma based DMAIC tool. Amit Yadav (Oct 2016) highlighted how the DMAIC methodology helped in achieving quality level in the automobile part industry. Successful implementation of six sigma had resulted in a reduction of DPMO from 68181 to 9090.9, and hence increased the sigma level from 2.99 to 3.86 with superlative solution. Varsha Karandikar (2014) effectively deployed the DMAIC approach for tackling issues like process variation, rejection and rework at a filter manufacturing industry. Prominent root causes were identified through the Cause and Effect Diagram, FMEA etc.

Punyisa Kuendee (2017) had used seven QC tools for minimising delay in delivery problem. The cause and effect diagram was used to find out the root causes of this problem. Using the Pareto chart cleared that the coordination problem was a major cause. Rahul Singh (2014) discussed how an application of Six Sigma i.e. the DMAIC methodology provides a framework for better quality and productivity improvement in a manufacturing enterprise. Ibrahim Alhuraish (2015) highlighted the effective combination of Six Sigma and Lean manufacturing in various industries of France. Nilda Tri Putri (2018) used Lean Six Sigma tool for reducing quantity of defects produced by cement bags company by identifying the sources and root cause analysis of non-conformance products. P. A. Marques (2016) proposed an integration framework where the life cycle stages inherent to a Lean or Six Sigma project can be systematically linked to the applicable causes and sub-clauses of ISO 9001:2015. Sushil Kumar (2016) stated that a Six Sigma tool-Taguchi approach is applied for defect reduction; further for analysing various significant process parameters of the casting process at a foundry in automobile casting industry. After reviewing the above papers, it was concluded and understood that the DMAIC is a core part of Six Sigma quality initiative. It refers to a data-driven improvement cycle used for improving, optimizing and stabilizing business processes and designs.

3. Case Study Objective

In this paper, the author(s) conducted case study in an electronics manufacturing company that assembles mobile phones. A DMAIC has been presented in this work. This is the approach to reduce defects in soldering in a mobile phone manufacturing company. DMAIC is used to kill the root causes of defects in the process thus improves quality. The objectives of this study are as given below:

- The proposed study focuses on reducing market Annual Failure Rate (AFR) and thereby reducing customer complaints.
- The presented study shows how the DMAIC methodology helps in reducing soldering defects quantity.
- The study also helps in reducing soldering defects ppm from 3800 to 200.

4. Research Methodology

- Define: This phase is the basis for all succeeding stages as it determines the needs of customers. Specific roles and responsibilities are assigned to different members of the team. A project charter is prepared.
- Measure: All the processes that influence customers are highlighted and defects generated in the same are determined.
- Analyse: This phase compels the reader to a question- WHY? It leads to the root cause of the defects in the process. The objective of this phase is to understand what causes the defects to generate.
- Improve: The objective of this phase is to brainstorm, devise and implement methods to bring defects under control.
- Control: This phase ensures that the new process after implementation of changes has all of its key variables under acceptable range. This is achieved using various tools like control charts, statistical process control (SPC), check-sheets etc.

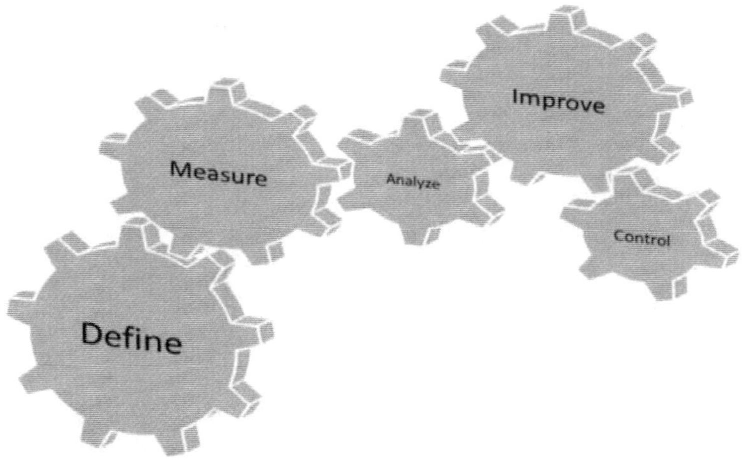

Figure 1. DMAIC Process

4.1. Define

During the VOC (Voice of Customer) analysis, it was found that almost 20 percent of the feature phones returned from the market due to the functional failure of LCD and camera. The cause of this problem was mainly bad processes followed during these drag soldering stages. The opportunity here was to create a standard process for soldering stages to help curb defects by 100%.

Table 1. Project Charter

S. No	Role	Department
1	Mentor	Quality HOD
2	Champion (Leader)	PQC In-charge
3	Team Members	Maintenance, Repair, Production, Process Engineering, Quality (Incoming, Process, Outgoing, Market), Reliability

To make sure that the VOC was translated properly to the CTQ (Critical to Quality), a cross functional team (CFT) was raised which comprised members each from incoming quality, production, process quality, outgoing quality, process engineering, market quality, maintenance, repair and training department. A PFMEA (Process Failure Mode Effect Analysis) meeting was held with a goal of reducing the defects of drag soldering from 3800 ppm to 1000 ppm (approx.) in the phase 1 and then further to 200 ppm in the phase 2.

Table 2. DMAIC Parameters

Phase	Objectives	Relevant Tools
Define	Define the end client Define its CTQ requirements Develop the high-level map of core processes involved	Project Charter CTQ Chart VOC Analysis
Measure	Measure the performance Map the magnitude of the problem	Data Aggregation Advance Excel
Analyze	Identify the root cause of the problems Prioritize opportunities to improve	Cause and Effect Diagram Histograms Pareto Chart
Improve	Innovative processes Develop and implement improvement plans	Brainstorming DOE (Design of experiment) FMEA (Process)
Control	Prepare Check sheets Control and standardize the improvements	Check sheets Control Charts Measurement System Analysis (MSA)

4.2. Measure

This phase involved collecting and measuring performance data to estimate the process capability and sigma level of the current system. Raw data was collected from all the three shifts for a period of 12 weeks. Histograms showing trends of defectives were plotted against all the three shifts.

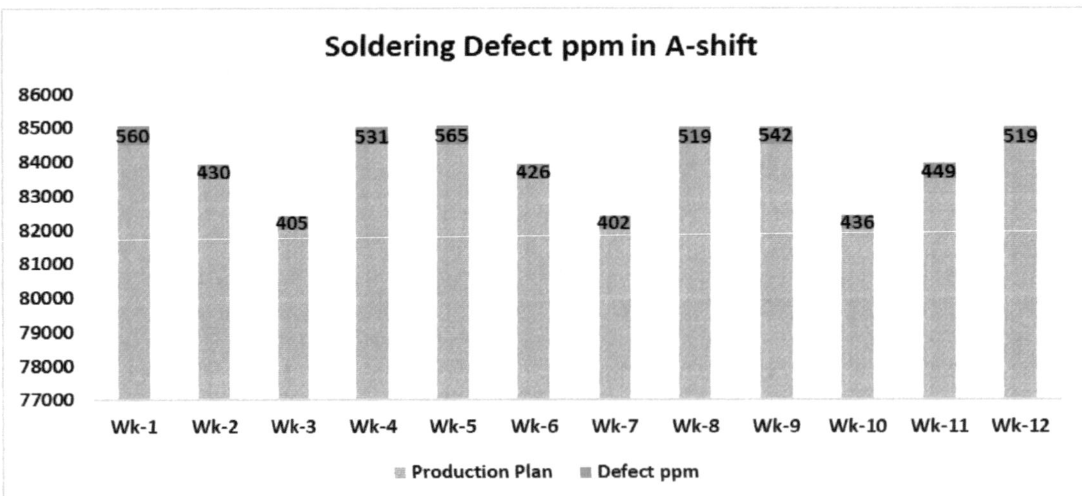

Figure 2. Soldering Defect ppm in A-shift

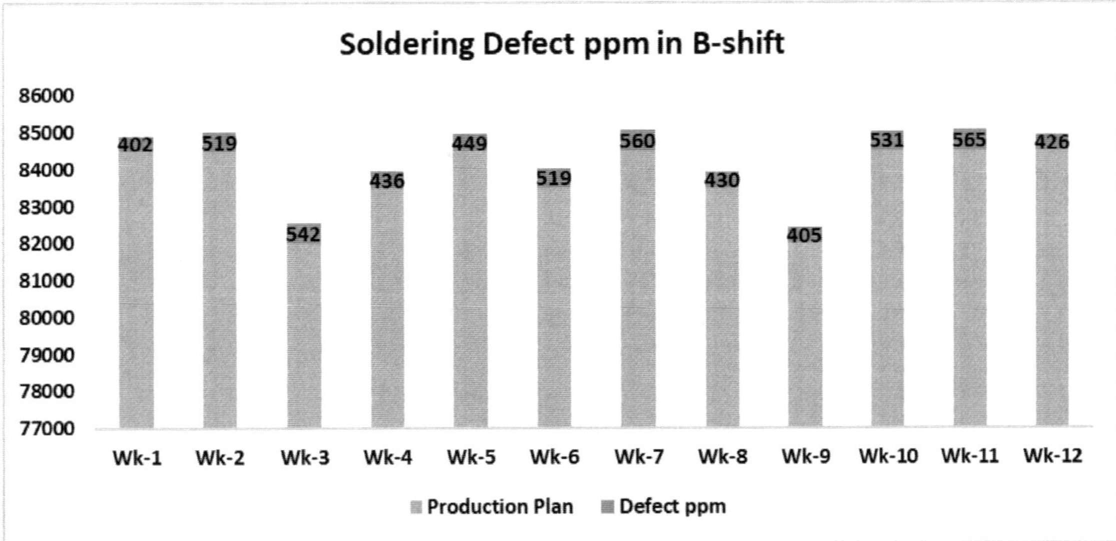

Figure 3. Soldering Defect ppm in B-shift

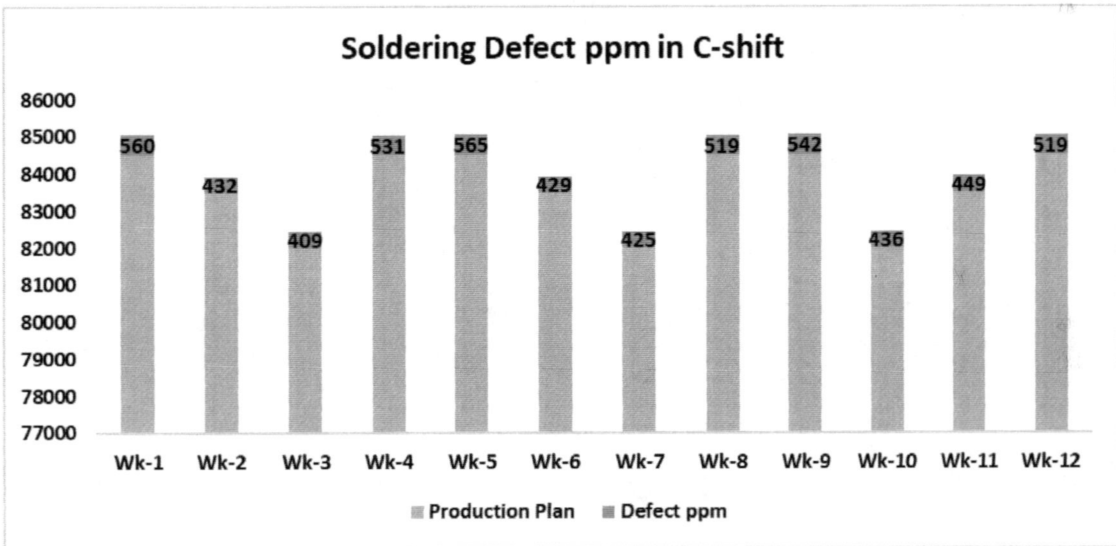

Figure 4. Soldering Defect ppm in C-shift

4.3. Analyse

The Analyse phase called for identifying and determining all factors that affected the process or procedure in the handling of the product from the beginning to the customer and lead to defects in LCD, LED, vibrator motor, mic, speaker, Bluetooth, camera etc.

For this purpose, GEMBA walks on the shop floor were conducted by team members shift-wise, line-wise, operator-wise. Various problems were identified and measured in the measure phase, leading to the non- conformance to the product were grouped in accordance with the 4M's (Machine, method, material and man) efficiency and were found to be as follows:

4.3.1. Machine

- Improper or unsuited temperature was being used during the soldering stages. When/ If the temperature at the bit end was lower than what was required, FPC's tinning would take more time for melting than the conveyor speed, resulting in low quality soldering. In contrary to this, when the temperature was on the high end, it would result in the burning of an FPC.
- Calibration was yet another cause which led to unsuitable temperature of soldering.
- Unsuitable bits were being used for soldering. There was no specification regarding which soldering bit was best for point-soldered and drag-soldered.
- Temperature controlled soldering station was not adapted. In an absence of the TCSS, there was usually a temperature loss during soldering of large FPCs.

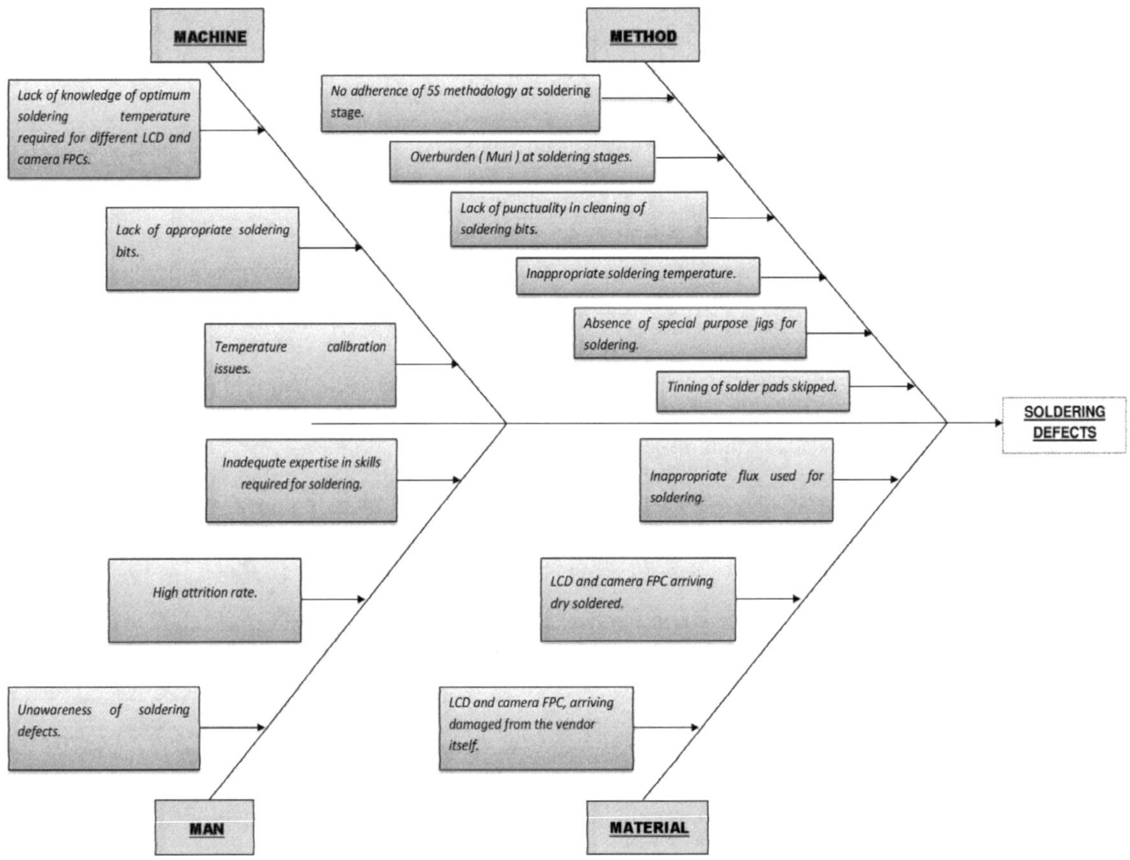

Figure 5. Cause and effect diagram analysis

4.3.2. Method

- No jigs were being used at soldering stations. This resulted in an unwanted movement while the PCB getting disoriented when soldering iron was dragged over them.
- Tinning of soldering pads was skipped.
- Huge variation in work content at different stages led to line imbalance problem. This further resulted in a less cycle time for soldering at soldering stages.
- There was no standard cleaning and sorting method adapted to various stages on assembly line. This further led to mixing of good and not good components.

- Low quality sponge was being used for cleaning of soldering bits. Further there was no instruction for method of cleaning and periodicity of cleaning at various soldering stages.
- Improper lighting (low luminous intensity) and magnifier were provided. There had been cases of improper lighting and hence low luminous intensity at the stages.
- Lack of soldering check stages on the assembly line.

4.3.3. Material
- Unsuited flux and soldering wire were being used.
- Damaged FPCs received from the vendor itself. There was no check point of checking FPC damage while unpacking LCDs and cameras on assembly line before soldering.

4.3.4. Man
- The operators lacked proper skills and expertise required for soldering.
- The attrition rate among operators was high which led to recruitment and training of new operators. This again made the process vulnerable to various defects.
- There was unawareness of knowledge of identification of defects among operators. This can be again related to training of operators.

4.4. Improve
In this phase, different problems were found during the analyse phase, which were targeted and improved using various actions as listed below: -
- The soldering temperature was increased to the range of 370-400 degree Celsius from 300-350 degree Celsius.
- Calibration was yet another cause which led to unsuitable temperature of soldering.
- As far as the use of soldering bits is concerned to all thin wired components like microphone, speaker, LED, Bluetooth wire and vibrator motor, round type bit was employed. For components with FPCs like camera and LCD, K series type bit was employed.
- To curb the issue of temperature loss during soldering of FPCs as in an LCD and camera, a normal soldering station was upgraded to electronics based on the TCSS embedded with a temperature sensor that keeps the tip temperature steady.
- PCB-specific jigs were designed and used for soldering. This resulted in the prevention of unwanted movement and the disorientation of the PCB when soldering iron was dragged over them.
- Another stage dedicated to tinning of solder pads was placed on assembly line whose sole work content was to supply PCBs to soldering stations after properly tinning their FPCs.

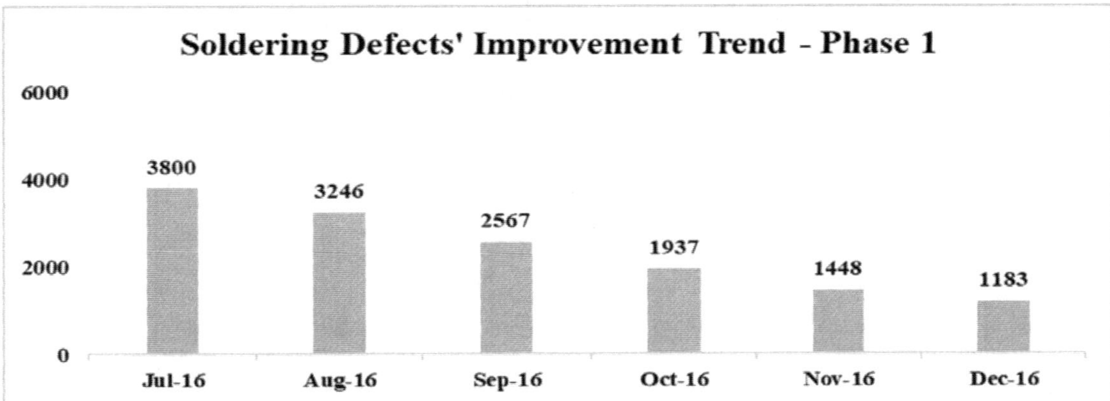

Figure 6. Soldering Defects' Improvement Trend – Phase 1

- Various Industrial Engineering tools like PMTS and ECRS were used to study the work content in order to distribute or eliminate them and hence bring cycle time within takt time.
- The well proven 5S methodology was introduced, adapted and taught at various levels in an organization. Also, mandatory 5S audits were planned after every couple of days.
- High quality brass and sponge were used for cleaning the soldering bits. Furthermore, there was no instruction in cleaning, method of cleaning and periodicity of cleaning at various soldering stages.
- Luminous intensity was improved from 600-700 Lux to 800-900 Lux.

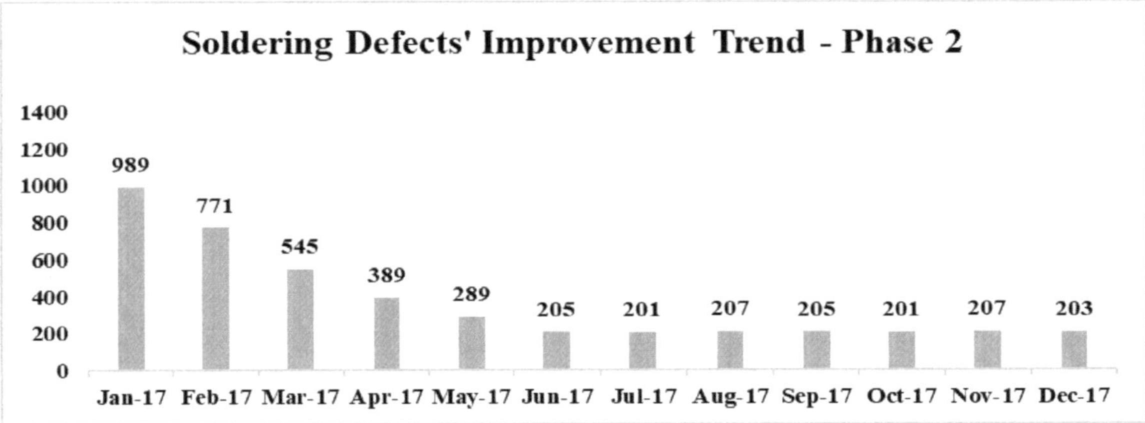

Figure 7. Soldering Defects' Improvement Trend – Phase 2

- A dedicated stage for checking the defects of soldering like dry soldering, solder balls, solder bridges was seated just after the soldering stages on the assembly line. In addition to this, the stage was provided with a magnifying glass so as to have proper and zoomed visuals of the defects.
- Now solder wire with flux 3.3% was made to be used.
- A full-fledged check sheet was developed containing check points that would otherwise get skipped during sampling of the PCBs from a vendor.
- A proper training and review program were developed for soldering operators.
- Appreciation and awards were distributed for achieving milestones in soldering operation.
- An in-house training school was established where training of soldering and in-depth knowledge of defects were imparted to assembly line workers.

4.5. Control

This phase required controlling or sustaining all the improvements in processes as adopted in the improvement phase. In other words, the purpose was to monitor the improvements that have been made and to assess the results that have been achieved through implemented kaizens. Moreover, a separate training school was developed, where operators are trained through the following activities:

- Soldering Manuals written in layman's language provide detailed information on all processes related to soldering with illustrations.
- Bulletin boards carrying pictures of correct and incorrect ways of performing soldering were installed on walls.
- High power magnifiers were installed for assembly line workers for better understanding of defects.

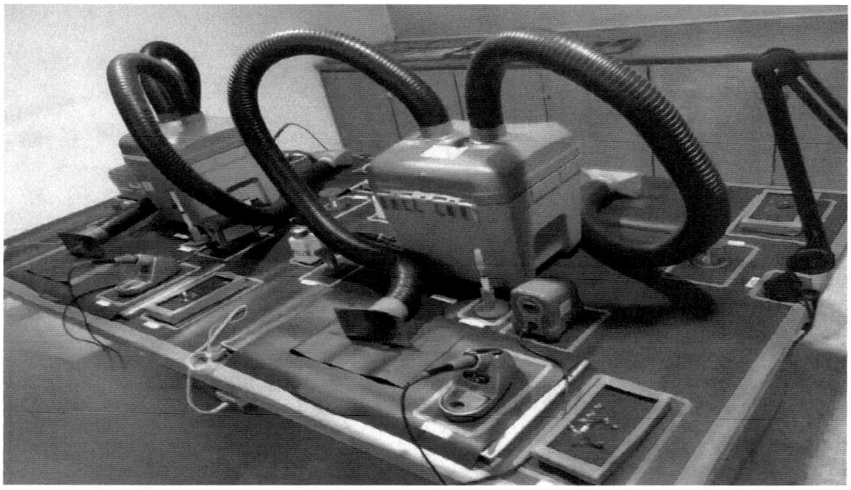

Figure 8. Workbench for soldering activity

5. Conclusion

This paper explains the DMAIC tool in its entirety and uses the same to counter the problems of soldering. Based on the results of this paper it can be seen how the DMAIC tool was effectively used to eliminate waste by reducing soldering defects from 3800 ppm to 200 ppm with a purpose to approach a six-sigma quality level in soldering processes. No doubt Lean Six Sigma is a challenging yet effective tool and not limited to the type and size of the industry. Whether small or large scale, it can be used in any type of industry, be it medical, cement, automobile industry, etc. It is not a standalone tool and hence uses combination of various tools viz. DMAIC, PFMEA, Quality control, you name it which makes it a robust package for improving process and hence deliver customer satisfaction par excellence. Having said that it can be concluded that the implementation of six sigma calls consumes time, effort, money and ground-breaking improvements; however, it still is a process improvement tool worth giving a shot.

Acknowledgement

We would like to express our deep gratitude to Professor Dr. Richa Sharma (Amity University, Noida) for her valuable and constructive suggestions during the development of this research work. We appreciate her willingness to give us her valuable time so generously. I would also like to thank Harsha Goel, Vipan Singla, Shushil Kumar, Satish Kumar Singh, Navneet Singh, Arun Bhatt and Arun Nagar for their valuable support.

References

[1] Alhuraish, I., Robledo, C., & Kobi, A. 2015, October. The effective of lean manufacturing and six sigma implementation. In 2015 International Conference on Industrial Engineering and Systems Management (IESM) (pp. 453-460). IEEE.

[2] Al-Aomar, R., Al-Mansouri, A., & Al-Laban, M. 2017, October. Reducing the interruption of power distribution: A six sigma application. In 2017 2nd International Conference on Knowledge Engineering and Applications (ICKEA) (pp. 226-230). IEEE.

[3] Amit Sahay and Ekta S Mehta, 2014 " A Practical Approach to achieve Six Sigma – Illustration through Case Study," International Journal of Current Engineering and Technology, pp 41-46.

[4] Coetzee, R., Van der Merwe, K., & Van Dyk, L. 2016. Lean implementation strategies: how are

the Toyota Way principles addressed?. South African Journal of Industrial Engineering, 27(3), 79-91.

[5] Galli, B. J. 2018. What risks does lean six sigma introduce?. IEEE Engineering Management Review, 46(1), 80- 90.

[6] Gupta, N. 2013. An application of DMAIC methodology for increasing the yarn quality in textile industry. IOSR Journal of Mechanical and Civil Engineering (IOSR-JMCE), 6(1), 50-65

[7] Karandikar, V., Sane, S., Sane, S., Jahagirdar, S., & Shinde, S. 2014. Process Improvement in a Filter Manufacturing Industry through Six Sigma DMAIC Approach. International Journal of Current Engineering and Technology, 4(4), 2546-2556.

[8] Kuendee, P. 2017, April. Application of 7 quality control (7 QC) tools for quality management: A case study of a liquid chemical warehousing. In 2017 4th International Conference on Industrial Engineering and Applications (ICIEA) (pp. 106-110). IEEE.

[9] LARIS, I. 2015. Evaluation of the Operational Performance in Implementing Lean Manufacturing and Six Sigma.

[10] Mahato, S., Dixit, A. R., & Agrawal, R. 2016, March. Process excellence in IT sector in an emerging economic scenario. In 2016 3rd International Conference on Recent Advances in Information Technology (RAIT) (pp. 636-640). IEEE.

[11] Marques, P. A., Meyrelles, P. M., Saraiva, P. M., & Frazão-Guerreiro, F. J. 2016, December. Integrating lean six Sigma with ISO 9001: 2015. In 2016 IEEE international conference on industrial engineering and engineering management (IEEM) (pp. 894-898). IEEE.

[12] Navas, R. K. B., Akash, R. P., Sathish, G., & Azharudeen, J. M. 2016, Decembe. Six Sigma in Education: Examination Result Analysis Using Six Sigma-A Case Study. In 2016 IEEE 4th International Conference on MOOCs, Innovation and Technology in Education (MITE) (pp. 245-250). IEEE.

[13] Putrì, N. T., Hidayat, B., Yusof, S. R. M., Fatrias, D., & Indrapriyatna, A. S. 2018, March. Reducing bag defects using six Sigma methodology: A case study in a cement industry. In 2018 7th International Conference on Industrial Technology and Management (ICITM) (pp. 11-16). IEEE.

[14] RUS, A. M., ZERBES, M. V., KIFOR, C. V., RUS, C. G., & GRECU, V. 2014. PROCESS IMPROVEMENT USING DMAIC METHODOLOGY IN AUTOMOTIVE INDUSTRY. CASE STUDY. Academic Journal of Manufacturing Engineering, 12(4).

[15] Singh, R., & Kumar, S. 2014. A Basic Approach to Six Sigma in SMEs to improve Production, Quality & Management System. International Journal of Current Engineering and Technology, 4(6), 4238-4241.

[16] Yasamin, E., Matin, M., & Brun, A. 2014. DMAIC Six Sigma definition and an analytical implementation: car industry case studies.

[17] Yadav, A., & Sukhwani, V. 2016. Quality improvement by using six sigma DMAIC in an industry. International Journal of Current Engineering and Technology, 41-46.

Angular Flame Hybrid Stove with Auto Align Base Frame

Apurva Soni, Praveen Kumar Maduri, Kushagra Singh, Rishabh Maurya

Galgotias College of Engineering and Technology, Greater Noida, India

Email- apurvasoni1836@gmail.com

Abstract- Liquified petroleum gas, a non-renewable source of energy is used on a large scale as a fuel in cooking. Increasing amount of fuel consumption year after year will lead to the depletion of the fuel. In this article, author proposes to combine more than one energy for cooking purposes. Combining electromagnetic induction technique with the conventional liquified petroleum gas burners will remove the dependency on one kind of energy making it available for more time. The proposed smart stove will consume much less space and provide better cooking experience while keeping the non-renewable energy conserved and ensuring the safety of the user.

Key Words: Liquified petroleum gas, non-renewable energy, fuel, electromagnetic induction, smart stove, safety.

1. INTRODUCTION

Liquified petroleum gas (LPG) is used as cooking fuel all over the world. LPG consists of propane and butane gases which are abstracted from crude oil. Crude oil is a non-renewable source of energy which is being imported from other countries in India. India itself has a very large market for LPG as all the conventional gas stoves use LPG, to reduce the usage of the fuel electric cooktops came into existence but the power consumption was high in these cooktops, to reduce the power consumption and making cooking much safer induction stoves [**1**] were introduced. In a country like India where villages are still in the process of development and where literacy rate is not 100% removing conventional gas stove with induction stove is not possible to overcome this obstacle a gas stove with both induction and gas burner [**2**] is introduced but the main drawback of this design is that it occupies much space and as both the burners can work at same time double the energy is being wasted. In India where population is growing at a high rate and technology is getting smaller day by day the available space for a specific task is reducing.

The system proposed in this article consists of a single burner which works on both LPG and induction i.e. Electricity. By using both the energies in one stove we are reducing the size of the stove and as only one mode can work at one time energy is not getting wasted. the holes of the burner [**3**] are designed in such a way that no hot spot is seen on the utensil which makes cooking process faster and thus reducing the consumption of non- renewable source of energy.

To reduce the consumption of Liquified petroleum gas, electronic and induction stoves were invented but consumers are much dependent on domestic burner which uses liquified petroleum gas only to cook food. Completely removing gas stoves on which the whole population of country is dependent on for cooking is not a good idea. Besides replacing, both gas stove and electromagnetic induction stove are combined together in a single burner to reduce space and reducing the consumption of one kind of non-renewable energy.

To make the burner safe for user use an LPG sensor [**4**], metal detector, thermistor and Node MCU is used to make it smart and safe for the user as the user can monitor the system [**5**],[**6**].

2. PROBLEM STATEMENT

Population is increasing at a very high rate thus the demand for basic needs is also increasing. Food is the basic requirement of human body and cooking it is must to kill bacteria and other germs in food.

To cook food LPG is being used at a very high scale about 75% of crude oil from which LPG is extracted is being imported from other countries.

As this cooking fuel Liquified petroleum gas, is stored in closed cylinders there are times in our daily life where we are short of liquified petroleum gas which leads to unavailability of food and inconvenience in our daily life routine. People keep spare induction stoves to avoid such situations but it consumes more space and money just for solving an inconvenience.

Generation today is working towards preserving our mother earth by using less amount of non-renewable energy but due to one obstacle or other we are always one step behind for example: old age people have difficulty in understanding new induction stoves while youngsters easily understand and uses the technology here we have the technology gap which can be removed by combining both the techniques thus making it usable by old and young both and along with that using least amount of Non - renewable energy.

In restaurants where food is being prepared at a very large scale just to keep the food hot a lot of LPG is wasted making food costly and exhausting a non-renewable source of energy.

3. DETAILS OF THE SYSTEM

Working of hybrid induction smart gas stove has many processes to have a better understanding of the processes they are given below in detail: -

1. Heating of utensils

In this hybrid induction smart stove, a switch is provided which has two operating modes. First – Induction mode, second - Gas mode. When the switch is turned to gas mode, a signal is received by microcontroller which commands the metal frame to move upwards to a length of 1.5cm. Motor is connected to the vertically upward moving frame ***(fig: 1)*** which rotates and moves the frame vertically upwards which is used to place the utensil while cooking on gas stove. The frame which is used to place the utensil for cooking is connected to an open circuit ***(fig: 2)***. When the utensil is placed on the frame the circuit gets closed giving signal to microcontroller which in return gives command to the valve to open for the flow of LPG. Parallelly microcontroller commands the gap igniter to ignite the gas coming out of valve, gap igniter used 5v dc to ignite the LPG. The holes through which the gas is coming out is not cut in circular mode but are cut in horizontal mode in a shape of rectangle***(fig:3)***. There is a tube through which the gas comes out through rectangular holes which are inclined at an angle to increase the area which is exposed to the flames. There are 4 such lines from where gas can come out which can be controlled with a knob. The LPG coming out through horizontal holes deduces the time consumed to cook food thus reducing the consumption of LPG gas while working on Gas mode.

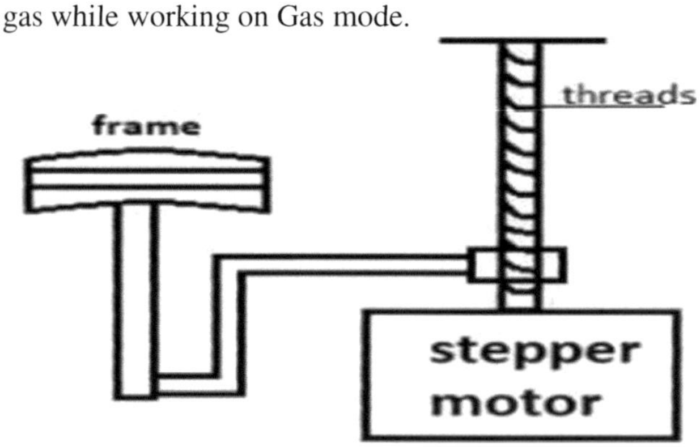

fig:1 mechanism to pop up frame [7]

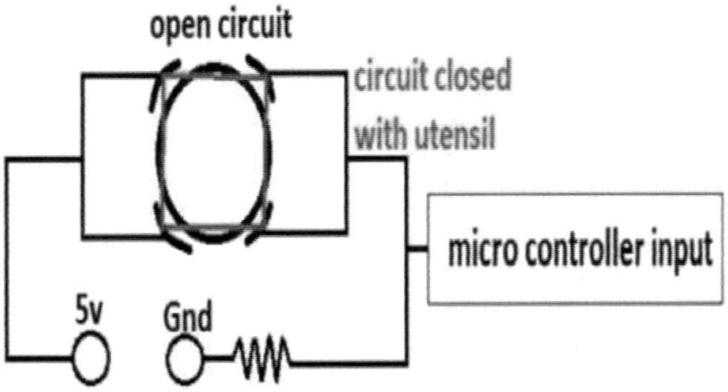

fig:2 proposed by author

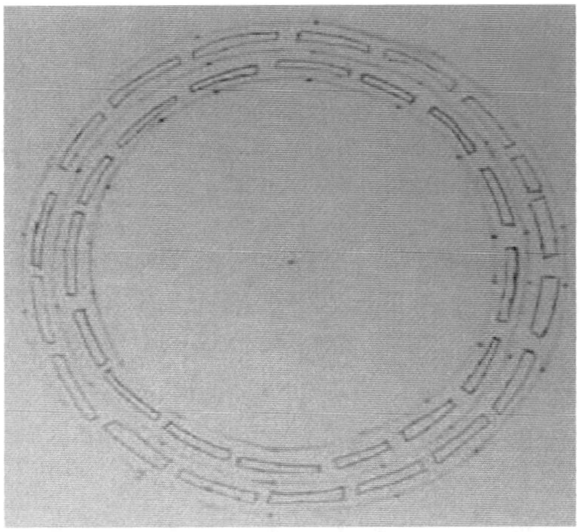

fig:3 proposed by author

When the switch is turned to induction mode the frame which is connected to motor moves vertically downwards giving space to place the utensil on cook top on the copper coil which is isolated with the help of ceramic plate. As soon as we place the utensil on cooktop due to electromagnetic induction eddy currents will be produced which will start heating the utensil for cooking. There are two coils which are under the ceramic plate which can be used separately depending on the size of the utensil. We can use utensil of any size on the cook top as we are using two coils to produce electromagnetic induction.

A thermistor is placed in the centre of the plate to keep the temperature under control. A voltage-controlled oscillator is used to control the eddy current which produces heat in the utensil that is, a voltage- controlled oscillator is used to control the temperature of the utensil in which cooking takes place. There is a fan connected to micro controller, when there is rise is temperature which is detected by the thermistor the fan starts and reduce the temperature of the coils.

2. AUTOMATIC SENSING PROCESS

In hybrid induction gas smart stove sensors are connected to microcontroller to make the stove safe for the users. MQ-5 the liquified petroleum gas sensor is used to sense any leakage of LPG, when MQ-5 detects any leakage microcontroller commands the gas valve to close and send a message on phones with the help of Node MCU [**5**]. Thermistor is used to detect the temperature of the induction copper coil, when temperature rises above a limit microcontroller sends a command to fan to start which reduces the temperature of the coil. Metal detector is used to detect on which coil the utensil is placed thus saving energy from being wasted, coil on which metal is detected is turned on and produces heating effect in the utensil. When an area of more than 50% is covered by the utensil only then the current in the coil moves thus producing heat in the utensil. Node MCU is connected to the micro controller to give user a wireless control of the knob to adjust the flame level among *low, medium, high, full.* To make the knob move automatically a servo motor is used connected to the knob which can be used manually and through mobile phones. The whole system is connected through Wi-Fi when the system is not connected then the motor will work like a simple knob to change the flame intensity [**6**].

3. SOURCE OF ENERGY

Cooking is highly dependent on Liquified petroleum, which with time is getting depleted thus to avoid any crisis we are integrating two types of energy together that are electricity and liquified petroleum gas. This burner is different from any burner till date available in market as it can work on both liquified petroleum gas and electricity that too on same burner without using up much space in the kitchen. Factor which makes it unique is whenever there is unavailability of one energy it will continue to work not producing any inconvenience in daily life. There are times if we want to switch from one energy to another, we have to change the whole burner but, in this burner, we can use it on electricity and gas both or either one of them without changing the burner.

4. WIRELESS CONTROLLING OF SMART STOVE BURNER

Node MCU is connected on micro controller *(fig:4)* which on the one end is connected to mobile phone and on the other with stepper motor to adjust the flame of the burner when on gas mode. This gives wireless control of burner to user making the stove smart and safer as the flames can be adjusted from a distance**.**

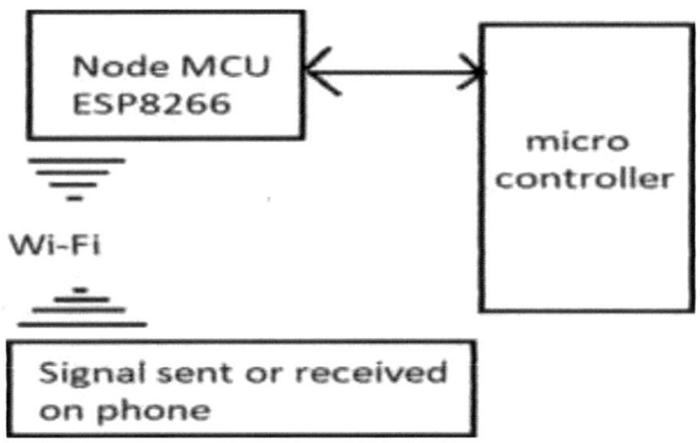

fig:4 Node MCU connections

5. VERICAL MOVEMENT OF FRAME

The gas stove needs air to function properly and a stand to place utensil. In conventional stoves a fixed metal frame is given on which utensil is placed but, in this invention, ***fig:1*** we are using movable 4 pillar like structures arranged in a circular pattern to give the utensil stability. Metal frame is moving vertically up and down to provide space for gas stove and induction stove as we need plain surface for induction stove while we need air between the utensil and gas in Gas stove. This mechanism is introduced to make the system convenient for the user. The frame is connected to a metal piece which is moving on a pole which has threads on it. One end of the thread pole is fixed and the other end is connected to motor, when motor rotates in clockwise direction the frame moves vertically up while when the motor moves in anticlockwise direction the frame moves vertically downwards. Motor is connected to microcontroller which commands the motor to move when the switch is turned to Gas mode.

6. MOVEMENT OF LPG GAS

when the switch is on the gas mode and utensil is placed on the frame open circuit gets closed then only microcontroller commands the gas valve to open. Microcontroller commands flame to burn at low mode. If the user wants to cook at any other mode or increase the intensity of the gas flame it has to be done manually by rotating the knob. To increase the intensity of the flame 4 valves are connected separately to different rings from which gas comes out. After the gas moves from the valve holes are provided in the pipe at a regular distance for the flow to air in LPG without air burning of LPG does not take place thus for proper burning and flame exposure holes for air are provided. The pressure of gas is so high that no leakage is observed through these holes.

7. HORIZONTAL HOLES OF BURNER

In hybrid induction gas smart stove horizontal holes are used besides circular holes which would increase the contact surface of the flame with the utensil and thus preventing heat spots on the utensil. Horizontal holes will cover better area than the circular holes when utensil is placed over flames. As the area covered by the flame is more the heat distribution is more and faster as compared to the conventional burners [**8**].Holes have a metal rectangular piece attached to it inside the pipe so that the gas which comes out is inclined at an angle to cover maximum area and provide better experience in cooking. The figures given below ***(fig-5)*** prove that the gas coming out of inclined horizontal opening will cover more area of a utensil than the gas coming out of circular holes cover.

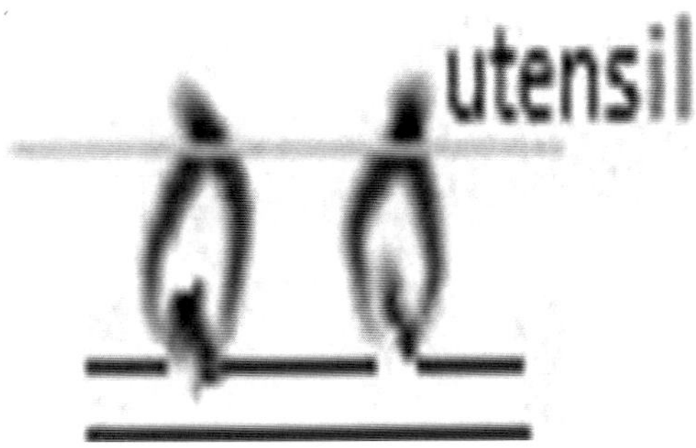

Fig:5 Circular holes

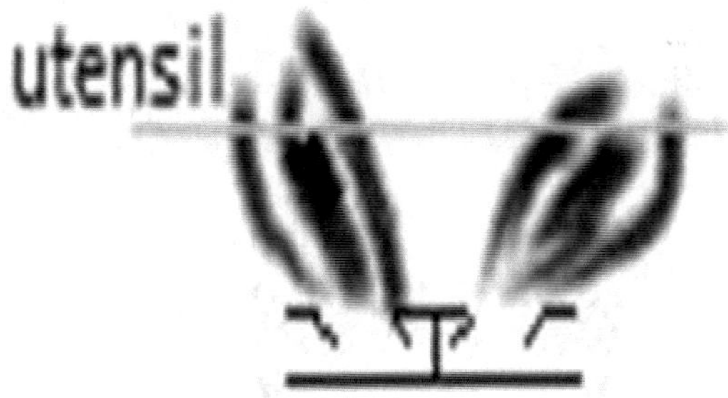

Fig:6 Horizontal Holes (proposed by author)

4. BLOCK DIAGRAM

a) Block diagram of burner:

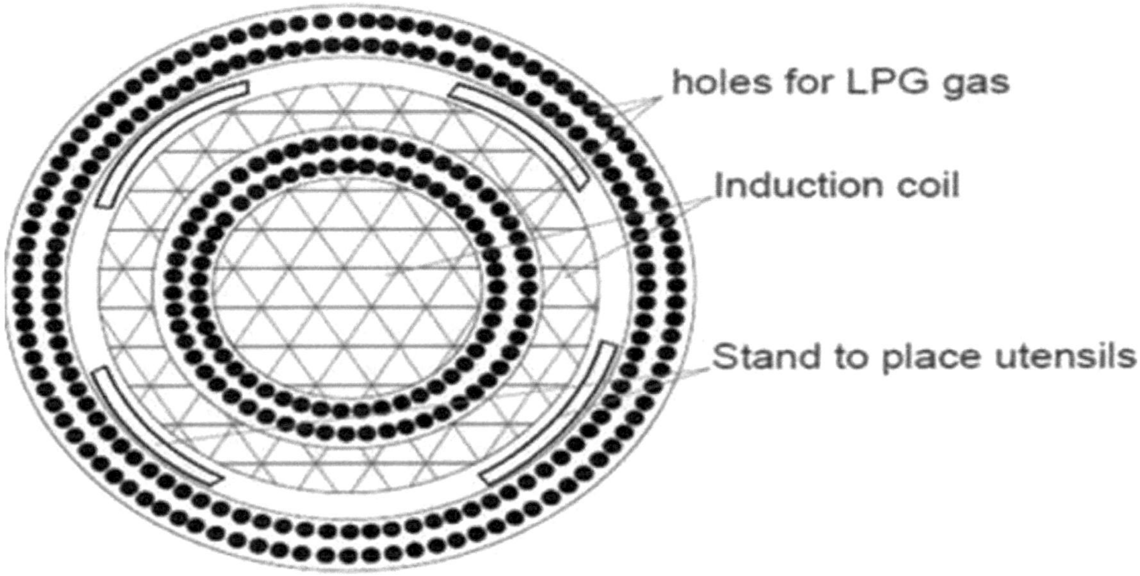

b) Block diagram of System:

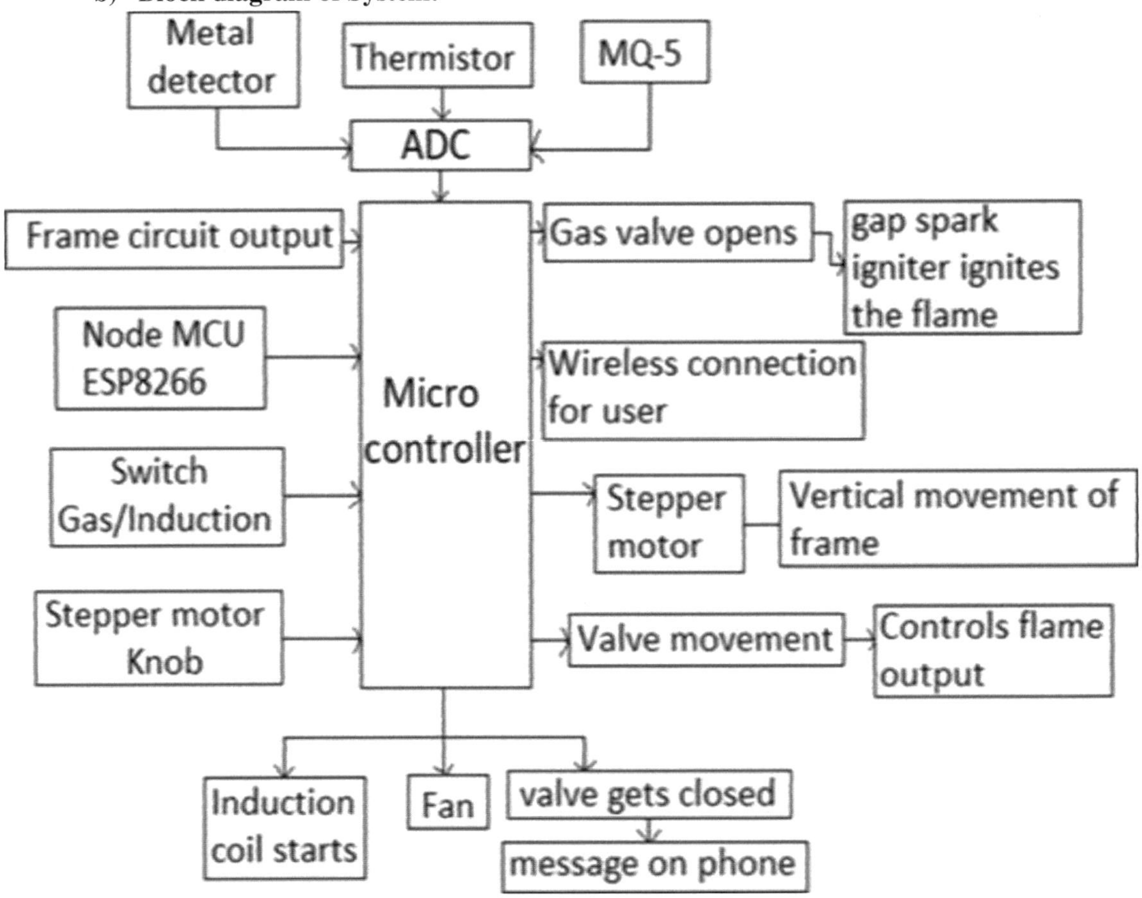

c) Block diagram at receiver end:

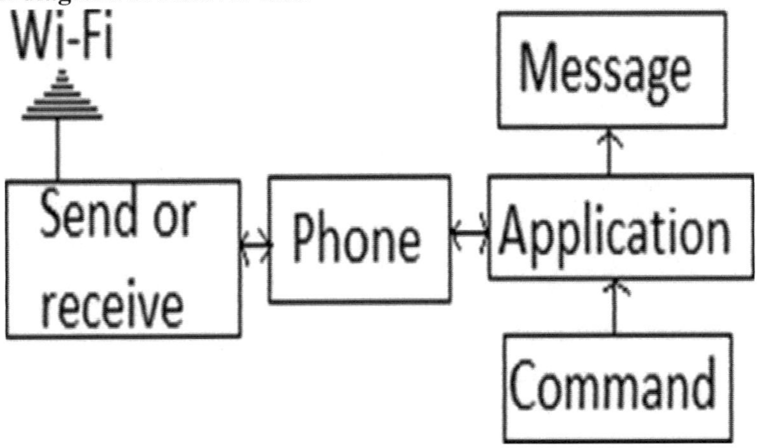

5. CHALLENGES AND FUTURE SCOPE:

The future work will be improving the system by joining it with machine learning for example: machine will learn from the continuous use of the burner when you reduce the flame or for which food item you use gas stove and induction stove, it can learn from the users daily routine and become smart day by day. One can also connect induction stove commands to the user phone making it ever easier to cook.

6. CONCLUSION

By implementing this hybrid induction smart gas stove not only space occupied by stoves reduced but it was very helpful in times when any one kind of energy either electricity or liquified petroleum gas were not available. Stitching modes from LPG to induction mode is very easy. The whole system is user friendly and very convenient to use.

7. REFERENCES

[1] Yanwu L, Cheng K.W.E, Chan K.W, Sun Z.G & Zhao S.W 2009 Development of a commercial induction cooker *3rd International Conference on Power Electronics Systems and Applications* PESA 2009 1 **- 3**.

[2] Chheti R, Chhoedron D, Tilak S & Robinson Duane A. 2017 *Analysis* on Integrated LPG Cook Stove and Induction Cooktop for Cooking Purposes in Bhutan *International Journal of Science and Research* 1071706 **23-26.**

[3] Surange J.R, Patil N.K, Rajput A.V, 2014, Performance Analysis of Burners used in LPG Cooking Stove-A Review. *International Journal of Innovative Research in Science, Engineering and Technology,* ISSN: 2319 – 8753

[4] Hussain A. A, Halah Y. A 2016 Electronic Design of Liquefied Petroleum Gas Leakage Monitoring, Alarm, and Protection System Based on Discrete Components *International Journal of Applied Engineering Research* ISSN 0973-4562 **11**, Number 19 pp. 9721- 9726.

[5] Anandha K S, Nair D, Rakesh K, Sampath K, Nair G.S 2017 IOT Based Smart Gas Monitoring System, IOSR *Journal of Electrical and Electronics Engineering* (IOSR-JEEE) e-ISSN: 2278-1676, p-ISSN: 2320-3331, **PP 82-87**.

[6] Saleque, A, Jahan S, Talukdar S, Islam Md, Azmir Md, Rashedin, Md. 2019. Development of Smart Cooking Stove: Harvesting Energy from the Heat, Gas Leakage Detection and IoT Based Notification System *International Conference on Robotics, Electrical and Signal Processing Techniques (ICREST)* **10.1109**/ICREST. 8644117.

[7] Nenni, 2018. Pop Up Camera Design Patent on Vivo Apex**. 10.31219**/osf.io/ph8n9.

[8] Khan M.Y, Saxena A 2013 Performance of LPG Cooking Stove Using Different Design of Burner Heads. I*nternational Journal of Engineering Research & Technology* (IJERT) ISSN: 2278-0181.

IOP Conf. Series: Materials Science and Engineering **748** (2020) 012029 doi:10.1088/1757-899X/748/1/012029

Valve Controlled Automatic Opto-Hydroponics System

Kushagra Singh, Praveen Kumar Maduri, Apurva Soni, Rishabh Maurya
Galgotias College of Engineering and Technology, Greater Noida, India

Email- kushagrasingh2707@gmail.com[1]

Abstract - In the growth of plants, various geological parameters like soil, air and water have been greatly affected by the increase in level of pollution and climate change. Degradation of soil, harmful gases in the air and loss of various macro nutrients are some common problems affecting the growth of plants. The only way that the plants could survive in a polluted environment is growing them in a well-managed and monitored greenhouse system. In this paper, a system is demonstrated which uses hydroponics technique and an artificial led light (Blue Red Pink) system to grow plants. The system is an automatic setup of hydroponics and led lights integration various sensors to manage the physical environment inside the greenhouse. System also proposes a hydroponics tube which is created for the seeds to directly grow on the tubes without shifting.

Keywords- microcontroller, sensors, automatic, led light, management, plants, hydroponics

1. INTRODUCTION

This study focuses on making an automatic opto-hydroponics system [1] which consists of two parts. The first part includes the hydroponics system in which plants are grown with a special hydroponic solution which contains all the nutrients required by plants. The second part is the artificial grow lights which mimics the light of the sun utilized by the plants to grow and perform photosynthesis. This light emits a particular wavelength favorable and best suited for the plants. The whole system is controlled, monitored and managed by the microcontroller setup which effectively controls all the physical parameters responsible for the growth of the plants [2]. The systems truly mimic the weather conditions like appropriate humidity, temperature and transpiration rate which the help of exhaust fans [3]. The system is also capable to manage the time period or the photo period of led grow lights automatically with the help of micro controller and timers. This system is best suited for areas where there is little or no light and also in areas where the soil is not supportive for the growth of plants. They are able to manage pH and nutrients to make sure plants are getting the exact nutrients they need. The systems are closed and recycle the water that is not used by plants. The ability to grow indoors allows farmers to control temperatures and lighting schedules to improve plant production [4]. Systems can be designed to make use of vertical space and increase planting density. Hydroponics also allow us to create farms in locations where soil conditions are too poor to support farming, or space is limited and a farm otherwise couldn't exist.

2. WHAT IS HYDROPONICS AND GROW LIGHTS

Hydroponics is growing plants without soil. The roots are grown in a nutrient solution that is dissolved in water. It was not formally called hydroponics until 1940 when William F. Gericke adopted the word from the Greek. Gericke was one of the first to grow plants in what he called "litter" where the plants were suspended above a tank of water containing a nutrient solution.

Hydroponic growing is an extremely efficient way to provide your plants with the food and water they need, when they need it, in the amounts they need which allows the plants to be as healthy as possible. The energy the plants save by taking nutrients up through their roots allows them to focus their energy on producing fruits and flowers. This allows the plants to grow faster and be ready to harvest sooner.

Hydroponic gardening systems can be set up to recycle water and nutrients, greatly reducing the resources necessary to grow food. Hydroponic way of growing plants can be useful in eliminating the need for herbicides and pesticides. It promotes an overall awareness of our environment and can even help to protect the environment by reducing traditional agricultural pollution [5].

IOP Conf. Series: Materials Science and Engineering **748** (2020) 012029 doi:10.1088/1757-899X/748/1/012029

A grow light/plant light is an artificial light source, this could be an electric light (mainly led lights) designed to stimulate plant growth by emitting a light which is appropriate for photosynthesis.

Grow lights can be used in various places where there is either no naturally occurring light, or where additional light source is required for the plants to grow well. For example, in the winter months when the available hours of daylight may be insufficient for the desired plant growth, lights are used to extend the time the plants receive light.

In sufficient amount of light can lead to long and spindly growth of plants which is not fruitful. Grow lights can either help to provide a light spectrum similar to that of the sun, or to provide a spectrum that is more favorable to the needs of the plants being cultivated.

Outdoor conditions are mimicked with varying color, temperatures and spectral outputs from the grow light, as well as varying the intensity of the lamps.

Factors affecting the growth of plants with grow lights- the photo period required by the plants, specific ranges of spectrum, luminous efficacy and color temperature.

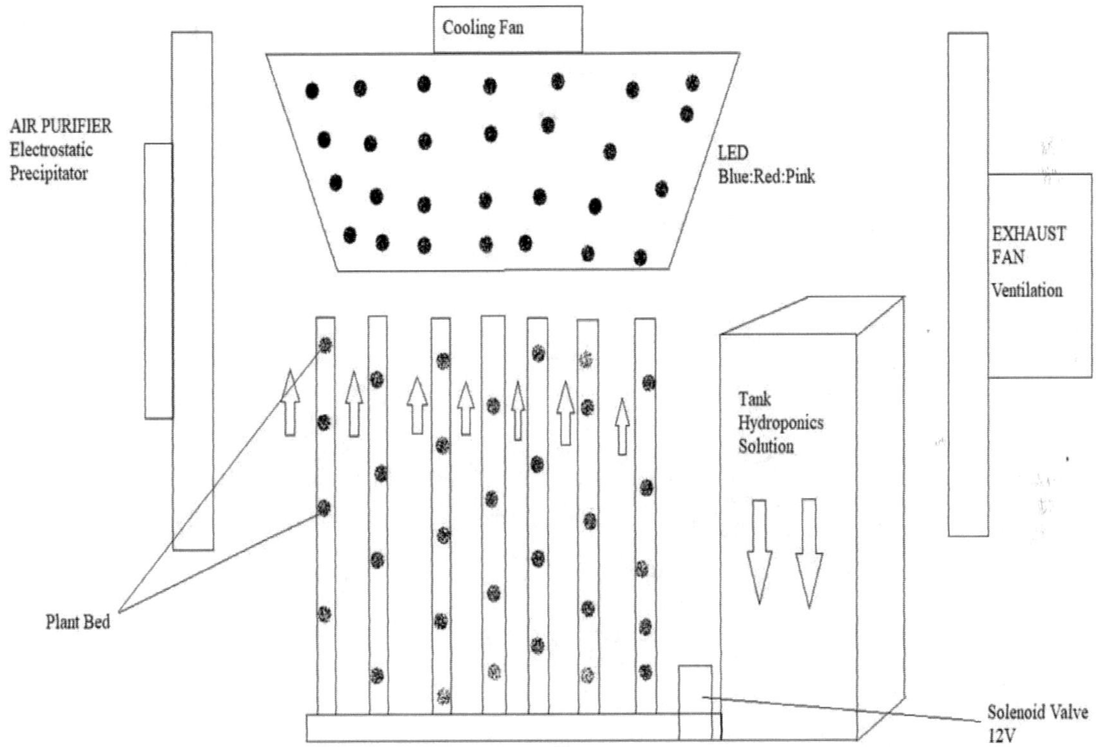

Fig 1: Hydroponics and Led syetem (Proposed Map)

3. METHODOLOGY

Automatic Opto-Hydroponics system is an automatic growth system for plants which provides artificial environmental conditions in a greenhouse system which is capable of managing, monitoring and configuring the artificial environment [6]. This setup consists of the following operating sections.

a. Automatic Hydroponics Setup

This setup is the combination of traditional hydroponic system and the advance automatic micro controller setup.

i. Humidity Sensor and Temperature Sensor (DHT22)

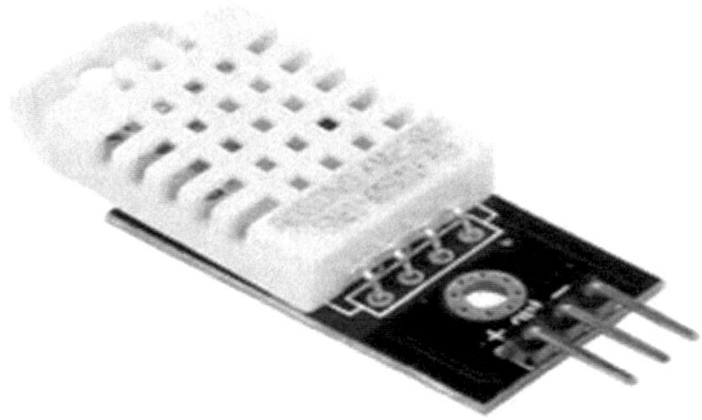

The DHT22 sensor consists of a humidity sensing component, an NTC temperature sensor (or thermistor) and an IC on the backside of the sensor. The term "NTC" means "Negative Temperature Coefficient", which means the resistance decreases with an increase in the temperature. To measure the humidity, it uses the humidity sensing component which has two electrodes with moisture holding substrate between them. So, as the humidity changes, the conductivity of the substrate changes or the resistance between these electrodes' changes. The change in resistance is further measured and processed by the IC which makes it ready to be read by a microcontroller. For measuring temperature this sensor uses an NTC temperature sensor or a thermistor.

A thermistor is a variable resistor that changes its resistance with a change of the temperature. The sensor is made by the fritting of semi conductive materials like ceramics or polymers to provide larger changes in the resistance with small changes in temperature.

Temperature Range	-40 - 125 °C / ± 0.5 °C
Humidity Range	0 - 100 % / ± 2-5%
Sampling Rate	0.5 Hz (one reading every two seconds)
Body Size	15.1mm x 25mm x 7.7mm
Operating Voltage	3 - 5V
Max Current During Measuring	2.5mA

ii. pH sensor Module

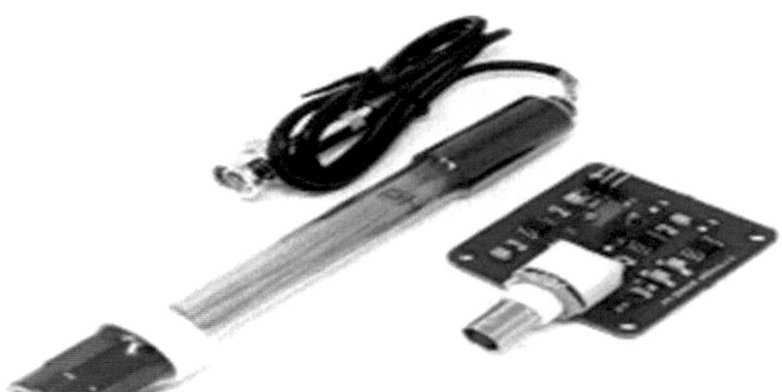

IOP Conf. Series: Materials Science and Engineering **748** (2020) 012029 doi:10.1088/1757-899X/748/1/012029

pH can be defined as the numeric representation of gram concentration in any solution. It varies between 0 to14. It is the logarithmic measurement of moles of hydrogen ions per liter of solution. The solutions having a pH value between 0 to 7 are acidic solutions with a large concentration of hydrogen ions whereas solutions having a pH value between 8 to 14 are basic solutions with small hydrogen concentration. The solutions having a pH value of 7 are neutral solutions. pH measurement gives the measure of alkalinity or acidity of a solution.

pH meter basically works on the fact that an interface of two liquids produces an electric potential which can be measured. In other words, when a liquid inside an enclosure made of glass is placed inside a solution other than that liquid, there exists an electrochemical potential between the two liquids.

Firstly, the electrode is placed inside the beaker filled with a solution whose pH is to be measure. The glass bulb which is fixed at the end of the measurement electrode consists of lithium ions doped to it which makes it act as an ion-selective barrier and allows the hydrogen ions from the unknown solution to migrate through the barrier and interacts with the glass, developing an electrochemical potential related to the hydrogen ion concentration. The measurement electrode potential thus changes with the hydrogen ion concentration. The reference electrode potential doesn't change with the hydrogen ion concentration and provides a stable potential against which the measuring electrode is compared. It consists of a neutral solution that is allowed to exchange ions with the unknown solution through a porous separator, thus forming a low resistance connection to complete the whole circuit. Potential difference between the two electrodes gives a direct measurement of the hydrogen ion concentration or pH of the system and is first pre-amplified to strengthen it and then given to the voltmeter.

U = EpH – Eref

EpH – Voltage potential of the measurement electrode

Eref – Voltage potential of a reference electrode

The pH is calculated based on the Nernst's equation which states that change in the total potential for every change in pH is

U = -kTpH

k- Boltzmann's constant, T- temperature.

-equivalent per litre of hydrogen ion

In the hydroponic setup, pH sensor continuously monitors the pH of the hydroponic solution to provide the chemical free growing environment [7].

VOLTAGE (mV)	pH value	VOLTAGE (mV)	pH value
414.12	0.00	-414.12	14.00
354.96	1.00	-354.96	13.00
295.80	2.00	-295.80	12.00
236.64	3.00	-236.64	11.00
177.48	4.00	-177.48	10.00
118.32	5.00	-118.32	9.00
59.16	6.00	-59.16	8.00
0.00	7.00	0.00	7.00

IOP Conf. Series: Materials Science and Engineering **748** (2020) 012029 doi:10.1088/1757-899X/748/1/012029

iii. Electrical Conductivity Sensor

Some of the root area factors like the salinity of the soil, irrigation water systems or fertilizer solutions can have a significant effect on the health, growth and quality of the plant if not monitored carefully [8]. Low and High salt levels could lead to the damage of plants. Hence, an electrical conductivity sensor in a hydroponics system helps to alert the salinity of the solution. In liquids, we often use the reciprocal of resistance, which is conductance, as a measurement. The water conductivity can reflect the level of electrolytes present in the water. The sensor is specially designed for the microcontroller to interface directly and give the output [9].

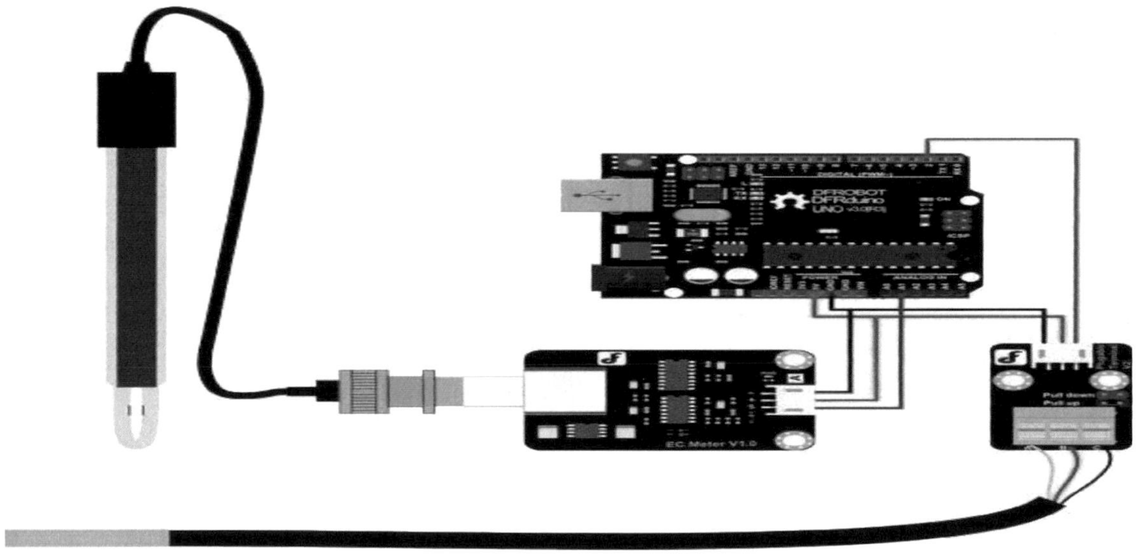

iv. NPK Sensor using optical transducer and microcontroller

In this technique of measuring the nitrogen, phosphorous and potassium content in the medium, three optical transducers are used.

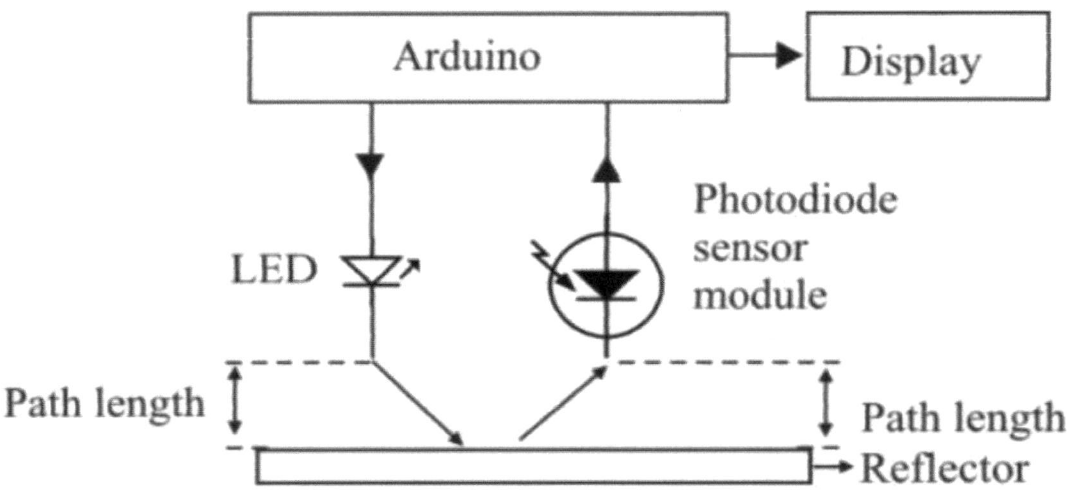

These optical transducers are used as a detection sensor with the help of Arduino microcontroller [10][11].

v. Solenoid Valve

A two-way solenoid valve is used at the outlet of the tank containing a hydroponics solution. In two-way valves are shut-off valves with one inlet port and one outlet port. When the solenoid valve is in de-energized condition, the core spring, assisted by the fluid pressure, holds the valve seal on the valve seat to shut off the flow. When the valve is in an energized state, the core and seal are pulled into the solenoid coil and the valve opens. Hence, the electromagnetic force is greater than the combined spring force and the static and dynamic pressure forces of the medium.

vi. Electrostatic precipitator and Exhaust fans

Exhaust fans are used to keep the greenhouse environment ventilated and remove the oxygen during the time of photosynthesis and at the time of respiration co2 is blown out of the green house. Electrical filters are used to purify the air so that plants could easily absorb them and grow well in a purified air. Polluted air can damage the leaves of the plants and can make them unhealthy.

vii. Grow Lights

LED grow lights are composed of multiple individual light-emitting diodes, usually in a casing with a heat sink and built-in fans. In this system, combination of Blue Red and Pink lights is used for the effective growth of plants.

IOP Conf. Series: Materials Science and Engineering **748** (2020) 012029 doi:10.1088/1757-899X/748/1/012029

viii. Tank

Tank contains the organic hydroponics solution N-3.90, P-0.51, K-1.47, Ca-0.01, S-0.12, Mg-0.018, Zn-0.026, Mn-0.0003, Cu-0.0005, B-Trace, Mo-0.003, +/- Herbal and Enzymatic properties.

ix. Hydroponics Tube and Seed Bed

Hydroponics tube consists of three main sections. The inner most is hydroponic solution flow path through which the organic hydroponic solution will flow with the downward opening which act as a drip with the help of gravity. The middle portion consists of sponge to store the solution. The outer portion has a seed bed on the upper side to rest the seeds and so that they can grow there. There are also two openings on the above for the exchange of gases from roots as well.

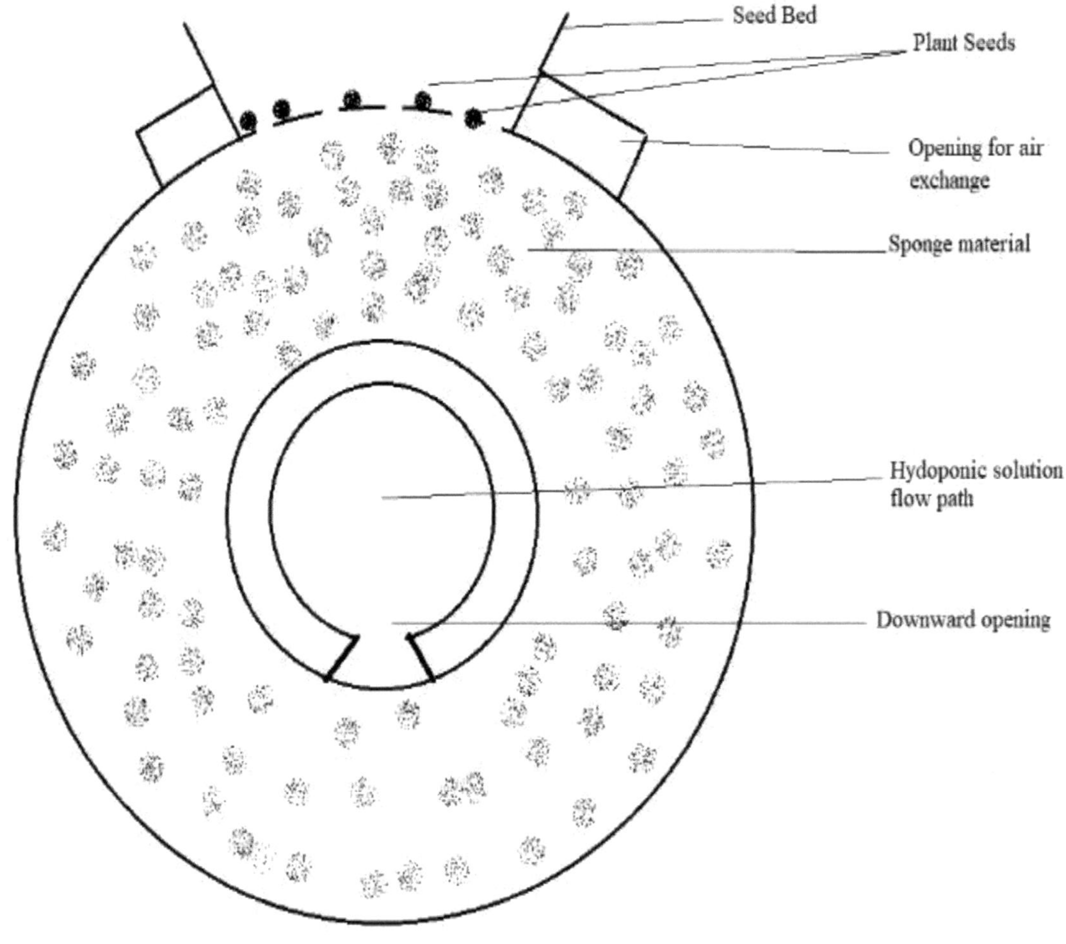

Fig 2: Cross section of Hydroponics Tube for plant growth (Proposed Diagram)

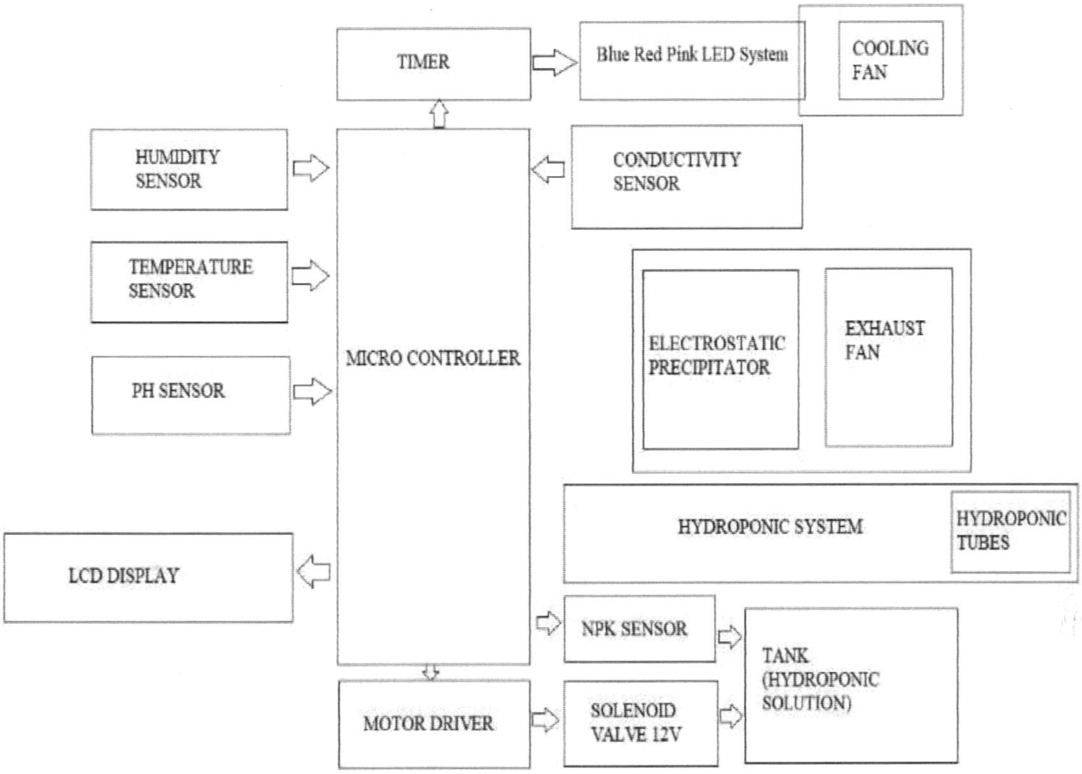

Fig 3: Full Block diagram

4. CONCLUSION/ RECOMMENDATION

Based on the implementation of automatic opto-hydroponics it can be said that our system is capable of healthiest growth of plants with minimum resources and maximum output. Aim of the system was to build cost effective and efficient setup to provide the best environmental conditions in an artificial setup.

Hydroponics is an environment-friendly and profitable technology. It has been promoted by the various governments and non-governmental organizations for its benefits in terms of food security. High-cost is the only major drawback in technology. However, a lot of R&D activity is being carried out globally to reduce the cost involved in technology.

Hydroponics involves the process of growing plants by using mineral nutrient solutions in sand, gravel, or liquid, without using soil. Due to the increasing success rates associated with the commercial hydroponics industry and the increasing difficulty of growing crops on soil, the hydroponics market is expanding exponentially. Along with the market expansion, manufacturers of hydroponic equipment are focusing on the development of new efficient systems.

Hydroponics is currently ahead of its time in India. The cost of produce coming out of a hydroponic unit currently is much higher than the cost of one coming out of traditional units. As such the market for hydroponic produce is limited to largely Metro's and a few tier 1 cities for now.

However, with constant innovation and adoption of newer techniques of growing, the cost differential is reducing. So, we will see progressive farmers adopting the technologies faster in coming years. But still the large part of farming in India will remain traditional due to cost considerations and also technology limitations.

5. REFERENCES

a. Albery W, John B.GD, Haggett, Robert Svanberg L. ,1985 ,The development of sensors for hydroponics. *Biosensors* 1.4, 369-397.

b. Pitakphongmetha J, Boonnam N, Wongkoon S, Horanont T, Somkiadcharoen D, & Prapakornpilai J,2016, Internet of things for planting in smart farm hydroponics style. *Computer Science and Engineering Conference (ICSEC)*

c. Moreno D. A, López-Berenguer C, Martínez-Ballesta M. C, Carvajal M, & García-Viguera C,2008, Basis for the new challenges of growing broccoli for health in hydroponics, *Journal of the Science of Food and Agriculture* **88.8** : 1472-1481.

d. Helmy H, Mahaidayu MG, Nursyahid A, Setyawan TA , Hasan A 2017. Nutrient Film Technique (NFT) Hydroponic Monitoring System Based on Wireless Sensor Network 2017 *IEEE International Conference on Communication, Networks and Satellite (Comnetsat)*

e. Ferentinos K. P, Albright L.D, 2003,Fault detection and diagnosis in deep-trough hydroponics using intelligent computational tools, **Biosystems Engineering 84.1** , 1330.

f. Gagnon V., Maltais-Landry G, Puigagut J, Chazarenc F, & Brisson J, 2010,Treatment of hydroponics wastewater using constructed wetlands in winter conditions, *Water, Air, & Soil Pollution* **212.1-4** 483-490.

g. Saaid M.F ,Sanuddin A, Megat A , Yassin M 2015. Autoimatated pH Controller System for Hydroponic Cultivation 2015 IEEE **9**th International Colloquim

h. Ferentinos, Konstantinos P, Louis D. A, and Norman R. S, 2000, Modeling pH and Electrical Conductivity in Hydroponics using Artificial Neural Networks, *IFAC Proceedings* Volumes **33.19** 173-178.

i. Kaewwiset T and Yooyativong T ,2017, Estimation of electrical conductivity and pH in hydroponic nutrient mixing system using Linear Regression algorithm. *International Conference on Digital Arts, Media and Technology (ICDAMT).*

j. Sihombing P, Karina N. A, Tarigan, J. T, & Syarif, M. I. 2018, Automated hydroponics nutrition plants systems using arduino uno microcontroller based on android, *Journal of Physics*: Conference Series. Vol. **978.** No. 1. IOP Publishing.

k. Masrie M, Aizuddin Rosman S, Sam R and Janin Z , Proc. of the 4th IEEE International Conference on Smart Instrumentation, Measurement and Applications (ICSIMA) 28-30 November 2017

Optimization of Process Parameters on CNC milling machine for Mild steel IS 2062:2011 E250 Gr. A with AlTiN coated tool insert in wet condition

Miss. Sumaiyya S.Nadaf[a], Prof. Mrs. Minakshi Y.Shinde[b]

[a]Student,Mechanical enginneering department, Karmveer Bhaurao Patil College of Engineering, Satara, Maharashtra, India,
[a]nadaf.sumayya@gmail.com
[b]Asst.Professor ,Mechanical enginneering department, Karmveer Bhaurao Patil College of Engineering, Satara, Maharashtra, India, [b]minakshi.shinde@kbpcoes.edu.in

Abstract. Now a days manufacturing industries wants to increase the quality of product and reduce manufacturing cost. The aim of this project work is to reduction of energy consumption during the machining phase, reduce surface roughness and increase material removal rate by optimization of face milling process parameters of mild steed IS 2062:2011 E250 Gr A with Taguchi orthogonal array. 27 experimental runs based on L27 orthogonal array of Taguchi method were performed. Spindle speed, feed rate and depth of cut this process parameters are optimized with multiple performance characteristics such as surface roughness (Ra), Material removal rate (MRR) and power consumption. The machining operation performed on VMC by using coolant with 63 mm cutter diameter. The analysis of variance (ANOVA) is also applied to identify the most significant factor using Minitab software. The analysis can be done on AlTiN insert coating. At the last conformation test were performed.

Keywords: CNC face milling, ANOVA, Surface roughness, MRR, Power consumption, Mild steel IS 2062:2011 E 250 Gr. A.

1. Introduction

Optimizing the process parameters has become a priority in manufacturing area. Milling process is widely used for machining variety of shapes by feeding the workpiece against the rotating multiple tooth cutter. G.Tamil Kumaran et.al [01] studied the optimal face milling parameter of EN 19/EN31 steel material with PVD and CVD coated tools on Fanuc series vertical CNC milling ANOVA technique used. Total of 27 experiments conducted at 3 levels for input variables cutting speed, feed rate and depth of cut. They found that for high depth of cut with increasing in cutting speed MRR increases.by machining PVD insert better surface finish was produced. Balinder Singh et al (2014) [02] used Taguchi method for optimization of process input parameter on CNC vertical milling machine with EN24 steel as workpiece material and carbide end mill as cutting tool.L27 orthogonal array with control parameters as cutting speed, Feed rate and depth of cut that the response parameters are surface roughness and material removal rate. Output analyze by Minitab software and S/N ratio obtained by Taguchi methodology. For surface roughness smaller-the-better type S/N ratio was applied and for material removal rate (MRR) larger-the-better type S/N ratio was applied. Avinash A. Thakre et al (2013) [03] work includes effects of various milling parameters with 3 levels of cutting parameters. For face milling operation experiments were conducted using Taguchi technique L9 orthogonal array on 1040 MS material on CNC vertical milling machine using carbide inserts. By experimentation results found that surface roughness parameter was mainly controlled by coolant flow it is approximately 60%.better surface finish is obtained contribution with optimal coolant flow rate .The second most important contribution was spindle speed (=22%). Ritesh Kumar Singh et al (2018) [04] Experiments carried out on face milling operation n Mild steel E250 (IS 2062:2011) material with 3 levels of input parameters with output parameters are material removal rate and surface roughness used to formulate combined objective function (COF) and Genetic algorithm (GA) used to optimize face milling parameter of L18 array. Cemented carbide milling cutter of 50 mm diameter with 5 numbers of cutting edges. Milon D Selvam et al (2012) [05] Use of Taguchi technique and genetic Algorithm (GA) for minimizing mild steel material with three zinc coated carbide tools inserted into face miller of 25 mm diameter.it is found that The surface roughness by Taguchi technique is 0.975 µm with 4.308% error form the predicted value and from genetic algorithm is 0.88 µm with 4.625% error from predicted value.

2. Material

This work is on Mild steel IS 2062:2011 E 250 Gr. A is one of the most commonly used construction materials. The Material is very strong and can be made from readily available natural materials. It has as a high resistance to breakage and is especially desirable for construction due to its good weldability and machinability. Dimension 40 x 38 x 21 mm was used for machining using face milling process. There are many applications of IS 2062 mild steel material.it is used in manufacturing of gears, forgings, fasteners, bearings, keyways, pinons etc.

Table 1. Composition of material.

Elements	(C)	(Mn)	(Si)	(S)	(P)
Measured Values (%)	0.23	0.55	0.18	0.034	0.042

3. Methodology

The Taguchi philosophy provides two tenets one is reduction in variation (improved quality) of a product or process represent a lower loss to society and second is proper development strategy can intentionally reduce variation.

In this work "Smaller-the-better" performance characteristics used for Surface roughness and power consumption. "Higher-the-better" performance characteristic used for Material removal rate.

The equations for calculating signal to noise ratio are as follows;
For smaller-the-better performance characteristics than;

$$\frac{S}{N} = -10 \times \log \times [\frac{1}{n}] \sum_{i=1}^{n} Y_i^2 \text{ ---(Eq. 1)}$$

For higher-the-better performance characteristics than;

$$\frac{S}{N} = -10 \times \log \times [\frac{1}{n}] \sum_{i=1}^{n} \frac{1}{Y_i^2} \text{ ---- (Eq. 2)}$$

4. Experimental setup

This work is on Mild steel IS 2062:2011 E 250 Gr. A (40 x 38 x 21 mm) material using face milling process. This work is carried out on VMC machine named as HURCO VM20i with cutter of 63 mm diameter with insert having coating of AlTiN. In this work three machining parameters such as spindle speed, feed rate and depth of cut and three performance parameters such as Surface roughness (Ra), Material removal rate (MRR) and Power consumption are used for further experiments. The photograph of experimental setup used for the experimental runs is shown in the Figure 1.

Fig. 1. CNC milling machine

The experiment was carried out in wet condition by using HICUT BIO 150 coolant. Experimentation is based on Taguchi design of experiment L27 orthogonal array is considered for experimentation.

Table 2. Process parameter and their levels.

Process parameter	Level 1	Level 2	Level 3
Spindle speed (rpm)	1000	1400	1800
Feed (mm/min)	600	800	1000
Depth of cut (mm)	0.2	0.3	0.4

5. Result and discussion

All 27 experiments run for AlTiN coated insert. The surface roughness was measured by Mitutoyo SJ210 surface roughness tester. The material removal rate of the workpiece is measured by ratio of the difference between weight of the workpiece before machining and weight of the workpiece after machining to the machining time that is achieved. In this power consumption AC three phase is measure by taking reading with non-contact type of clamp by holding cable between the clamps.

5.1. Surface roughness

Table 3. Experimentation results by using AlTiN coated insert.

Expt.No	Level 1	Level 2	Level 3	Surface Roughness (µm)	(S/N) ratio
1-1	1000	600	0.2	0.1301	17.71445
2-1	1000	600	0.3	0.1402	17.06504
3-1	1000	600	0.4	0.1612	15.8527
4-1	1000	600	0.2	0.1905	14.4021
5-1	1000	600	0.3	0.2295	12.78435
6-1	1000	600	0.4	0.2413	12.34885
7-1	1000	600	0.2	0.2725	11.29267

8-1	1000	600	0.3	0.3009	10.43156
9-1	1000	600	0.4	0.3401	9.367867
10-1	1400	800	0.2	0.0857	21.34038
11-1	1400	800	0.3	0.1101	19.16425
12-1	1400	800	0.4	0.1507	16.43773
13-1	1400	800	0.2	0.2016	13.91019
14-1	1400	800	0.3	0.2304	12.75035
15-1	1400	800	0.4	0.2689	11.40818
16-1	1400	800	0.2	0.2985	10.50111
17-1	1400	800	0.3	0.3200	9.897
18-1	1400	800	0.4	0.3600	8.87395
19-1	1800	1000	0.2	0.1272	17.91026
20-1	1800	1000	0.3	0.1666	15.5665
21-1	1800	1000	0.4	0.1800	14.89455
22-1	1800	1000	0.2	0.2223	13.06121
23-1	1800	1000	0.3	0.2619	11.63729
24-1	1800	1000	0.4	0.3088	10.20645
25-1	1800	1000	0.2	0.3412	9.33982
26-1	1800	1000	0.3	0.3750	8.519375
27-1	1800	1000	0.4	0.3810	8.3815

ANOVA for Surface Roughness-

Effect of process parameters on surface roughness for smaller-the-better performance characteristics at various levels is shown in Figure 2.

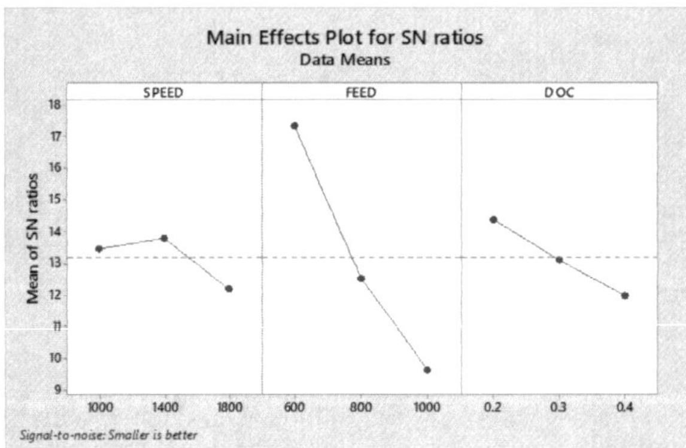

Fig. 2. Main effect plot for S/N ratio of surface roughness (μm)

Figure 2 indicates surface roughness is directly proportional to feed rate and depth of cut. It shows that increase in feed rate and depth of cut there is increase in surface roughness. As spindle speed increases surface roughness decreases and then increases.

The ranking of process parameters is enumerated in Table 4 for obtaining the minimum surface roughness the parameters should be set as level 2 of speed, level 1 of feed rate, level 1 of depth of cut.

Table 4. Response table for S/N ratio of Surface roughness (μm)

Level	Spindle Speed (rpm)	Feed rate (mm/min)	Depth of cut (mm)
1	13.473	17.327	14.386
2	13.809	12.501	13.091
3	12.169	9.623	11.975
Delta	1.641	7.705	2.411
Rank	3	1	2

The ranks indicate the relative importance of each factor to the response. The ranks and the delta values (Table 4) for various input parameters for surface roughness (Ra) shows that feed rate and depth of cut has the greatest effect on surface roughness (Ra) and least effect of spindle speed was seen.

Table 5. Analysis of variance for means of surface roughness (μm)

Source	DF	Adj SS	Adj MS	F-Value	P-Value	% Contribution
Spindle Speed (rpm)	2	0.008987	0.004493	21.07	0.000	4.580
Feed rate (mm/min)	2	0.167787	0.083893	393.41	0.000	85.517
Depth of cut (mm)	2	0.015162	0.007581	35.55	0.000	7.727
Error	20	0.004265	0.000213			2.173
Total	26	0.196201				100

It is observed that increase in feed rate and depth of cut, it produces a rough surface. Feed rate contributes 85.51 % the largest as compared to the other process parameters, followed by depth of cut with 7.72 %, spindle speed with 4.58 %. All speed, feed rate and depth of cut are the significant parameters.

5.2. For Material removal rate

Table 6. Experimentation results by using AlTiN coated insert.

Expt.No	Level 1	Level 2	Level 3	Material Removal Rate (gm/min)	(S/N) ratio
1-1	1000	600	0.2	9.9800	19.98265
2-1	1000	600	0.3	11.9760	21.56627
3-1	1000	600	0.4	14.0894	22.97789
4-1	1000	600	0.2	14.9700	23.50447
5-1	1000	600	0.3	21.6434	26.70653
6-1	1000	600	0.4	27.8512	28.8969
7-1	1000	600	0.2	17.1086	24.66431
8-1	1000	600	0.3	24.6083	27.82164
9-1	1000	600	0.4	31.9361	30.08565
10-1	1400	800	0.2	11.9760	21.56627
11-1	1400	800	0.3	14.9700	23.50447
12-1	1400	800	0.4	15.9680	24.06505
13-1	1400	800	0.2	17.1086	24.66431
14-1	1400	800	0.3	24.9501	27.94145
15-1	1400	800	0.4	32.3677	30.20224

16-1	1400	800	0.2	19.9600	26.00325
17-1	1400	800	0.3	28.7425	29.1705
18-1	1400	800	0.4	37.4251	31.46327
19-1	1800	1000	0.2	13.3067	22.48142
20-1	1800	1000	0.3	16.3309	24.26024
21-1	1800	1000	0.4	18.4246	25.308
22-1	1800	1000	0.2	19.9600	26.00325
23-1	1800	1000	0.3	28.7885	29.1844
24-1	1800	1000	0.4	37.7198	31.5314
25-1	1800	1000	0.2	23.9521	27.58687
26-1	1800	1000	0.3	35.2236	30.93669
27-1	1800	1000	0.4	43.0793	32.68537

ANOVA for Material removal rate -

Effect of process parameters on material removal rate for larger-the-better performance characteristics at various levels is shown in Figure 3

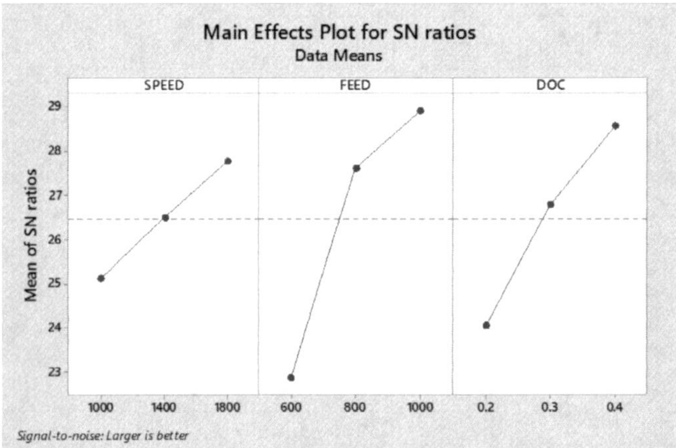

Fig. 3. Main effect plot for S/N ratio of Material removal rate (gm/min)

Above Figure 3 indicates that the level 3 of feed rate gives the maximum material removal rate. Figure 3 show that material removal rate is directly proportional to the speed, feed rate and depth of cut. As these process parameters increases MRR is increases. At level 1 of each process parameter shows minimum MRR that is MRR decreases with low speed, low feed rate and low depth of cut.

For obtaining the maximum material removal rate the parameters should be set as level 3 of speed, feed rate and depth of cut.

Table 7. Response table for S/N ratio of Material removal rate (gm/min)

Level	Spindle Speed (rpm)	Feed rate (mm/min)	Depth of cut (mm)
1	25.13	22.86	24.05
2	26.51	27.63	26.79

3	27.78	28.94	28.58
Delta	2.64	6.08	4.53
Rank	3	1	2

The ranks indicate the relative importance of each factor to the response. The ranks and delta values indicate that Depth of cut has greatest effect on Material Removal Rate and is followed by Feed rate and Spindle Speed in that order.

Table 8. Analysis of variance for means of Material removal rate (gm/min)

Source	DF	Adj SS	Adj MS	F-Value	P-Value	% Contribution
Spindle Speed (rpm)	2	218.2	109.082	12.18	0.000	10.100
Feed rate (mm/min)	2	1083.1	541.558	60.49	0.000	50.138
Depth of cut (mm)	2	679.8	339.906	37.96	0.000	31.469
Error	20	179.1	8.953			8.290
Total	26	2160.2				100

The MRR is increases with increases in feed rate, so that the contribution of feed rate is largest with 50.13 % as shown in Table 8. The contribution of depth of cut is 31.46 % and spindle speed is 10.10 %. The spindle speed, Feed rate and depth of cut has contribution for MRR and it is significant parameter.

5.3. Power consumption

Table 9. Experimentation results by using AlTiN coated insert

Expt.No	Level 1	Level 2	Level 3	Power consumption (KW)	(S/N) ratio
1-1	1000	600	0.2	1.9031	-5.58933
2-1	1000	600	0.3	1.9769	-5.92002
3-1	1000	600	0.4	2.5507	-8.13336
4-1	1000	600	0.2	1.9315	-5.71801
5-1	1000	600	0.3	2.3860	-7.55342
6-1	1000	600	0.4	2.6132	-8.34359
7-1	1000	600	0.2	2.0849	-6.38176
8-1	1000	600	0.3	2.5223	-8.03609
9-1	1000	600	0.4	2.6643	-8.51189
10-1	1400	800	0.2	2.1000	-6.44439
11-1	1400	800	0.3	2.2212	-6.93197
12-1	1400	800	0.4	2.6586	-8.49335
13-1	1400	800	0.2	2.0905	-6.40539
14-1	1400	800	0.3	2.4200	-7.67663
15-1	1400	800	0.4	2.6814	-8.56727
16-1	1400	800	0.2	2.3121	-7.28032
17-1	1400	800	0.3	2.5905	-8.26773
18-1	1400	800	0.4	2.7382	-8.74938
19-1	1800	1000	0.2	2.1474	-6.63827

20-1	1800	1000	0.3	2.5450	-8.11399
21-1	1800	1000	0.4	2.6984	-8.62231
22-1	1800	1000	0.2	2.3575	-7.4494
23-1	1800	1000	0.3	2.4712	-7.85822
24-1	1800	1000	0.4	2.7836	-8.89236
25-1	1800	1000	0.2	2.4428	-7.7578
26-1	1800	1000	0.3	2.6700	-8.53039
27-1	1800	1000	0.4	2.8291	-9.03302

ANOVA for Power consumption-

Effect of process parameters on power consumption for smaller-the-better performance characteristics at various levels is shown in Figure 4

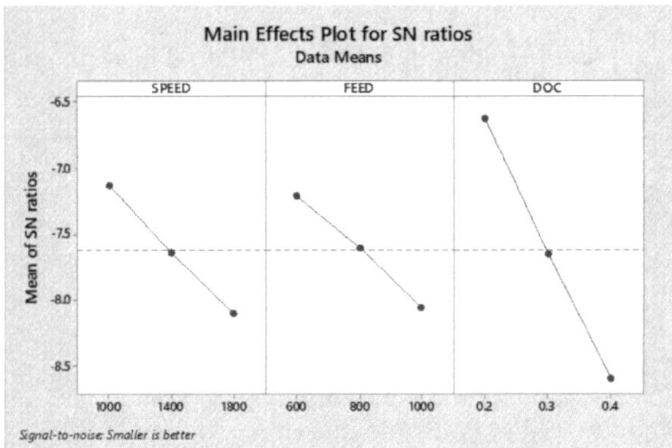

Fig. 4. Main effect plot for S/N ratio of power consumption (KW)

Figure 4 indicates power consumption is directly proportional to spindle speed, feed rate and depth of cut. It shows that increase in spindle speed, feed rate and depth of cut there is increase in power consumption.
The ranking of process parameters is enumerated in Table 10. For obtaining the minimum power consumption the parameters should be set as level 1 of speed, level 1 of feed rate, level 1 of depth of cut.

Table 10. Response table for S/N ratio of power consumption (KW)

Level	Spindle Speed (rpm)	Feed rate (mm/min)	Depth of cut (mm)
1	-7.132	-7.210	-6.629
2	-7.646	-7.607	-7.654
3	-8.100	-8.061	-8.594
Delta	0.968	0.851	1.965
Rank	2	3	1

The ranks indicate the relative importance of each factor to the response. The ranks and the delta values shows that depth of cut and spindle speed has the greatest effect on power consumption (KW) and least effect of feed rate.

IOP Conf. Series: Materials Science and Engineering **748** (2020) 012030 doi:10.1088/1757-899X/748/1/012030

Table 11. Analysis of variance for means of power consumption (KW)

Source	DF	Adj SS	Adj MS	F-Value	P-Value	% Contribution
Spindle Speed (rpm)	2	0.2970	0.148522	18.92	0.000	14.891
Feed rate (mm/min)	2	0.2347	0.117372	14.95	0.000	11.767
Depth of cut (mm)	2	1.3056	0.652817	83.18	0.000	65.463
Error	20	0.1570	0.007848			7.872
Total	26	1.9944				100

It is observed that increase in spindle speed, feed rate and depth of cut, it produces maximum power consumption (KW). Depth of cut contributes 65.46 % the largest as compared to the other process parameters, followed by spindle speed with 14.89 %, feed rate with 11.76 %. All speed, feed rate and depth of cut are the significant parameters.

6. CONCLUSION

Results of surface roughness (Ra) we conclude that minimum surface roughness was obtained as 0.0857 µm and maximum as 0.3810 µm. Surface roughness is most significantly affected by feed rate, depth of cut and speed with 85.51 %, 7.72 % and 4.58 % contribution respectively.

Results of material removal rate (MRR) we conclude that maximum material removal rate is obtained as 43.0793 gm/min and minimum is 9.9800 gm/min. Material removal rate is most significantly affected by feed rate, depth of cut and speed with 50.13 %, 31.46 % and 10.10 % contribution respectively.

Results of power consumption (KW) we conclude that minimum power consumption was obtained as 1.9031 (KW) and maximum as 2.8291 (KW). Power consumption is most significantly affected by depth of cut, speed and feed rate with 65.46 %, 14.89 % and 11.76 % contribution respectively

References

[1] G. Tamil Kumaran, R. John Stephen "Optimization of machining parameters for face milling operation using ANOVA", IOSR Journal of Mechanical and Civil Engineering (IOSR-JMCE).

[2] Balinder Singh, Rajesh Khanna, Kapil Goyal and Pawan Kumar, "Optimization of input process parameter in cnc milling machine of EN24 steel", International Journal of Research in Mechanical Engineering & Technology, Vol. 4, Issue 1, 2014.

[3] A. A. Thakre, "Optimization of milling parameters for minimizing surface roughness using Taguchi's approach", International Journal of Emerging Technology and Advanced Engineering, vol. 3, no. 6, pp. 1–5, 2013.

[4] Ritesh Kumar Singh, Anil Kumar, Guru Sewak Kesharwani "Concurrent optimization and an experimental analysis of face milling operation parameters for optimal performance on mild steel work piece", International Journal for Research in Applied Science & Engineering Technology (IJRASET) ISSN: 2321-9653; IC Value: 45.98; SJ Impact Factor: 6.887,Volume 6 Issue VI, June 2018.

[5] Milon D. Selvam, Dr.A.K.Shaik Dawood,Dr. G. Karuppusami, "Optimization of machining parameters for face milling operation in a vertical cnc milling machine using genetic algorithm", An International Journal (ESTIJ), ISSN: 2250-3498, Vol-2, August 2012.

[6] Turgay kivak, "Optimization of surface roughness and flank wear using the Taguchi method in milling of hadfield steel with PVD and CVD coated inserts, "Measurement, ELSEVIER, Volume 50, Page no. 19–28, 2014.

[7] Kuldip Singh Sangwana, Girish Kanta "Optimization of Machining Parameters for Improving Energy Efficiency using Integrated Response Surface Methodology and Genetic Algorithm Approach", Science Direct, Procedia CIRP 61, 2017.

IOP Conf. Series: Materials Science and Engineering **748** (2020) 012031 doi:10.1088/1757-899X/748/1/012031

Exergoeconomic and enviroeconomic analyses of single slope solar desalination unit loaded with/without nanofluid: A comprehensive review

Desh Bandhu Singh[1], Gagan Bansal[1], Jeetendra Kumar Yadav[2*], Navneet Kumar[2], Sumit Tiwari[3], Anuj Raturi[1]

[1]Mechanical Engineering Department, Graphic Era (Deemed to be University), Bell Road, Clement Town, Dehradun, 248002, Uttarakhand, India

[2]Galgotias College of Engineering and Technology, Plot No 1, Knowledge Park II, Greater Noida, 201306, UP, India

[3]Mechanical Engineering Department, Shiv Nadar University, NH - 91, Tehsil Dadri, Gautam Buddha Nagar,Uttar Pradesh - 201314, India
Email:yjkiitd@gmail.com

Abstract. The exergoeconomic and the enviroeconomic analyses of basin type single slope distiller unit based on solar energy can play an important role for mitigating the contemporary problem of acute shortage of drinking water throughout the world particularly developing and under developed countries. The technology of solar distiller unit has been known to humankind since the 16th century; however, it has not been preferred for purifying water on a large scale till date because of low potable water output, high initial investment, and non-availability of a technician at the local level. This technology is simple in nature and has the potential to mitigate the drinking water shortage; however, it has not been fully explored so far. A lot of configurations have been reported by various researchers since the development of solar distiller unit technology. This paper summarizes some of the important contributions reported so far on the basis of exergoeconomic and enviroeconomic parameters of single slope solar desalination units by incorporating the effect of nanofluid. A comparative study has been presented followed by a detailed discussion about exergoeconomic and enviroeconomic parameters of a single slope solar desalination unit loaded with/without nanofluid. It has been observed that the application of water-based nanofluid in the distiller unit results in the enhancement of performance. At last, the future scope has been presented.

1. Introduction

The design and analysis of basin type single slope solar desalination unit on the ground of exergoeconomic and enviroeconomic parameters can play an important role in mitigating the contemporary issue of shortage of fresh water throughout the globe. The access of water to human beings is limited to less than 1 percent and even this small amount of water is getting polluted day-by-day due to various anthropogenic activities. The basin type single slope solar desalination unit can work in passive as well as active modes. In the case of passive mode of operation of single slope solar desalination unit, external source of heat is not present for supplying heat to water in the basin; whereas, in the case of active mode of operation, an external source of heat in the form of solar collector or similar other device is present which supply heat to water in the basin. This review paper talks about the analysis of a single slope solar desalination unit on the basis of exrgoeconomic and enviroeconomic parameters.

2. Exergoeconomic and enviroeconmic parameters

Exergoeconomic parameter can be stated as the ratio of either exergy gain per unit uniform annual cost or exergy loss per unit uniform annual cost. If exergy loss is considered, then the objective is of minimization type. The exergy represents the quality and hence analysis of the system on the basis of exergy is a more realistic analysis. The exergy loss of various components is calculated and the component responsible for the highest exergy loss is targeted to minimize so that objective can be fulfilled. If exergy gain is considered, then the objective is to maximize the output. The exegoeconomic parameter correlates the exergy with the uniform annual cost. enviroeconomic analysis controls the pollution of the environment by giving motivation for reducing pollutants and encourages the use of renewable energy technology which does not create pollution. The enviroecomic parameter is also known as environmental cost. It is calculated by computing the amount of carbon mitigation on the basis of exergy. This amount is multiplied with the cost of carbon on the international level to get environmental cost.

3. Nanofluid

If nano-particles in controlled amounts are mixed with fluid using some technology, then the resulting fluid is termed as nanofluid. ultimate physiological properties in it and shows remarkable progress in the field of nanotechnology. Nano-fluids are the result of the latest innovation which has extra-ordinary thermo-physical and optical properties and coined a well-known name i.e. nano-fluid for the foremost instance. Nano-fluid can be produced by the standardized mixture with the optimum concentration ratio in between working medium to nano-particles (NPs) having 1-100 nm sized NPs in the base fluid. In the case of water-based nano-fluid, nano-particles are mixed with water and similarly, other kinds of base fluids are also available which are used accordingly as per the application.

4. Basin type passive/active single slope solar desalination unit loaded with/without nanofluid

Working in the field of enviroeconomic and the exergoeconomic scrutiny of Single Slope Solar still (SSSS) distiller unit, several researchers have made an effective attempt in characterizing the setup and to develop the process for enhancing the quantity and quality of drinkable water in all atmospheric conditions. The use of nanofluids also plays a significant role in enhancing the heat and mass transfer in the available solar still.[1]. The effective comprehensive analysis containing the elaborative study of the application of nanofluids in solar still is discussed by Bait & Si-Ameur.[1] The study revealed that the application of nanofluids in various distiller units enhances several characteristic properties like thermal conductivity, heat transfer, stability etc. the size, shape, orientation etc. of nanofluids also plays an influential role in the characterization. [1].

Tiwari et al. [2]in 2015 effectively Characterized elaborative Exergoeconomic and enviroeconomic analyses in their laboratory with flat plate collector of partially covered photovoltaic active solar distillation system. The comparative study with various past research analysis were made.[2]. Energy calculation based on entropy was initiated. The authors conclude comparatively better exergy and overall thermal efficiency. The setup was innovatively designed as shown in Fig. 1.

FPCs partially covered active SSSS designed and Fabricated was designed and fabricated. The final results obtained have been likened with the results obtained by other researchers working in similar technology and applications. The results projected the daily thermal efficiency of the proposed model is comparatively not as much of the other two models. The daily exergy efficiency of the model under study is higher than the model in Fig. 2 but at the same time, it was a little lower than the model

IOP Conf. Series: Materials Science and Engineering **748** (2020) 012031 doi:10.1088/1757-899X/748/1/012031

in Fig. 3. Also, the exergy, the thermal energy, the overall efficiency and the energy efficiency evaluated on a daily basis of the proposed model are greater than the other two models[2].

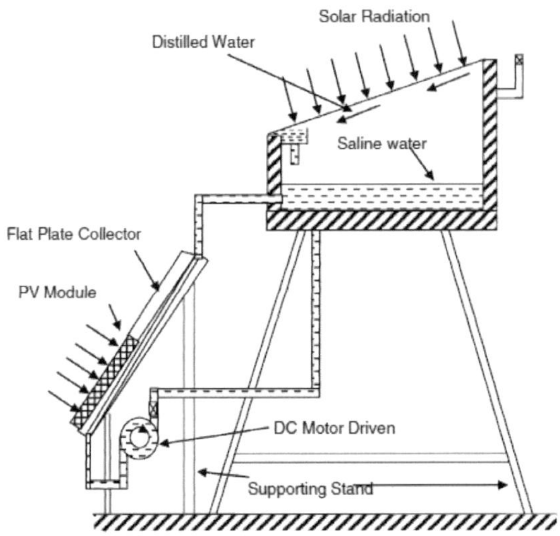

Figure 1. Side view of an active PVT solar distillation system [2]

Singh & Tiwari [3] gave the descriptive review on the Exergoeconomic parameters, enviroeconomic parameters and the productivity analyses of basin type N-PVT-CPC-SSSS. Figure 4 shows the effective Pectoral view of N-PVT-CPC-SSSS[3]. Several systems were analyzed and reviewed by Singh and Tiwari [4] and the revolution in the field of solar energy and economical energy was elaborated. The sustainable evaluation was developed with the experimental analysis of various solar collectors setups.

Figure 2. Hybrid (PV/T) active solar still [2]

The use of various nanofluids in SSSS setup with N identical PVTs were also characterized.[5]. Systematic and experimental study on single basin SSSS with different water nanofluids were also accomplished by Elango et al.[5] Different nanofluids like Al_2O_3, ZnO, Fe_2O_3 and SnO_2 which are water-based were experimentally (chemically) prepared and then one by one analyzed. Observations revealed that SSSS with water-based nanofluids gave higher exergoeconomic and enviroeconomic characteristics.[5] The cost of production, thermal analysis and the overall efficiency of the system gave optimum results. The interesting fact from the analysis says that 29.95% (almost one third) higher production rate was observed when the SSSS with aluminum oxide water-based nanofluids were used [5]. Panchal et al. gave the experimental analysis of heat pipe collector based SSSS that are properly coupled [6].

Figure 3. Integrated FPCs and DSSS[2]

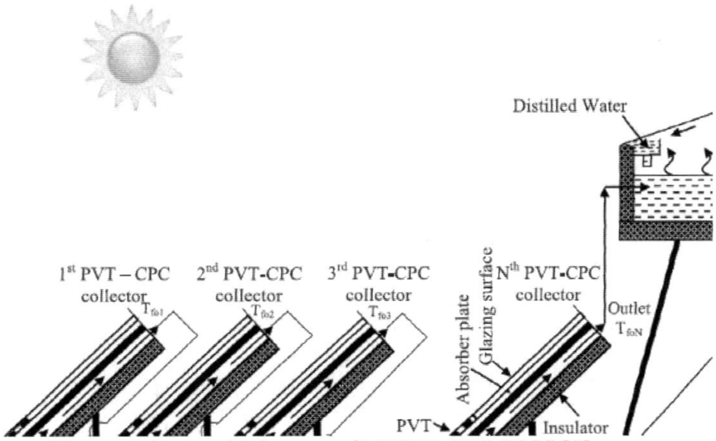

Figure 4. Pectoral Diagram of N-PVT-CPC-SSSS [3]

Ibrahim et al. [7] gave the Exergoeconomic analysis on SSSS for the cost analysis and converting the feasible results into optimizing outputs. The setups were developed and also the computational analysis was performed until the optimality [7]. The comparative diagnosis was done between the existing system and the developed system for the exergoeconomic and enviroeconomic characterizations. The outputs revealed that there is almost 36% reduction in the cost of exergy destructions and the fact is there is also a decrease of 45% in the exergoeconomic cost of fresh water soon after the optimization. One of the effective descriptions of solar still was simulated and optimized by Rashidi et al. on the volume of the fluid system. The entropy generation and flow of water-based nanofluid in the SSSS were simulated.[8] The productivity and improvement in SSSS with Aluminium oxide nanofluid which was under water-based atmosphere were identified. The entropy generation, solid volume fraction, productivity etc. of the SSSS with water-based nanofluids have been examined[8]. The generated Volume of Fluid model helps to simulate the condensation and boiling phenomenon of fluid in SSSS. The arithmetic results deliver the systematic outputs and showed that the productivity of SSSS is directly proportional to the solid volume fraction of nanoparticles. The productivity raised by almost five times as the solid volume fraction is raised between 0 to 5 percent. Also the average Nusselt number is raised by 18% with the solid volume fraction rise in the same range. Thermal entropy and viscosity showed different values at different geometric positions of the solar still. [8]. The entire fluid energy and exergy based analysis were carried out in a single slope solar still setup.

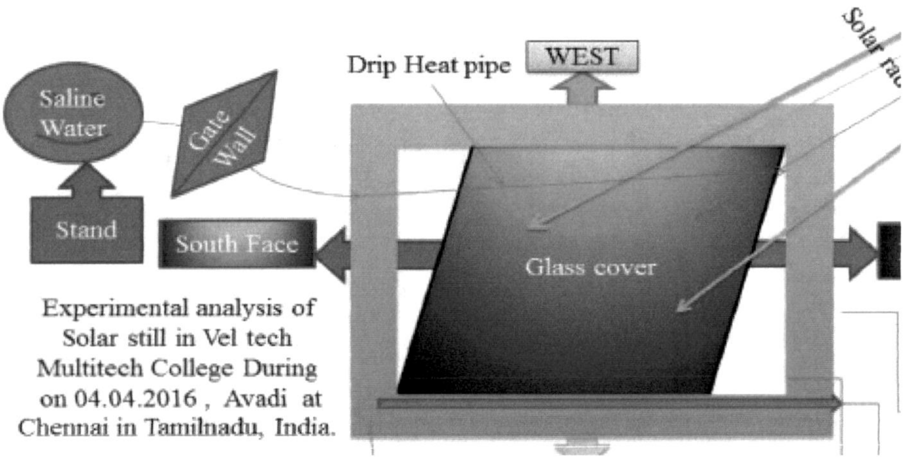

Figure 5. Setup of SSSS with different fin wick absorbing materials[11]

Rashidi et al. [9] optimize the stepped solar still cascaded with nanofluids. The series of optimization techniques and productivity enhancements were diagnosed. Here, the computational fluid dynamics approach was used to differentiate the RSM and the calculated results. The output shows a 2.1 % deviation in the comparative outputs. Also on adding the nanoparticle (range between 0 – 5%), the hourly productivity is raised by 22%[9]. The reviews showcase the effect of adding different particles to the existing system. Singh et al. [10] also revealed the Effect of enviroeconomic and exergoeconomic matrices on the existing life cycle and cost analysis of passive solar stills with single slope available with and without water-based nanofluids. [10] Environment conditions also play a vital role in the diagnosis of the exergoeconomic properties of the system. Therefore, when tested in the climatic conditions of Delhi in the month of May, the 0.144 kWh per unit cost based on exergoeconomic parameter has been identified to be 0.144 kW h/Rs which is slightly higher than double slope solar still having 0,137kWh/ Rs values under similar atmosphere. Shanmugan [11] worked on some of the miscellaneous materials having absorbing characteristics and performed the productivity enhancement analysis of solar still by the inclusion of PCM and various shaped Nanoparticles. The particles have different compositions based on suitability and availability. [11]. Very effective characterization was obtained on different compositions. Figure 5 shows the Setup of SSSS with different fin wick absorbing materials. Various combinations in the setup would be probably the solar still without fins, Materials variations of fin wick required in still, Calculation setup for solar intensity per second absorbed to a still etc.

Conclusions
It has been tried to review the effect of nanofluid on the performance of basin type double slope solar desalination unit. On the basis of present work, the following conclusions can be drawn:

i. The output of passive single slope solar desalination loaded with/without nanofluid is not appreciable and needs to be improved for making it popular as well as competitive.
ii. The demerits of low outputs of passive type single slope solar desalination unit can be overcome by active double slope solar desalination unit.
iii. The use of nanoparticles with water enhances the performance of a single slope solar desalination unit on the basis f exergoeconomic and enviroeconomic parameters.

Recommendations
The exergoeconomic and enviroeconomic parameters of solar desalination unit loaded with nanofluid can further be improved by optimizing the mass flow rate, number of collectors and water depth with the help of optimization techniques.

Acknowledgement
The team of current review work would like to thank the Graphic era deemed to be university and Galgotias College of Engineering and Technology for providing all sorts of amenities and infrastructure.

References
[1] Bait O and Si-Ameur M 2018 Enhanced heat and mass transfer in solar stills using nanofluids: A review *Solar Energy* **170** 694-722

[2] Tiwari G N et al 2015 Exergoeconomic and enviroeconomic analyses of partially covered photovoltaic flat plate collector active solar distillation system *Desalination* **367** 186-196

[3] Singh D B and Tiwari G N 2017 Exergoeconomic, enviroeconomic and productivity analyses of basin type solar stills by incorporating N identical PVT compound parabolic concentrator collectors: A comparative study *Energy Conversion and Management* **135** 129-147

[4] Singh D B and Tiwari G N 2017 Performance analysis of basin type solar stills integrated with N identical photovoltaic thermal (PVT) compound parabolic concentrator (CPC) collectors: A comparative study *Solar Energy* **142** 144-158

[5] Elango T *et al* 2015 Performance study on single basin single slope solar still with different water nanofluids *Desalination* **360** 45-51

[6] Investigation E *et al 2011* i a e m e **6340** 1-9

[7] Ibrahim A G M, Rashad A M, and Dincer I 2017 Exergoeconomic analysis for cost optimization of a solar distillation system *Solar Energy* **151** 22-32

[8] Rashidi S, Akar S, Bovand M, and Ellahi R 2018 Volume of fluid model to simulate the nanofluid flow and entropy generation in a single slope solar still *Renewable Energy* **115** 400-

410

[9] Rashidi S, Bovand M, Rahbar N, and Esfahani J A 2018 Steps optimization and productivity enhancement in a nanofluid cascade solar still *Renewable Energy* **118** 536-545

[10] Singh D B, Tiwari G N, Al-Helal I M, Dwivedi V K, and Yadav J K 2016 Effect of energy matrices on life cycle cost analysis of passive solar stills *Solar Energy* **134** 9-22

[11] Shanmugan S, Palani S, and Janarthanan B 2018 Productivity enhancement of solar still by PCM and Nanoparticles miscellaneous basin absorbing materials *Desalination* **433** 186-198

Design and Fabrication of low capacity Torque Transducer

Pooja Yadav[1], Meena Pant[1], Ankita Sheoran[1], Pawan Kumar Arora[2] and Harish Kumar[1#]
[1]National Institute of Technology Delhi, New Delhi, India
[2]Galgotias College of Engineering and Technology, G B Nagar, India
[#]E-mail: harishkumar@nitdelhi.ac.in

Abstract

The current work focuses on the development of a Torque Transducer capability o f 10 Nm and analysis of it using FEA software ABQUS V.2018, it also focuses on identification of important parameters that impacts the stress values like mesh type and global seed size, values of stresses obtained using both analytically and computational are compared and parameters for which most conforming results are obtained is identified. Effect of mesh size and type has been thoroughly studied and it is analyzed what mesh type and what mesh size gives results most closer to analytical values. Using the same analysis it is found out at what point maximum stress, strain values occur on the surface, this helps in ascertaining the locations where strain gauges could be mounted on the sensing elements. For analysis purpose model is prepared using Autodesk Fusion 360 software and on that FEA is performed. Further the designed model has been fabricated using series of CNC machining operations After the fabrication is complete, strain gauges are mounted on previously identified location to measure the strain / stress values.

Keywords: Torque transducer, Strain gauges, Stress/ Strain, Finite element analysis, ABAQUS

1. Introduction

In recent decades, demand for more reliable torque measurement with low uncertainty has been expanded [1]. Torque is a force that twists a structure. Different from axial loads that produce a uniform, or mean stress on the cross section, torque creates a stress distribution over the cross section. Torque leads to development of angular acceleration. It is a vector quantity. The direction of the torque vector is given by right hand rule; it is given by the expression.

$$T = F.r \tag{1}$$

Shafts transmitting a torsional load that is torques are very common in engineering applications like a motor, generator, automobiles, aircrafts, agricultural machinery etc. Are some of the few examples of torque application, where shafts of different materials, shape and types are being used. Shafts of uniform circular cross section are commonly used in these applications and they most of the times carry pure torsional loads, torques. If a load is applied to an engine shaft which is fixed, it generates reaction torque on the shaft [2]. The two equal and opposite torques at the end of the shaft will deform it. Torque being an important physical quantity its calculation has always an important for various analyses. Torsion theory gives relationship between torque and other

IOP Conf. Series: Materials Science and Engineering **748** (2020) 012032 doi:10.1088/1757-899X/748/1/012032

related quantities and constant for cylindrical body and that equation is referred to as torsion equation. Torsion equation is given as,

$$\frac{T}{Ip} = \frac{\tau}{r} = \frac{G\theta}{l} \tag{2}$$

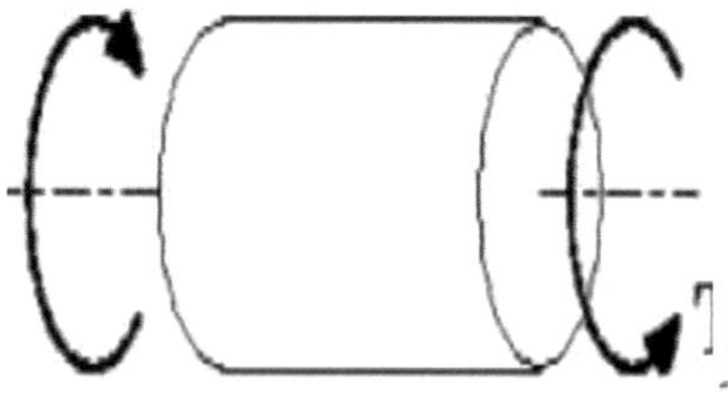

Figure1.1 Twisting moment on cylindrical object

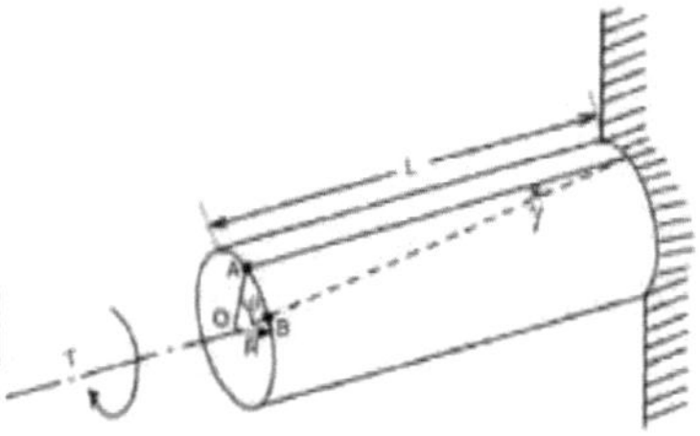

Figure 1.2 Cylindrical shaft subjected to torsion

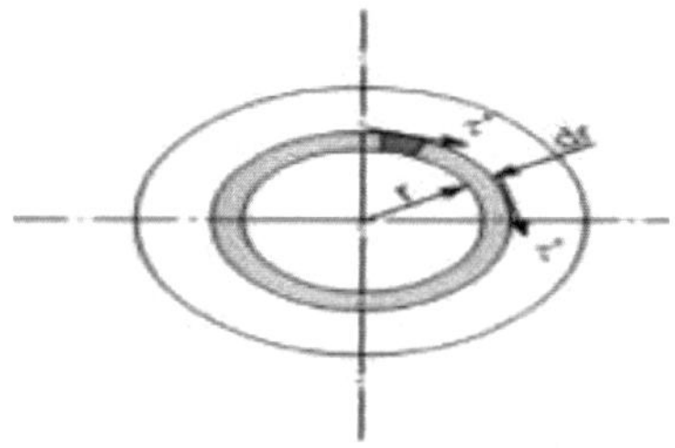

Figure 1.3 Cross- section under Torsion

2. Design analysis of transducer

For a circular shaft which is rigidly clamped at one end and twisted from other end, relationship between various involved parameters is well established by torsion theory[3]. Torque transducer are generally made up of steel, which is suitable for most of the applications, other material that could be considered for fabrication of torque transducer is high strength Al alloy like Al7075, which is better suited for low strength transducers. Sensing element is of square cross section, on which strain gauges are mounted to sense the deformation produced [4]. It undergoes deformation under the influence of external torque that is with help of strain gauges mounted on it. Relationship between torque and shear stress for rectangular cross-section (according to St. Venant's theory) is given by following analytical expression based on which dimension cross-section is found out expression is:

$$T = \frac{x^2 y^2 \tau}{3y + 1.8x} \tag{3}$$

T = Torque applied, τ = shear stress induced due to applied torque, x and y = length of sides of cross-section, for Square cross-section x = y, so the equation becomes

$$T = 0.208 x^3 \tau \tag{4}$$

For design of sensing element, it is required to find out the dimension of square cross- section, for T= 10 Nm and x = 0.009, τ = 65.95 MPa.

3. Finite Element Analysis

3.1 CAD model

On the basis of calculations and suggested design following drawing of transducer was prepared in Autodesk 360 software using the same software CAD model of the drawing was prepared.

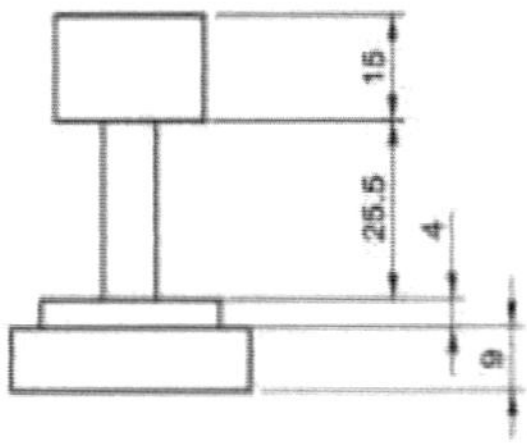

Figure 3.1(a) front view

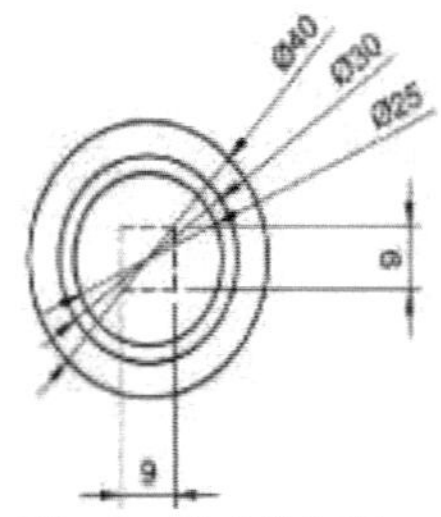

Figure 3.1(b) Top view

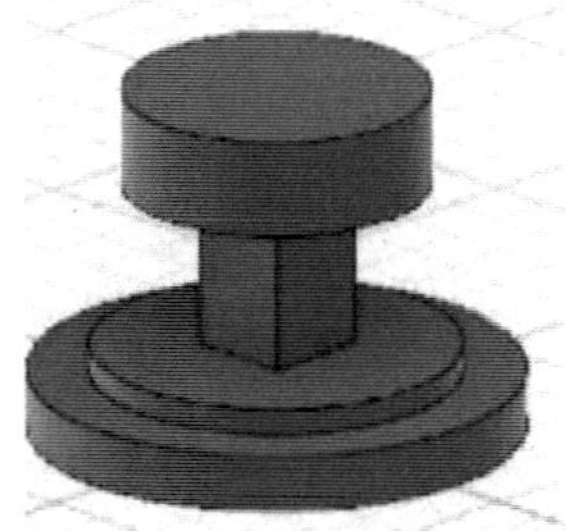

Figure 3.1 (c) CAD model

After preparation of the CAD model, finite element analysis of the drawing was carried out using ABAQUS CAE and following results were obtained. On applying suitable boundary conditions and constraints and adopting standard procedure for FEM, different mesh types and mesh sizes were analyzed and following results were elaborated using it.

3.2 Tetrahedral mesh type

Shear stress distribution on sensing element under external torque of 10 Nm shows that maximum values of stress are obtained on middle region of each side of sensing element.

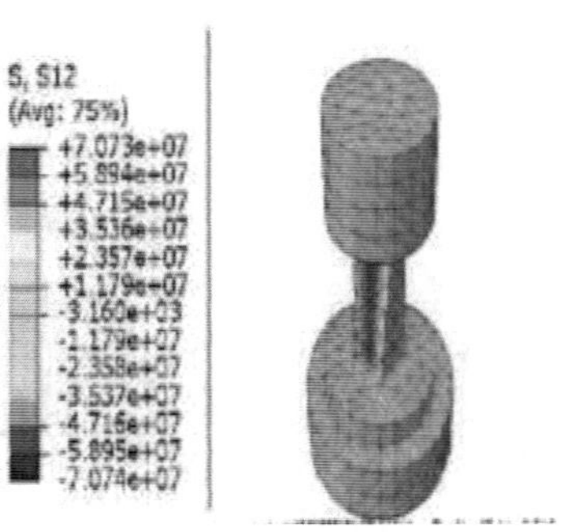

Fig.3.2 (a) shear stress for tetrahedral mesh (global seed size 0.004)

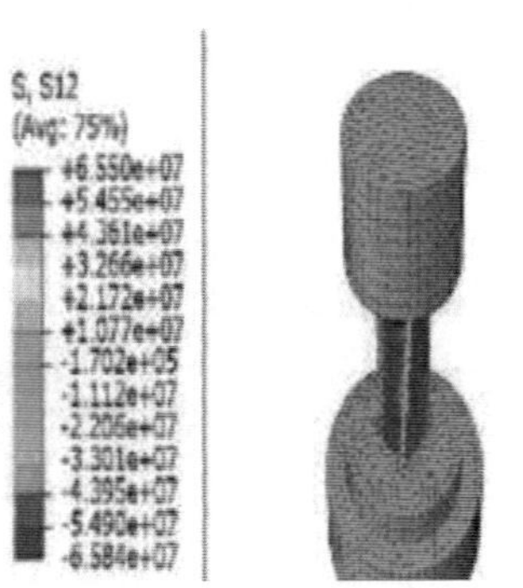

Fig.3.2 (b) shear stress for tetrahedral mesh (global seed size 0.002)

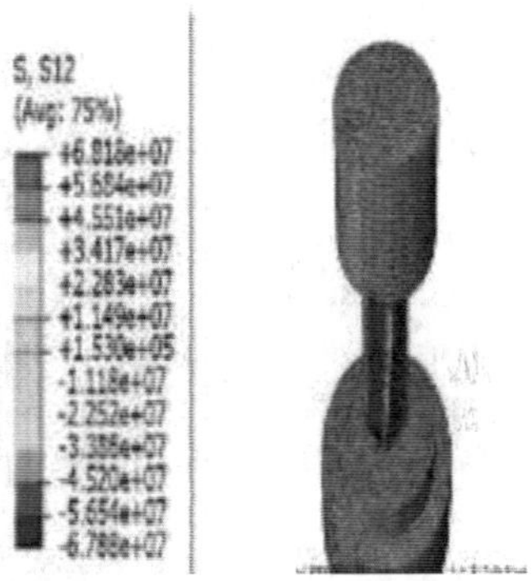

Fig.3.2 (c) shear stress for tetrahedral mesh (global seed size 0.001)

For tetrahedral mesh type, shear stress values for all the three considered mesh sizes i.e. 0.004, 0.002 and 0.001 are 70.73 MPa, 65.5 MPa, 68.18 MPa respectively which are suitable for safe design factor of safety being 2.

3.3 Hexagonal mesh type

Shear stress distribution for this mesh type follows the same pattern as tetrahedral mesh type i.e. it is maximum on middle region and decreases towards side, for same external torque value and global seed size of 0.004 shear stress values obtained were 27.2MPa.

IOP Conf. Series: Materials Science and Engineering **748** (2020) 012032 doi:10.1088/1757-899X/748/1/012032

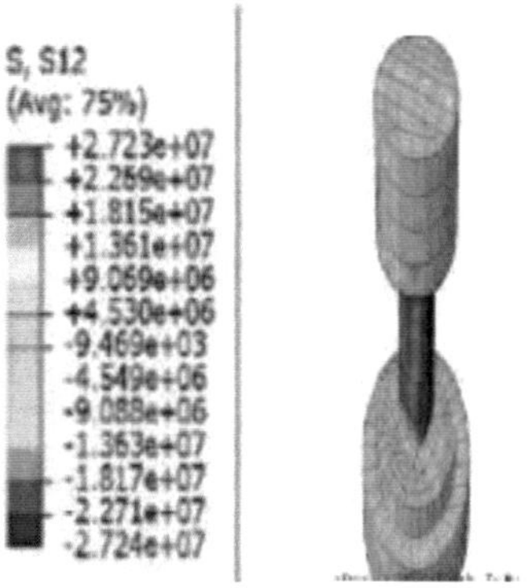

Fig.3.3(a) shear stress for hexahedral mesh (global seed size 0.004)

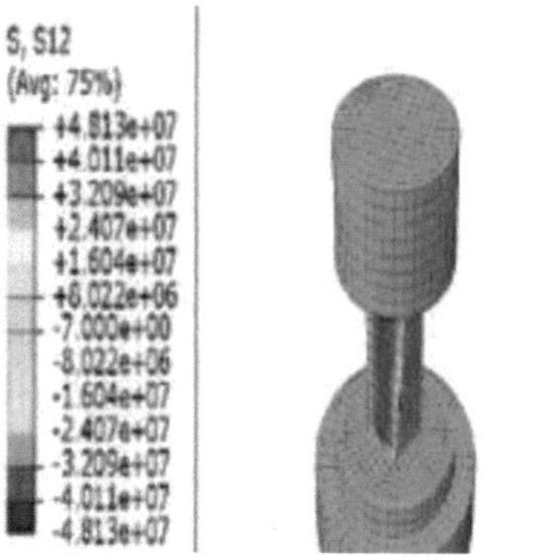

Fig.3.3 (b) shear stress for hexahedral mesh (global seed size 0.002)

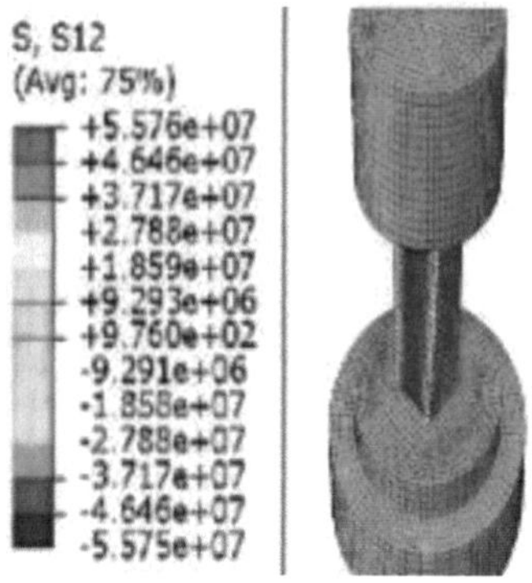

Fig.3.3 (c) shear stress for hexahedral mesh (global seed size 0.001)

For hexagonal mesh type shear stress values obtained using FEM analysis are not conforming with analytical results and as mesh is being refined values tend to get closer to analytical results, closest results were obtained for global seed size of 0.001 which was 55.76 MPa. On studying the FEA results obtained for the two considered mesh type that is tetrahedral and hexagonal it was concluded that out of the two, tetrahedral mesh gives results closest to analytical result this is because of close packing factor of tetrahedral mesh type. Closest results were obtained for global seed size of 0.002 of tetrahedral mesh. Analytically found value is 65.25 MPa.

4. Strain gauge

A strain-sensitive material is the material whose resistance varies with instantaneous strain developed over its surface. They can be either metallic or can be a semiconductor. When a material such as this is stretched, length of it increases and thereby decreasing its cross- section which leads to an increase in its electrical resistance. Change in its resistance is measure of its strain values. Thus a strain gauge is an electrical unit which uses change in electrical resistance to measure strain. Strain gauge are utilized to measure strain on am object, it consist mechanical inputs like force, pressure, weight etc. Into change in electrical resistance that could be measured. It is a sensor whose resistance changes with applied force strain gauge is widely used electrical measurement technique, for mechanical quantities. Strain can be tensile or compressive, thus strain gauge can be used to identify both expansion and contraction. Strain can be generated due to external load or stress in the body. Strain may be caused due to pressure, torque, heat, structural changes of the material. Under certain conditions by measuring the strain we can evaluate the quantity causing the strain. This is widely used for experimental stress analysis. It not only evaluates stress on the part but also predicts its safety and endurance. For shock or vibration instrumentation, resistance strain gauges may be used, it might be the active element in commercial or special purpose transducer.

On application of stress on metal conductor its resistance changes, as resistance is a function of length and cross-section area, resistance of conductor is also changed due to change in resistivity of the conductor, it is called as peizoresistive property so strain gauges arc also known as peizoresistive gauges. On application of tensile force on metallic wire of length l its length increases, if L_1 is the initial length of wire and L_2 final length of wire then strain is given by equation 5,

$$\in = \frac{L_2 - L_1}{L_1} \tag{5}$$

Further its diameter decreases, since resistance of the conductor is inversely proportional to length, change in resistance due to change in length can be measured and calibrated, thus strain gauges is used to measure force and other parameters like displacement, stress etc. Relationship between input and output parameters is expressed by term gauge factor or gauge gradient, Gauge factor (GF) is defined by equation (6)

$$GF = \frac{\frac{\Delta R}{R}}{\frac{\Delta L}{L}} \tag{6}$$

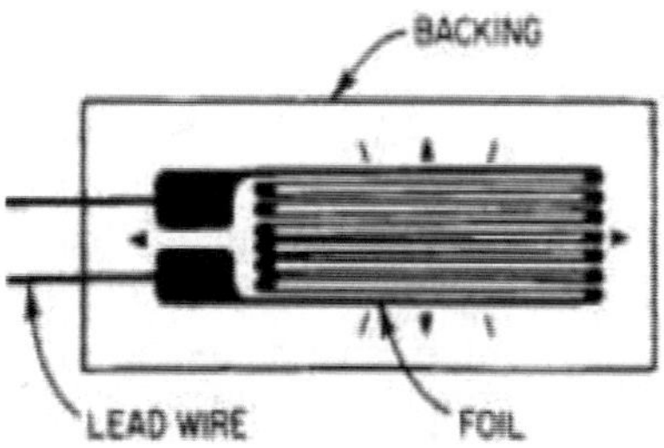

Figure 4.1 strain gauge constructions

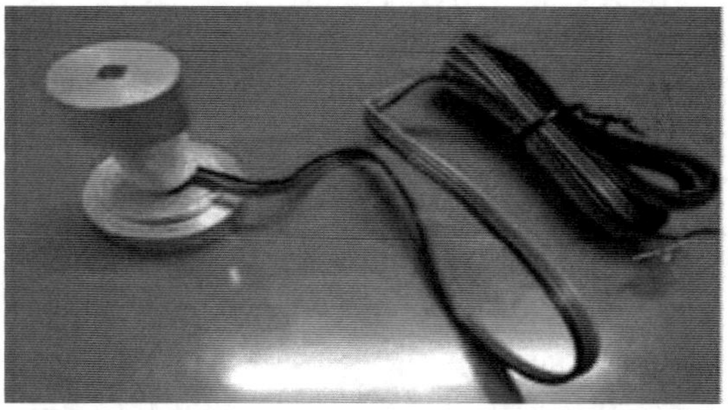
Figure 4.2 final strain gauged transducer

For the wires of strain gauge to have sufficiently high electrical resistance (60-350 ohm) they need to be very thin, so it is difficult to handle, so in order to facilitate ease of handling it needs to be planted on a carrier medium like paper or plastic Backing material is used to provide ease of handling, Required amount of lateral resistance is provided so that it could be shortened without buckling, both compressive and tensile stresses could be measured. Connection terminals are provided, a protective coating is applied over strain gauge.

For measuring the value of stress/ strain developed on the sensing element strain sensor have been used. Position of fixing the sensors have been ascertained using Finite Element Analysis done earlier, for the purpose of fixing the sensor the surface should be dirt free and it should be fixed at an angle of 45° on the surface taken from horizontal surface. This is the point of maximum stress/strain concentration. It is important to properly do the curing of the model. The connections are made according to wheat-stone Bridge.

5. Results and discussion

Torque transducer of capacity 10 N -in has been designed using torque and shear stress relationship for square cross-section obtained from St. Venant's investigation, Here dimensions of the sensing element is 9*9 mm value of shear stress induced for 10 N-m has been found which is equal to 65.949MPa. Computational results obtained though FEA have been found conforming with analytical values. Out of two different mesh types and three different global seed sizes used for analysis purpose the results for tetrahedral mesh and global seed size of 0.002 is found to be closest to analytical values. FEA analysis also helped in Guiding out regions of maximum stress/strain on the sensing element. After the fabrication being done of the designed part using CNC machining processes, Strain gauges were required to be affixed at pre-determined locations on the sides of sensing elements. These should be dirt free for affixing the strain gauges and suitable curing and post curing is carried out. Strain gauges are affixed in order to sense the values of stress or strain generated at points of maximum shear stress as shown by analysis on the ABAQUS.

Conclusion

Presented work dealt with design and development of torque transducer of capacity 10 Nm, finite element analysis was carried on using ABAQUS software and it was found out to be close to analytical results. After the finite element analysis of the model and ascertaining positions of maximum strain or stress, strain gauging was done on that position. Shear stress and strain values were found to be maximum at the centre of sides of the cross- section, where strain gauge needs to be fixed for measurement of strain values. Torque transducer being important for calibration of various torque measurement techniques, there is requirement of development of other torque transducers of nominal capacities. Presented work could be redesigned to prepare torque transducer of increased capacities moreover in the present work sensing element is of square cross-section this can be modified to circular parabolic or any irregular shape and similar analysis could be performed on it. Based upon the analysis new design could be fabricated and calibrated using standard calibration procedures. Metrological Characterization of the designed model needs to be done against standard torque machine.

Reference

[1] Kumar H., Kumar A., Gupta S. (2011), Design studies and testing of a torque transducer, Indian Journal of Pure & Applied Physics, pp. 653-656.

[2] Warnock F.V., Benham P. P. (1965), *Mechanics of solids and strength of materials,* Sir Isaac Pitrnann, London.

[3] Diddens D., Reynaerts D., Brussel H.V. (1995), Design of a ring shaped three axis micro force / torque sensor, *Sen and Actut A*, pp. 225-232.

[4] Madni A. M., Hansen R. K., Vuong J. B. & Wells. R. F. (2005), A differential capacitive torque sensor (DCTS), *Proceedings of' IEEE Sensors*, pp. 1286-1289.

[5] Zhang. J., Chen G., Wang W., & Sun H. (2016), Design and application of strain brushless torque sensor, *Icamia.* pp. 1-5.

[6] Aghili F., Buehler M. & Hollerbach J. M. (2001), Design of a hollow hexaform torque sensor for robot joints, *International Journal of Robotics Research*, pp. 967-976.

[7] Hoehn. K., Olsson A., & Arkwright J. W. (2017), High Capacity Torque and Compression measurements using fibre optic sensors, *International conference of fibre – optic and photonic sensors for industrial and safety applications*, pp. 39-44

[8] Kim I. M., Kim. H. S. & Song J. B. (2012), Design of joint torque sensor and joint structure of a robot arm to minimize crosstalk and torque ripple, *International conference*

on ubiquitous robots and ambient intelligence, pp. 404-407.

[9] Nazar M., Islam U., Cheng P., Oelinann B. (2018), Torque sensor design considering thermal stability for harsh industrial environments, *International conference on sensing technology,* pp.83-86.

[10] Sorli M., Pastorelli S.(1995), Six axis reticulated structure force / torque sensor with adaptable performances, *Mechatronics*, pp. 585-601.

[11] Hon Y., Zeng D., Yao J., Kang K. & Zhao Y. (2009), Optimal design of a hyperstatic stewart platform based force / torque sensor with genetic algorithms, *Mechatronics*, pp.199-204.

[12] Zhiwen Y., Zhibin F., Weina H.(2015), Torque measuring device structural design and experimental analysis for drive shaft, *China Measurement &Test*, pp. 120- 123.

[13] Jinquan H., Xiaoping W., Hui Y.(2015), Design of high-precision torque measurement system based on STM32, *Transducer and Microsystems Technologies*, pp. 99-101.

[14] Baolin Y., Ying Y. ,Zeitiing L., Sliiqhig G.(2011), Study on design principle of strain torque sensor, *Machinery Design & Manufacture*, pp.13-15.

[15] Chunmei Z., Zhaoxia W., Xiaofeng H. (2009), Design and study of torque measurement system." *Machinery Design & Manufacture*, pp 30-32.

AUTHOR INDEX

Institute of Physics
Dirac House, Temple Back
Bristol BS1 6BE UK

ISSN: 1757-8981
ISBN 978-1-7138-2657-6
9 781713 826576

CURRAN ASSOCIATES INC.
proceedings.com